Marine Gas Turbines

Ocean Engineering: A Wiley Series

Michael E. McCormick   Ocean Engineering Wave Mechanics
John B. Woodward   Marine Gas Turbines

MARINE GAS TURBINES

JOHN B. WOODWARD
Professor
Department of Naval Architecture
and Marine Engineering
University of Michigan

A WILEY-INTERSCIENCE PUBLICATION

JOHN WILEY & SONS, New York–London–Sydney–Toronto

Library of Congress Cataloging in Publication Data

Woodward, John B
Marine gas turbines.

(Ocean engineering, a Wiley series)
"A Wiley-Interscience publication."
Includes bibliographies.
1. Marine gas-turbines. I. Title.

VM740.W6 623.87′23′3 74-31383
ISBN: 0-471-95962-6

Printed in the United States of America

10 9 8 7 6 5 4 3 2

SERIES PREFACE

Ocean engineering is both old and new. It is old in that man has concerned himself with specific problems in the ocean for thousands of years. Ship building, prevention of beach erosion, and construction of offshore structures are just a few of the specialties that have been developed by engineers over the ages. Until recently, however, these efforts tended to be restricted to specific areas. Within the past decade an attempt has been made to coordinate the activities of all technologists in ocean work, calling the entire field "ocean engineering." Here we have its newness.

Ocean Engineering: A Wiley Series has been created to introduce engineers and scientists to the various areas of ocean engineering. Books in this series are so written as to enable engineers and scientists easily to learn the fundamental principles and techniques of a specialty other than their own. The books can also serve as text books in advanced undergraduate and introductory graduate courses. The topics to be covered in this series include ocean engineering wave mechanics, marine corrosion, coastal engineering, dynamics of marine vehicles, offshore structures, and geotechnical or seafloor engineering. We think that this series fills a great need in the literature of ocean technology.

MICHAEL E. McCORMICK, EDITOR
RAMESWAR BHATTACHARYYA, ASSOCIATE EDITOR

November 1972

PREFACE

As this book is written, the gas-turbine engine has yet to find its equilibrium position in the marine field vis-à-vis its older competitors. Over the 20 years or so preceding, many experimental installations have been made; most of these passed after briefly seeming to be important. More lately, service installations—that is, nonexperimental—have been made in both military and commercial vessels. Some of these perhaps will be unsuccessful in that the initially enthusiastic owners will eventually regret their choices, while some will probably serve through long and successful careers. Meanwhile, the gas turbine is developing rapidly: efficiency is increasing, the ability to burn commercially feasible fuels is being extended, and engines of greater horsepower are being introduced. It, therefore, seems highly likely that the probability of attractive application will rise and that more applications will therefore be encouraged, all resulting in an increasing penetration of the marine field by the gas-turbine engine. In consequence, the marine engineering profession will increasingly need to be familiar with this type of machinery. To such need is this book addressed.

In similar fashion, the gas turbine is advancing into other ocean-engineering services, particularly for the offshore-petroleum industry. This application is also treated here.

In attempting to meet the needs that the above suggest, I have two purposes. One is to introduce the engineering student to some of the concepts of the marine gas turbine, and indeed of the gas turbine in general. Toward this purpose, some rather elementary material is included, particularly in the first four chapters. I seek to give this student an idea of what the gas-turbine engine is, how it works, what the principal parameters of significance are, and to show him a sampling of vessels in which the engine has found application. But it is not intended that this reader, or any other, will learn here how to design such engines, or to analyze them more than superficially. His needs in this area can be better served by more specialized literature.

The second purpose is to outline to the practicing design engineer or the more advanced student the problems encountered in designing a marine power plant centered on the gas-turbine engine. Although the gas turbine can be offered by its builder as an almost self-contained plug-in power module, there is much the ship designer—the marine engineer—must know about and provide for a successful plant. Therefore, I attempt here to acquaint the designer with these problems and, by illustrations from many sources, show him their typical scope. On the other hand, it should be stated that this is not a design manual, nor a handbook. Rapid change, if nothing else, precludes my attempting such a thing. A designer, confronted with an actual project, should obtain his working data from up-to-date sources provided by the supplier of his engine.

Many sources are to be acknowledged as contributing material to this book. In a basic way, A. O. White's chapter on gas turbines in *Marine Engineering* (Society of Naval Architects and Marine Engineers, 1971), the U.S. Navy's design manual, NAVSHIPS 0941-038-7010, and the *Gas Turbine Handbook* (Gas Turbine Publishing Company, Tom Sawyer, editor) form a background. Many individuals and companies have contributed specifically: A. O. White and J. C. McMahon of General Electric (Schenectady), Paul Dupuy of General Electric (Cincinnati), Chester Stasiowski of Turbo Power & Marine, T. B. Lauriat of Avco-Lycoming, Frank Vibrans of Nickum & Spaulding, and Charles Miller of the Naval Ship Engineering Center contributed significantly with personal information and provided data from their organizations. Literature of the American Society of Mechanical Engineers, Society of Naval Architects and Marine Engineers, Society of Automotive Engineers, Institute of Marine Engineers, and the Society of Naval Engineers was also a rich source. Literature of the first two was especially helpful. Among the ASME papers, I particularly commend the several authored by Eugene Weinert and his colleagues at the Naval Ship Engineering Center, Philadelphia Division.

The University of Michigan supported this work in several ways. First, I was given a sabbatical leave to begin the project. Subsequently the University acquiesed while I used, at its expense, enormous amounts of paper, reproductive services, and typist's labor. One might say that the typist, Constance Goodsell, was doing only what she was paid for, but the quality of her work and her intelligent coping with the rough spots were well above the commonplace norms.

Cost of manuscript preparation was supported in part by Newport News Shipbuilding, and this help is gratefully acknowledged.

John B. Woodward

University of Michigan
October, 1974

CONTENTS

NOTATION

CHAPTER 2 NOTATION

C_p	constant pressure specific heat
C_v	constant volume specific heat
h	enthalpy
k	ratio of specific heats, C_p/C_v
Q_A	heat added
Q_F	heating value
R	pressure ratio
T	temperature
v	specific volume
w_A	air rate
w_F	fuel rate
w_N	net work

CHAPTER 3 NOTATION

A	flow area, or heat transfer area
a	correction factor; see Eq. 3.17
B	depth of flow passage
C	velocity of fluid, or product, w × c (w = weight flow rate)
c	specific heat
D	wheel diameter
E	energy rate
e	base of natural logarithms
H	enthalpy, or heat transfer unit
h	enthalpy, or surface heat transfer coefficient

k	thermal conductivity
$\dot{m}$	mass flow rate
P	pressure
ΔP	pressure difference
Q	torque
r	radius
T	temperature
ΔT	temperature difference
U	linear wheel velocity, or overall heat transfer coefficient
V	volume flow rate
w	velocity of fluid relative to wheel, or weight flow rate
x	tube wall thickness
α	nozzle angle; see Figure 3.11
β	vane or fluid direction angle; see Figure 3.2
ϵ	utilization of energy, or heat exchanger effectiveness
ρ	fluid density
ω	angular velocity of wheel

Subscripts

A	axial component
c	cold
F	fast
h	hot
i	inside
L	large
m	mean
max	maximum
min	minimum
o	outside
P	products
R	reactants
S	slow
u	tangential component
1	entrance
2	exit

CHAPTER 4 NOTATION

BHP	brake horsepower
C_1	velocity of jet from turbine nozzle

N	rev/min
N_3	power turbine rev/min
NPT	power turbine, rev/min
P	power
Q	torque
r	a subscript denoting rated conditions
rev/min	revolutions per minute
SFC	specific fuel consumption (=fuel rate)
T_{am}	ambient temperature
T_o	ambient temperature
T_8	exhaust gas temperature
U	linear turbine speed (= rotational speed × radius)
W_8	exhaust gas flow

CHAPTER 5 NOTATION

bhp	brake horsepower
B_P, B_U	Taylor propeller coefficients
C_Q	see Eq. 5.12
C_T	see Eq. 5.13
D	propeller diameter
ehp	effective horsepower
g_0	unitary conversion constant
H	propeller pitch
H_E	see Eqs. 5.15 and 5.16
H_L	see Eqs. 5.15 and 5.16
H_P	see Eqs. 5.15 and 5.16
J	see Eq. 5.4
K_Q	see Eq. 5.2
K_T	see Eq. 5.3
L	vessel length
$\dot{m}$	mass flow rate
N	rotational speed, rev/min
n	rotational speed, rev/sec
P	power
P_B	brake horsepower
P_D	delivered horsepower
P_E	effective horsepower
P_S	shaft horsepower
P_T	thrust horsepower
PC	propulsive coefficient

Q	torque
R	resistance
S	thrust
shp	shaft horsepower
T	thrust
t	thrust deduction
V	speed
V_A	speed of advance
V_J	see Eqs. 5.15 and 5.16
w	wake fraction
δ	Taylor advance coefficient
η	efficiency
η_H	hull efficiency
η_o	open-water propeller efficiency
η_P	propeller efficiency
η_R	relative rotative efficiency
θ_P	see Eq. 5.13
ρ	density of water

Subscript 0 implies rated or base conditions.

CHAPTER 6 NOTATION

a.c.	alternating current
d.c.	direct current
ehp	effective horsepower
Hz	cycles per second
J	propeller advance coefficient
K_Q	propeller torque coefficient
K_T	propeller thrust coefficient
N	rev/min
P_E	effective horsepower
P_S	shaft horsepower
Q	torque
r	reduction gear ratio; subscripts d, gen, gt, for diesel, generator, gas turbine, respectively
shp	shaft horsepower
T	thrust
V	vessel speed

Subscript 0 implies rated or base condition; prime (′) denotes a ratio with the subscripted quantity, for example, N' is N/N_0.

CHAPTER 7 NOTATION

AH	ahead circuit breaker (from Figure 7.12)
D, Dx1	switch contacts (from Figure 7.11)
FCU	fuel control unit (from Figure 7.13)
I	moment of intertia of transmission system
K_Q, K_T	propeller torque coefficient and thrust coefficient
m	mass of hull
M	switch contacts (from Figure 7.11)
n	shaft speed
N_{HP}	gasifier rev/min (from Figure 7.6)
Q	torque
Q_d	propeller torque
Q_f	friction torque
R_T	hull resistance
SW GR	switch gear (from Figure 7.12)
t	time
T_F	combustor temperature (from Figure 7.6)
T_P	propeller thrust
T_X	exhaust temperature (from Figure 7.6)
V	vessel speed
$x_1 \cdots x_2$	logical conditions (from Figure 7.4)
α	power-turbine nozzle angle (from Figure 7.6)

CHAPTER 8 NOTATION

ASTM	American Society for Testing and Materials
C_s	value in centistokes (unit of kinematic viscosity)
D	diesel, used in designation of ASTM diesel fuels
F	degrees Fahrenheit
GT	gas turbine, used in designation of ASTM gas turbine fuels
t	time
u	fluid velocity
y	dimension as used in Figure 8.7
μ	viscosity
τ	shear force per unit area in fluid flow

CHAPTER 9 NOTATION

A	cross sectional area of duct
c_p	constant pressure specific heat

dB	decibels
d	hydraulic diameter of duct
f	friction factor
g	acceleration of gravity
g_o	unitary conversion constant
H	fluid head
J	mechanical equivalent of thermal energy
K	parameter f $l/$d
l	distance between points along duct centerline
m	mass flow rate
P	pressure
P	duct perimeter
R	gas constant, perfect gas equation
R	sound absorption, dB per unit of duct length
T	temperature
V	average velocity
W	sound power level
Z	elevation above datum
α	sound absorption coefficient
μ	viscosity
ρ	density
τ	viscous shear force per unit area
∇u	velocity gradient

CHAPTER 10 NOTATION

F_0	magnitude of exciting force
H	horizontal force
k	spring constant
M	moment
m	mass
N	rev/min
R	radius of gyration
T_R	rolling period
t	time
V	vertical force
W	weight
x	horizontal dimension (Eq. 10.2)
x	vertical displacement (Eqs. 10.4, 10.5)
y	vertical dimension
θ_R	angle of roll

$\dot{\theta}$	angular velocity
ω	frequency of vibration
ω_f	frequency of exciting force

CHAPTER 11 NOTATION

CFR	code of federal regulations
Cps	cycles per second (Hertz)
dB	decibels
GPM	gallons per minute
GR	reduction gear
Hz	hertz
JP4, JP5	jet engine fuels
NEMA	National Electrical Manufacturers Association
PPS	pounds per second
psig	pounds per square inch, gage
SFC	specific fuel consumption
μ	micron

CHAPTER 12 NOTATION

G.G.	gas generator, Figure 12.2
C.P.P.	controllable-pitch propeller, Figure 12.2
F	temperature, degrees Fahrenheit
gal/min	gallons per minute
hr	hours
M.E.	main engine, Figure 12.2
psig	pounds per square inch, gage
rev/min	shaft speed, revolutions per minute

CHAPTER 13 NOTATION

F, °F	temperature, degrees Fahrenheit
ft, ft^2, ft^3	feet, square feet, cubic feet
gal/min	gallons per minute
H_2	hydrogen
HeXe	hydrogen-xenon mixture
hp	horsepower
hr	hours

Hz, HZ	hertz
kw, KW	kilowatts
KW_e, KW_t	kilowatts electrical, kilowatts thermal
kw/ft^2	kilowatts per square foot
lb, lbs	pounds
lb/in^2	pounds per square inch
lb/kw	pounds per kilowatt
lb/kw-hr	pounds per kilowatt-hour
lb/sec	pounds per second
lb/shp-hr	pounds per shaft horsepower hour
mph	miles per hour
NEMA	National Association of Electrical Manufacturers
O_2	oxygen
psia, PSIA	pounds per square inch
Pu^{238}	plutonium 238 isotope
Q	heat flow rate
R, °R	temperature, degrees Rankine
rev/min, RPM	shaft speed, revolutions per minute
shp	shaft horsepower
ZrH	zirconium hydride
η	efficiency

Marine Gas Turbines

Chapter ONE

INTRODUCTION

1.1. BRIEF INTRODUCTION TO THE GAS-TURBINE ENGINE

I do not attempt here to give a full and intimate description of the marine gas turbine engine, but rather an introductory sketch for the benefit of the beginning student in marine engineering. That student is advised to supplement this text with other reading, especially the chapter on gas turbines in *Marine Engineering* [White (1971)]. And if he is interested in becoming a designer of these machines, then he must delve further into the mountain of technical literature that has accumulated since the gas turbine first became a viable prime mover during World War II.

For the balance of this section, let's speak in straightforward and elementary terms to describe the function and form of the gas turbine engine.

First, let us consider the turbine. A turbine is a wheel that absorbs kinetic energy from a fluid stream. It may be quite crude, as with the windmills and waterwheels of earlier technology, or highly sophisticated and efficient, as with contemporary steam and gas turbines. Its chief competition in converting energy into usable mechanical form is the cylinder-piston combination that is the heart of the various reciprocating engines, principally diesel and spark ignition. The advantages offered by the turbine are its relative compactness and the fact that the rotating output of the turbine is usually what is wanted to turn a wheel, propeller, generator, or pump. Its compactness is a consequence of the continuous nature of the process by which it handles its working fluid, quite in contrast to the batch process inherent with the cylinder-piston. The latter is handicapped by the need for repeated accelerations and decelerations of piston, valves, connecting rod, and of the fluid being handled. None of this occurs in the turbine; both rotation and flow are continuous, so that a given mass of metal, arranged artfully into a turbine wheel, can handle a much greater throughput of fluid than the comparable cylinder-piston. Rate of energy absorption, or power, is roughly proportional to the throughput, hence the attractive weight-per-horsepower characteristic of turbines. (A rotary engine, such as the Wankel, is intermediate in this respect between turbine and reciprocating engines.)

The preceding paragraph fails to mention several significant points, however. First, the rotational speed of a turbine is nearly always much too high for its load. For example, the most efficient speed of a ship propeller is likely to be in the neighborhood of 100 rev/min, while the most efficient speed of a turbine (either steam or gas) is likely to be at least several thousand rev/min. (The reasons for the high speed of the turbine are to be developed in Chapter 3.) The solution is usually a speed-reduction gear set between the turbine and its load. This device adds to weight, cost, and complexity of the turbine—but truly not very much—not enough to alter much the striking compactness that characterizes the turbine. And, indeed, most marine reciprocating engines also require reduction gears, though the speed ratio is seldom as great as with the turbine.

The second point can be much more significant. It concerns the origin of the fluid stream that is the input to the turbine. For the windmill and waterwheel, the answer is obvious, but these members of the turbine family have no place in marine propulsion. For the steam turbine, the answer is radically different. A massive and complex array of components is required to produce and deliver the fluid—steam at high pressure and temperature—to the turbine, and to remove it and process it for reuse. This array occupies the major fraction of space and weight for the propulsion plant and is nearly fatal to the steam-turbine concept. The internal combustion engines, especially diesel, have virtually eliminated the steam turbine from many sizes and types of vessels because the internal combustion process dispenses with most of this external plant; this allows them to reverse the compactness advantage of the turbine, or at least equal it, with comparable or better efficiency.

The gas turbine engine is a successful attempt to combine the compactness and simplicity (i.e., absence of reciprocating parts) first demonstrated by the steam turbine, with the advantages of internal combustion demonstrated most widely by the diesel engine. Now, it is impractical to have the combustion, and the compression of air that must precede combustion, accomplished within the passages of the turbine itself. (A diesel engine can do these things within its cylinder because its batch nature allows different processes to be assigned to different portions of the piston cycle.) Therefore, several major components are required by the gas turbine in addition to the turbine itself, but their physical magnitude does not approach that of the steam processing plant that must accompany the steam turbine.

The first major component is the *compressor,* which raises the pressure of air taken from the atmosphere by factors of perhaps five to twenty. It is either a centrifugal or axial-flow rotating device that receives its driving torque directly from the turbine, since it is always mounted on the same shaft. The second is the *combustion chamber* (or *combustor*), in which the air from the compressor is heated by the injection and burning of fuel, typically a clean, low-viscosity, petroleum oil. The air/fuel ratio (by weight) is likely to be 60:1 or higher, so that the product offered to the turbine is hot, high-pressure, air, slightly contaminated with the products of hydrocarbon combustion. This air is at high pressure, having suffered only a slight loss in pressure as it passed through the combustion process, and at a temperature usually in the 1500 to 2000 F range if the machine is running at rated power. It is then expanded through a nozzle to convert part of its thermal and pressure energy into kinetic energy; the resulting high velocity air—or "gas," if you prefer—stream then impinges on the blades of the turbine.

To summarize briefly, the basic gas turbine engine consists of a compressor and combustor whose combined function is to produce a high-pressure, hot, gas; a nozzle to expand the gas into a high-velocity stream; and a turbine to extract the energy from this fluid into a rotating shaft. Of course, upon looking at an actual machine you will observe it to be more complicated than this description indicates. For one thing, several auxiliary systems are essential, just as with any of the familiar types of engines. For example, fuel and lubricating oil must be supplied, there must be some means of starting the engine, and control and monitoring systems are required.

Another complication almost universally found in marine gas turbines is the use of two turbines in series, rather than the single one implied above. The first of these (the *compressor-drive, high-pressure,* or *gas-generator* turbine) functions solely to drive the compressor; the compressor and this turbine are, in fact, mounted on opposite ends of the same shaft. The second of these (the *low-pressure, free,* or *power* turbine) is mounted on a separate shaft, is driven by the gas exhausting from the other turbine, and functions solely to drive the output shaft of the engine. The principal advantage of this splitting of functions is that the torque characteristic of the engine is much more suited for mechanical drive applications, such as the ship propeller, than if both compressor and external load are driven by the same turbine.

The combination of compressor, combustor, and the compressor-drive turbine is usually called the *gas generator,* since its function is to generate the gas that drives the power turbine. The marine gas generator is often an aircraft jet engine that has been adapted to marine use, and is replaceable as a unit.

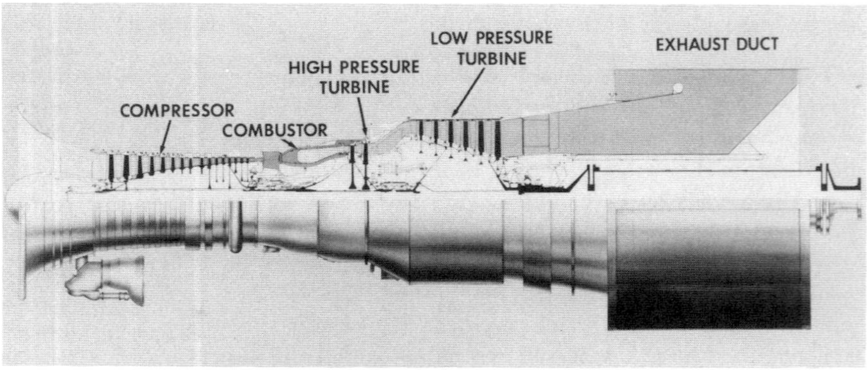

Figure 1.1. General Electric LM2500 gas-turbine engine.

Figures 1.1 and 1.2 illustrate the preceding discussion by cutaway and exploded views of the General Electric LM2500 marine gas turbine engine. It is an adaptation of the TF39 military and CF-6 commercial turbofan aircraft engines, and has a nominal rating of 25,000 horsepower. All parts mentioned to this point are shown, but without accessories, foundations, and ducts for air supply and exhaust.

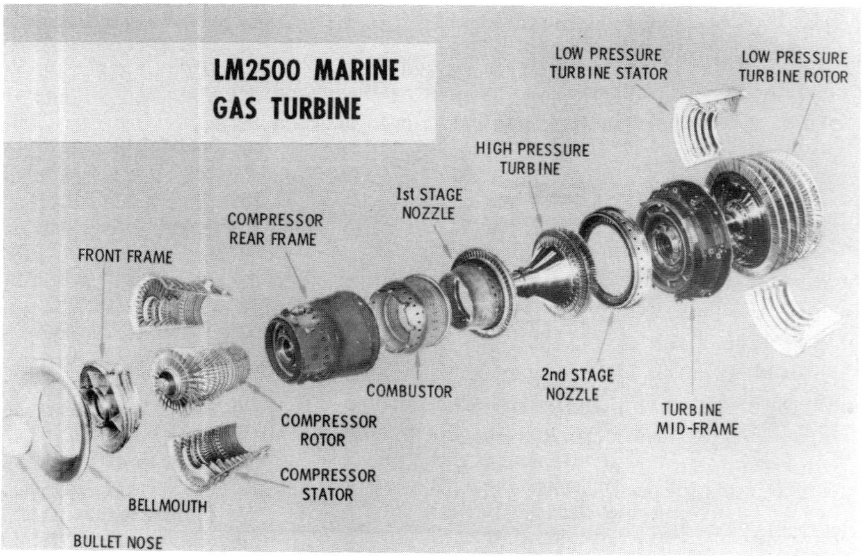

Figure 1.2. General Electric LM2500 gas-turbine engine—exploded view.

The student should also be familiar with several amplifications of the basic engine. These are illustrated in schematic form by Figure 1.3. Figure 1.3*a* shows the engine as discussed to now. Sketch *b* shows the addition of *regeneration,* accomplished by heat exchange between exhaust gas and the air leaving the compressor. This is a scheme for improving efficiency, since it enables part of the heating function to be accomplished by heat recovered from the exhaust. Sketch *c* shows the addition to part *b* of *reheating,* accomplished by a second combustor between the compressor-drive turbine and the power turbine. This scheme increases the output and also increases efficiency when combined with regeneration. Sketch *d* shows intercooling, a process of cooling the air as it is compressed. Its objective is to reduce the work of compression required, and it also improves efficiency of the engine when combined with regeneration. Sketch *e* shows both reheating and intercooling combined with regeneration.

Figures 1.3*d* and *e* also illustrate the *twin-spool* concept. Note that compression is accomplished by two compressors in series (C_1 and C_2). Each of the two is driven by its own turbine, C_2 being driven by T_1 and C_1 by T_2. The advantage is an increase in efficiency of the compression process, gained by allowing each half of the compressor to turn at its own best speed. Although the sketch doesn't indicate it, the shafts are concentric, that is, the shaft carrying T_2 and C_1 passes through the center of the shaft carrying T_1 and C_2.

1.2. CLASSIFICATION AND ILLUSTRATION OF ENGINES

Some classification of gas turbine engines has been implied previously, especially with respect to Figure 1.3. For example, one may speak of a simple engine (Figure 1.3*a*), of a regenerative engine (Figure 1.3*b*), and so forth. They may also be classified in several other ways. A common classification divides them into aircraft-derivative engines and industrial (or heavy-duty) engines.

An engine of the aircraft type has been adapted from a jet engine, usually by changing the materials of certain parts (or perhaps by coating them) to resist sea-salt corrosion and by addition of the power turbine to absorb the kinetic energy of the jet exhaust. An engine of this class, the General Electric LM2500, has been noted previously, and pictured in Figures 1.1 and 1.2. The outstanding characteristic of the aircraft-derived engines is their lightness. The LM2500 weighs 34,000 lb, including typical mounts, base, and enclosure, for approximately 1½ pounds per horsepower.

Figure 1.4 illustrates another aircraft-derivative engine, the Turbo Power &

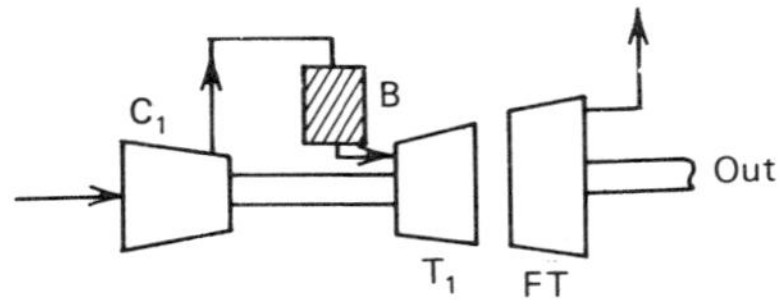

(a) Simple cycle

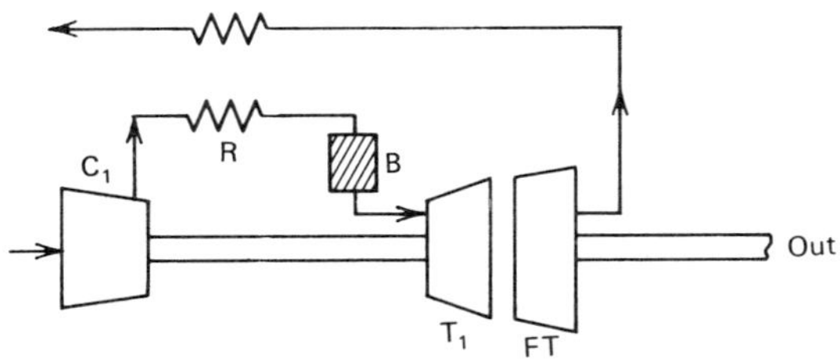

(b) Regenerative cycle

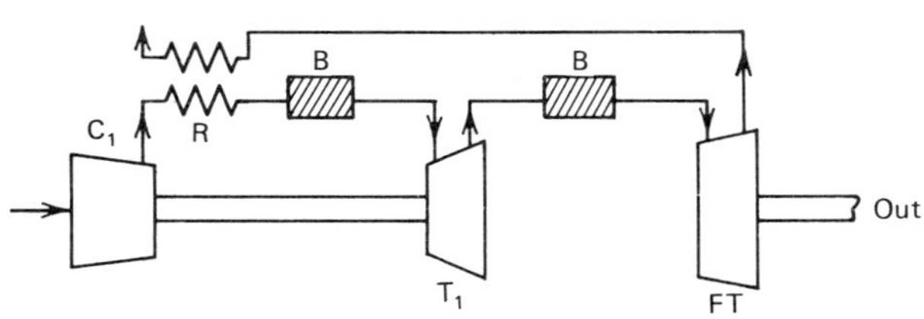

(c) Regenerative cycle with reheat

Figure 1.3. Several gas-turbine cycle arrangements. From *Marine Engineering*, 1971, published by the Society of Naval Architects and Marine Engineers.

Marine Systems (United Aircraft) FT4. Its rating and weight are of the same magnitudes as those of the GE engine.

The industrial-type marine gas turbine had its origins in applications such as gas pipeline pumping, where the light weight of aircraft service is not demanded. A typical example, the General Electric MS-5000 engine, is pictured in Figure 1.5. The rugged construction of the casing, in particular, is evident in this picture. This engine is rated in the neighborhood of 20,000 horsepower,

C_1 = Low press. comp.
C_2 = High press. comp.
B = Burner
T_1 = High press. turb.
T_2 = Low press. turb.
FT = Free power turb.
R = Regenerator
I = Intercooler

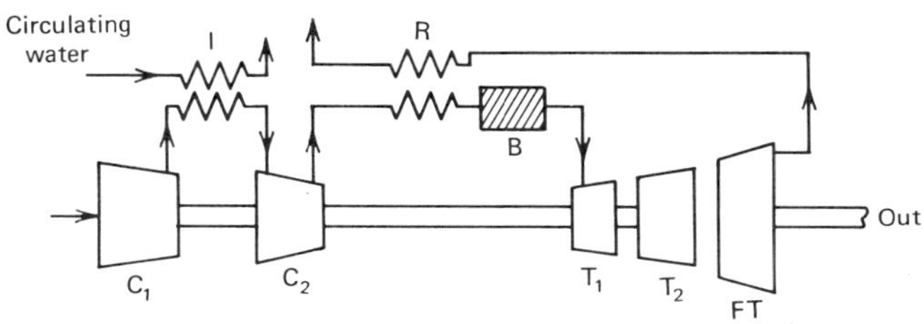

(d) Regenerative cycle with intercooling

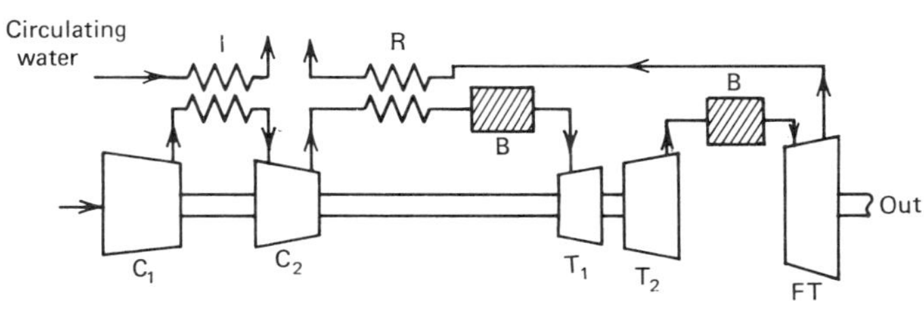

(e) Regenerative cycle with reheat and intercooling

and weighs in the neighborhood of 200,000 lb, including its base. With a regenerator, which is optional with this engine, the weight is approximately doubled.

Figure 1.6 illustrates by an exploded view an aircraft-derivative engine of a size and rating suitable for small vessels. It is the Avco Lycoming TF25, rated at 2000 horsepower continuous, and weighing approximately 1020 lb (not including the reduction gear shown in the figure). When assembled, this

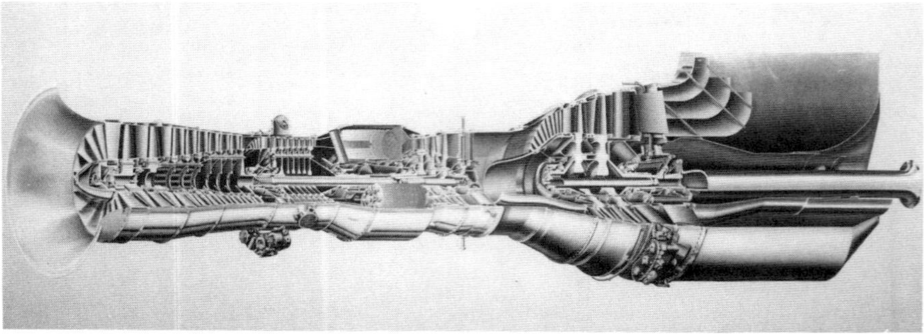

Figure 1.4. Turbo Power & Marine Systems (United Aircraft) FT4C engine.

engine is approximately 50 in. long, 41 in. high, and 30 in. wide. The basic engine is referred to as the "power-producer module" in Figure 1.6. This component contains the compressor, combustor, compressor-drive turbine, and power turbine. It is illustrated in Figure 1.7 by a sectional view.

Figure 1.8 illustrates an industrial engine that has been applied to small craft propulsion. It is the Ford Motor Company 420 engine, whose principal

Figure 1.5. General Electric MS-5002B engine. Compressor (foreground) and compressor-drive turbine shown.

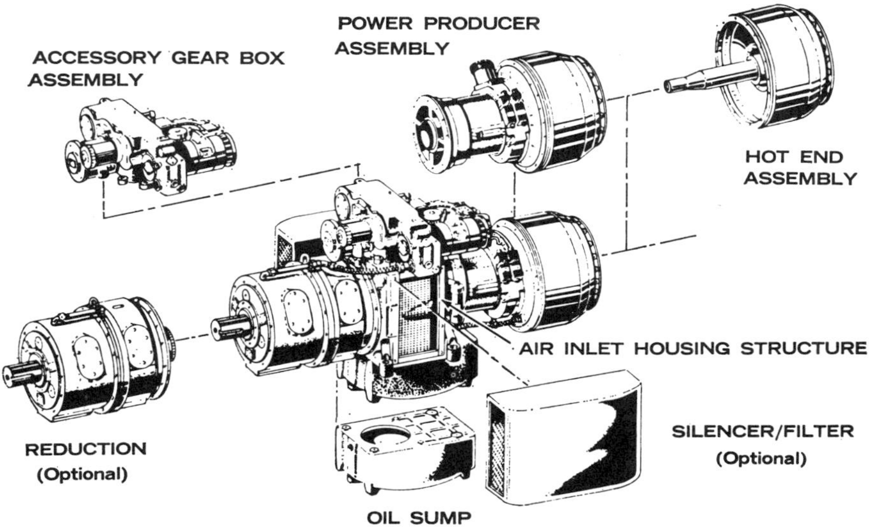

Figure 1.6. Avco-Lycoming TF25 engine, exploded view.

application is truck drive. The marine version is rated for 400 continuous horsepower, and weighs about 1700 lb. (Production of this engine ceased in 1973.)

1.3. THE MARINE ENGINE

The marine gas turbine, like most of the other engines used for marine propulsion, is usually an adaptation of a machine that was originally developed for some other purpose. The aircraft-derivative gas turbine is a prominent example. It is, therefore, instructive to discuss the special requirements facing a marine propulsion engine, since an understanding of these requirements is essential in deciding on the suitability of an engine, and in making a choice between different engines. This discussion is the major theme of this section, where it is offered in general terms, but with special references to the gas turbine engine when these are appropriate.

The requirements can be put into three categories: those demanded by (1) the load, (2) the environment, and (3) the service. The second of these should be subdivided into the burden placed by the environment on the engine, and the burden placed by the engine on the environment.

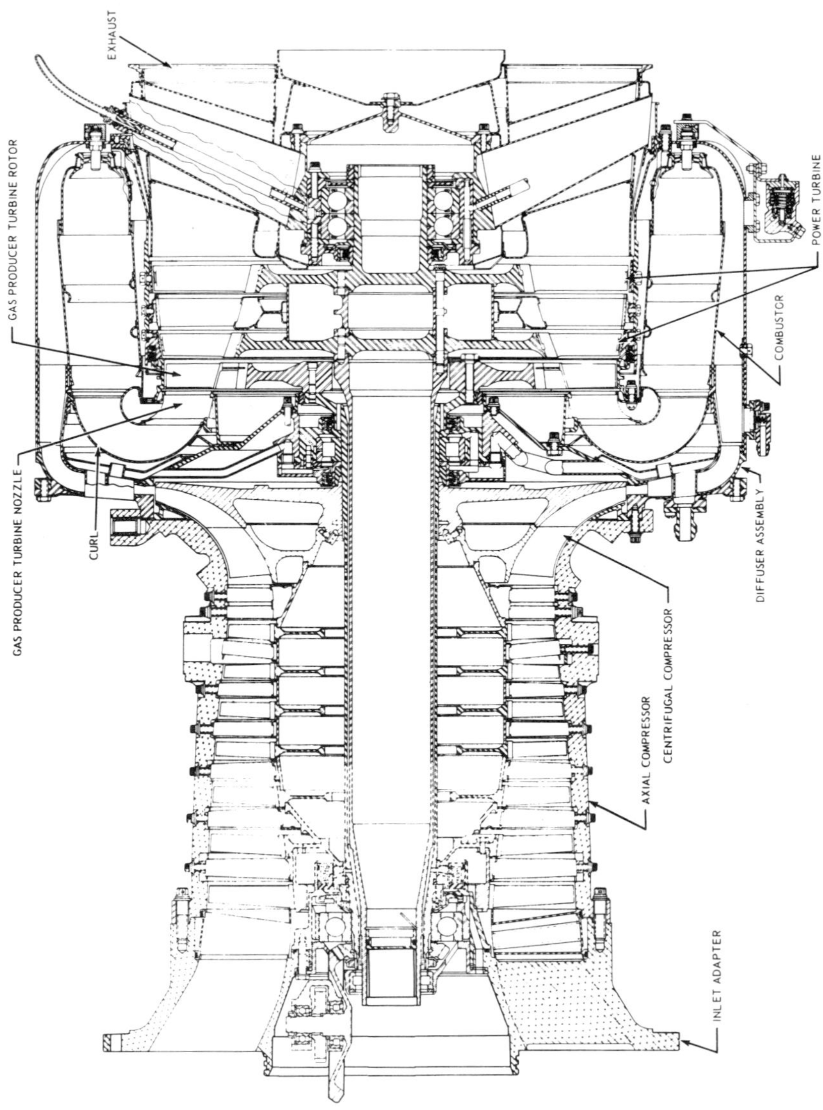

Figure 1.7. Avco-Lycoming TF25 engine, power-producer module.

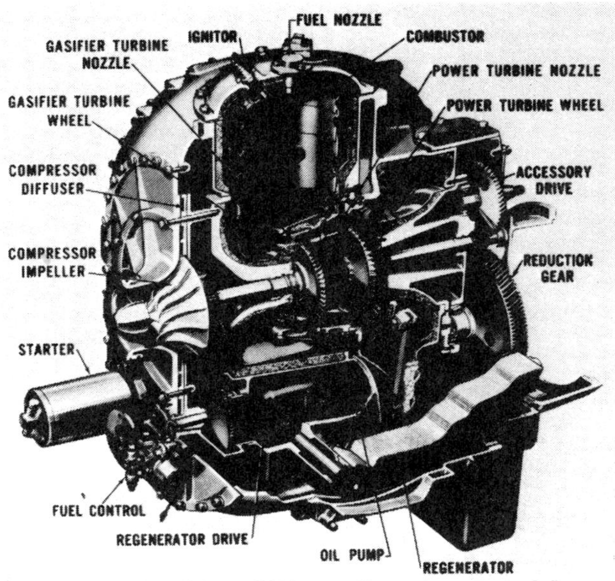

Figure 1.8*a*. no legend

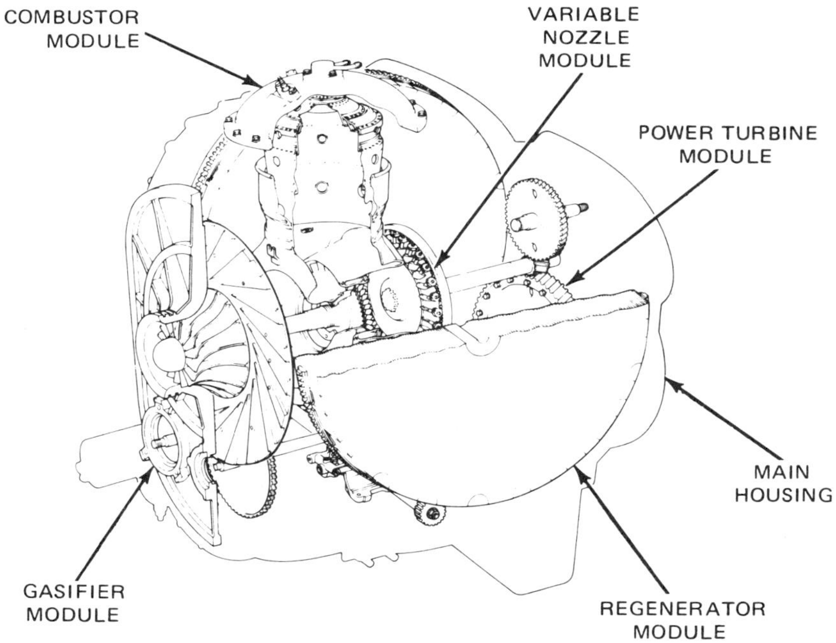

Figure 1.8*b*. Ford Motor Company gas-turbine engine.

A. Load Requirements

The load to be driven by a marine propulsion engine is usually a screw propeller, though it may also be a jet pump, or an electrical generator or hydraulic pump if nonmechanical power transmission is used. In any case, the engine is connected to its load by a rotating shaft. The power demand of the load is thus apparent to the engine as a demand for a certain torque at a certain rotational speed (in rev/min) , since shaft power is proportional to the product of these two quantities. Because the shaft must obviously turn at the same rev/min at both ends, and because the power transmitted must be the same at both ends (minor losses in bearing friction are being ignored here), both torque and rev/min must be the same for both engine and load. It is too much to expect that such disparate devices as a marine propeller and, say, a modified aircraft engine would absorb and produce the same torque at the same rev/min. Indeed, even if the engine were to be designed specifically for the propeller, obtaining an exact match in torque and rev/min at the desired propulsive power would demand large sacrifices in efficiency of one or both components. But, as noted in the first section, shaft-speed change, usually by mechanical gearing, is so common in all forms of driver-load matching that almost any reasonable combination of driver torque and rev/min can be matched to that of a marine propulsor in this way (or electrically, or hydraulically). The problem is thus "no problem," and the statement of a few sentences back that torque and rev/min must be the same for load and engine is true only when the gearing is regarded as being part of either the engine or of the load.

It may be said that the rev/min of any practicable engine can be changed to suit that of any propulsor, though the speed changer will cost something, of course. But this statement is made with reference to one operating point, the point at which maximum power is required. Because of the differing characteristics of engine and propulsor, the exact match is always disturbed when the propulsion plant operates at lower powers. If, however, the engine torque becomes greater than the load torque, the arrangement is satisfactory, since fuel input can be reduced to reduce the engine torque as necessary. In fact, engine torque capability must be greater, since an excess is essential to accelerate the load up to its maximum condition. Because the propeller torque-rev/min curve falls steeply as rev/min is reduced,* most engines meet this requirement. One exception is the single-shaft gas turbine. Its torque declines radically at low rev/min, and some means of reducing load torque until full rev/min is reached must be adopted with this type of engine. The two-shaft gas turbine, however, has excellent torque characteristics.

The requirement of the load must also be met during transient conditions:

* The same is true of pumps and is basically true of electrical generators, also, when the ultimate load is a pump or propeller.

all loads, marine and otherwise, must be accelerated and decelerated, started and stopped. A related point is the time taken to bring the engine from a cold condition to the condition at which it can produce useful power. The steam power plant, for example, takes several hours to bring from the cold to the operating condition, since steam pressure must be raised slowly, and piping and turbines must be carefully brought to operating temperature before power production can begin. It might be implied in consequence that several hours of warm-up time are acceptable for a marine engine, but often advantage can be gained from an ability to start and bear a load in a short interval of time. The gas turbine, especially the aircraft-derivative type, does not have the heavy metal parts that require careful warm-up, and can be brought from a cold start to full output in less than a minute. This characteristic, among others, makes it suitable as a boost engine in naval and coast guard service, that is, an engine that can be brought on the line quickly to give a large increase to the power normally supplied by a cruising engine.

The marine propulsion load changes rapidly in maneuvering situations, and especially when an emergency stop is required. The gas turbine is outstanding in meeting this requirement, since the inertia of its moving parts is small, and because its stored energy is small (essentially nil unless a regenerator is used). In this respect, it contrasts sharply with the steam propulsion plant, which stores a significant amount of energy in the steam contained in the boiler and piping. However, the inertia of hulls is so large compared to that of the machinery of any modern type that acceleration and deceleration are usually not governed by machinery response times.

The propulsion load must be reversed, that is, the vessel must have the capability of going astern. The most common method of achieving astern thrust is to reverse the rotation of the propeller. Only reciprocating engines, such as the diesel, are themselves reversible. Turbines, both steam and gas, are not reversible, so that some auxiliary method must be used. A separate turbine wheel, designed to drive in the opposite direction, can be built into the basic machine, or installed separately, or a reverse train can be incorporated into the reduction gearing. An alternative that has been most widely used with gas-turbine propulsion is reversal of propeller thrust through reversal of its pitch, rather than reversal of rotation. The use of this type of propeller (the controllable-pitch propeller, or "CPP") relieves the engine of the necessity to provide reversal.

B. Environmental Requirements

The marine engine must operate in an environment that is wet, salty, and fraught with motions of widely ranging magnitudes and frequencies.

Wetness manifests itself in several ways. The most obvious one is the presence of gross water, especially in the atmosphere (i.e., spray). Propulsion machinery is rarely exposed directly to spray, but combustion air must be taken from the spray-laden atmosphere close to the sea surface. If it is indeed the sea—not an inland fresh-water area—the spray bears salt that will be deposited in the engine with consequent corrosive effects and obstruction of air passages. This problem has not been generally noticeable among steam plants and among the several reciprocating types of internal combustion engines, but the gas turbine inherently consumes air at greater rates than its competitors, and so takes in greater amounts of sea salt—amounts great enough to cause corrosion, and to degrade performance through obstructing air passages in the compressor. Design to minimize the salt ingestion and operational practices to remove salt from the engine are important features of marine gas turbine technology.

Sea-humid air promotes corrosion of several common metals, so that parts exposed to this air must be of suitably resistant material, or properly protected by a resistant coating. The adaptation to marine use of engines of all types usually involves replacement of certain parts, or their coating, to achieve acceptable corrosion resistance.

An important aspect of the water problem is the possibility that fuel will become contaminated with it, with deleterious consequences to the engine. The gas turbine is perhaps more sensitive to this hazard than either the steam turbine or the diesel, since its hottest parts are exposed continuously to the combustion gas stream, and requires the greatest care in design and operation to ensure that fuel reaches the engine in a clean condition.

Motions are an inevitable part of the marine environment. A hull in rough water experiences three periodic rotations—rolling, pitching, and yawing—and three periodic translations—heaving, swaying, and surging. In severe weather, heaving and pitching accelerations can approach 1 g in magnitude, at least near the vessel ends. Vibratory motion is also likely. There are several sources of vibratory excitation, including the engine itself, but the one that is unique to the marine environment is the propeller. Vibrations originating at the propeller may be transmitted to the engine directly via the propulsion shafting, or indirectly via the hull.

Closely related to vibration is shock, which can be defined as a very sudden vibratory impulse. Its only likely origin is an explosion, and it, therefore, need be considered only for military vessels.

Motions of essentially zero frequency are also significant, these being of several types. There are the lists, heels, and trims that the hull occasionally experiences, and its deflections that may be caused by changes in loading, for example.

C. The Environment as Affected by the Engine

An engine produces several outputs that must be absorbed by its surroundings, both aboard the vessel and in the space in the vicinity of the vessel, producing several types and degrees of environmental impacts. This is true in all applications, but each, including marine, has its unique aspects.

It is inevitable that an engine reject heat to its surroundings, since the second law of thermodynamics renders unavailable a large fraction of the heat released from its fuel, and there are losses due to the inefficiencies of components. The rejected heat appears in both exhaust gases and in cooling water. The heat put into cooling water has come to be regarded as "thermal pollution," at least when present in large amounts. The gas turbine releases a greater fraction of its wasted heat as hot exhaust gas than does any of its competitors, and consequently rates excellent marks on the thermal pollution scene. However, marine power plants are typically so small and so mobile compared to the large electrical-generating plants that are suffering from this problem, that none of the marine engines are handicapped by their thermal emissions to the surrounding water.

Release of heat in the form of hot exhaust gas is not regarded as environmentally bad in the biological sense, but can be a fire hazard, both aboard the vessel and in the immediate surroundings. As an example of the latter hazard, small vessels with through-hull engine exhausts can be a threat to wooden piers. The through-hull exhausts of diesel and spark-ignition engines are always cooled internally by injection of sea water, thus removing this threat, but gas turbine exhausts are typically of too great a volume for such measures and must be designed with regard for hazard they may create.

More local thermal environmental problems may come from the heat transferred via convection and radiation from the surfaces of the engine. Since a marine engine usually must be installed in a small and enclosed space, this heat can be deleterious to other components in the space, as well as to personnel. Most marine engine installations, therefore, require combinations of insulation of the hot surfaces and ventilation of the engine compartment, to keep surrounding temperatures at a tolerable level.

The exhaust gas of any combustion engine in any application carries some constituents that have been recognized as harmful in the biosphere. If a petroleum fuel is being burned, unburned hydrocarbons, carbon monoxide, and several oxides of nitrogen—the components of automotive exhausts subject to legal strictures—are present, and there is sulfur dioxide if the fuel contains sulfur. Since the gas turbine employs a continuous combustion process, and usually burns low-sulfur fuels, it is perhaps the best of the marine engines in this respect. However, to this date (1973) marine-exhaust emissions have not been subject to legal limits. The major exception has been restrictions on visible smoke imposed by some harbor municipalities.

Noise generated by internal combustion engines is a problem in all applications, so that silencing must be employed, especially for noise escaping via the exhaust. In the marine application, the proximity of living spaces, or the necessity for crew members to enter the machinery compartment, means that sound isolation of the engine is often necessary. This is especially true with the gas turbine, since its intense high-pitched sound is intolerable to humans unless strongly attenuated.

Nuclear radiations should be mentioned as an environmental factor, though only for sake of completeness, for none of the combustion engines generate any such radiation.

And also as a measure of completeness, solid wastes generated by marine engines should be mentioned. An example is the ash produced by coal-fired boilers. This is not likely to be of concern in contemporary marine design.

D. Service Requirements

Requirements of marine service involve weight/volume characteristics, endurance, load profile, reliability, and maintenance practices.

Weight of propulsion machinery and the volume that it occupies compete with fuel capacity, payload, and living quarters. The least machinery, therefore, may be said to be the best machinery, for reductions in machinery weight or volume can be used to advantage for these other things, or will allow the vessel to be built with less displacement for fixed requirements for payload, living quarters, or endurance.

Endurance is usually designated by the distance the vessel can travel on one loading of fuel at a specified fraction of rated propulsive power. It is most directly related to the weight or volume that can be devoted to carriage of fuel, and, as implied in the preceding paragraph, is affected by the machinery weight and volume. The fuel rate (i.e., efficiency) of the engine is also significant to endurance, since the distance steamed* on a given weight of fuel is proportional to that efficiency.

Since efficiency has been mentioned here, it perhaps can command a few words of its own, though there is little uniquely marine in it, beyond the endurance consideration just given. Efficiency is, of course, the prime factor in determining the fuel cost of engine operation, and hence is an important concern in any engine application.

A load profile describes the lengths of time that an engine operates at

* "Steamed" is commonly used, as here, even though the engine may not involve steam at all.

various fractions of its rated power. These vary widely, depending on the type of vessel and its service, so that there is no single uniquely marine load profile. As one example, a tanker in typical long-haul service—Persian Gulf to Philadelphia, say—steams for several weeks continuously at full power, with only brief intervals of shutdown between voyages. In this respect, it resembles a base-load electric power station. On the other hand, the boost engine in a Coast Guard vessel may be cold for long intervals between brief intervals of full power, and, in this respect, it resembles a standby electric power station. Naval vessels may spend long intervals in port, or they may remain at sea for long intervals. The sea time is likely to require extended cruising at low propulsive power, with occasional short periods of full power during combat or exercises.

One significance of load profile is that it largely determines the hours of engine wear per hour of actual clock time and thereby indicates the effective life of the engine, or the intervals between maintenance events. It is also significant to endurance, since miles/shp-hr (shp = shaft horsepower) is greater at lower powers than at higher, and since efficiency of an engine is usually poorer at powers below the rated level.

Reliability of a vessel is the probability that it will accomplish its assigned mission. (The word "mission" has military connotations, but applies also to commercial vessels, for which the mission may be to complete a scheduled voyage on time.) It is affected by many factors, even such uncontrollable factors as weather and sea state. The condition of the machinery is obviously a factor, there being little likelihood of a mission being accomplished with an inoperative propulsion plant (though an analysis would have to consider partial failures, especially if the propulsion plant were multishaft or multiengine), and this condition is controllable by maintenance practices, and by the choices of machinery made in design. As applied to the machinery itself, reliability is the probability of nonfailure over a specified interval. The same could be said of an engine in any service—unreliability is welcome nowhere—but the values required, and the influences that determine the values obtainable, may be unique to the several possible marine applications.

Maintenance is the process of examinations and replacements that is required to keep the propulsion machinery operating in a satisfactory manner. (This should be contrasted to repair, which is the process of restitution following a failure.) In general, the maintenance process is determined by the type of machinery, but it is linked in several ways to the type of service. For example, the marine environment affects some maintenance practices, such as establishing the need to remove sea-salt deposits from gas turbine compressors. And as noted above, load profile strongly affects maintenance intervals. The requirements of the service also determine whether maintenance needs are significant. A minor replacement that required shutdown every 100 hr, say, might be a scarcely noticed factor in an aircraft engine application, but is likely to be unacceptable for a tanker on a 10,000-mile voyage.

And how do the service requirements influence the choice of the gas turbine as a marine propulsion engine?

The weight/volume requirements have been the strongest point in favor of the gas turbine, especially for the aircraft-derivative engine. For example, a typical 20,000 shp commercial ship machinery plant might weigh 100 tons if based on the aircraft gas turbine, 400 tons if based on an industrial gas turbine with regenerator, 700 tons if medium-speed diesel, and 800 tons if steam [based on example figures offered in Chapter I of *Marine Engineering* [Kane (1971)]]. The volume comparison runs in the same order, although the gas turbine advantages are less striking because of the large volumes occupied by inlet and outlet ducting, and by the regenerator when that device is present.

The figures quoted just above are without fuel. Fuel rates of contemporary gas-turbine and steam-turbine engines are nearly the same, with perhaps a slight advantage to the gas turbine. Ship propulsion diesels have about a 10–15% better fuel rate than the best gas turbines, and so regain at least part of the latter's weight advantage when fuel weight is considered, or can enjoy a correspondingly greater endurance.

In small-craft installations requiring only several hundred horsepower, weight of the machinery alone strongly favors the gas turbine over any competitor, but in this range of powers its fuel rate is at least 1.5 times that of a diesel engine. The weight advantage most definitely depends on the endurance required.

The gas-turbine engine is satisfactory for any reasonable load profile if its rating is chosen properly. This point must await the fuller discussion of a later chapter for elucidation.

Practically nothing can be said quantitatively about reliability, since hard information on this subject requires extensive collection of data under actual service conditions, and the gas turbine is too new in regular marine service for this to have been done. However, lengthy service of these machines in aircraft and industrial services has established a high level of reliability that should carry over into the marine field if the unique features of load, environment, and service are carefully accounted for in design and operation.

Each type of engine has its unique maintenance requirements, and these are usually fitted artfully into the service requirements. The aircraft-derivative gas turbine may require frequent replacement of its hottest parts, a task that could cause an unacceptable loss of productive time if done *in situ*. But because of the lightness and small size of this engine, the practice in the early gas-turbine ships has been to replace the entire gas generator section of the engine as a unit. This can be done within normal port shutdown times, even aboard ships that are scheduled for fast turnarounds in port. Maintenance is discussed more fully in a later chapter, but let this example serve for the present to illustrate that it is generally possible to fit maintenance requirements to service requirements.

1.4. THE MARINE DESIGNER'S TASK

The marine designer responsible for the propulsion plant has the fundamental task of choosing the type of machinery (diesel, gas turbine, steam, etc.). In some instances this task is almost trivial; for example, an owner's preferences may leave him no choice. In other instances, the choice of machinery may be imbedded in a thorough analysis of a large-scale transportation or military system, and thereby require judgement of factors far beyond the ship itself. And many cases fall between these extremes.

It is not the intention here, however, to treat machinery selection. Students who are interested in this process are advised to read Chapter I of *Marine Engineering* [Kane (1971)]. If especially interested in the naval application, they may gain some further insight by reading the first two chapters of Gillmer's *Modern Ship Design* [Gillmer (1970)]. And several technical papers have expounded on analyses of large-scale systems that have resulted in the choice of gas turbine propulsion; see Caldera et al. (1967) for an example describing a commercial system.

This book assumes that whatever the application may be, the gas-turbine engine has been chosen for propulsion, this engine having been found satisfactory when weighed against the criteria discussed in the preceding section, and against the economic and other criteria of the application. The question to be explored here is what must subsequently be done in design to ensure a well-engineered propulsion plant. At first glance, it might seem like very little, since the engine is to be purchased as a package, tempting the marine designer to adopt a "bolt it down, hook it up, and push the GO button" attitude. As I hope you suspect, this is an overly simplified approach. Actually, there is a large engineering effort needed in spite of the package nature of the propulsion engine. As discussed in the preceding section, the load, the environment, and the service make demands upon the engine, and the designer's task might, in consequence, be summarily described as ensuring that all of these demands are successfully met.

Let us then look at the engineering required, after the choices of machinery are made, putting the discussions of the previous section into particular terms. As before, the discussion is applicable to propulsion engines in general, but refers to the gas turbine where its features are unique.

Consider first the interaction between the engine and its load. In the strictly physical sense, this refers to the interaction between the output coupling of the engine and the shaft by which torque is transmitted to that load.

The shaft itself may present a design problem commanding major attention. Although a shaft appears to be an elementary component, it experiences thrust, bending, and torsion; it is subjected to corrosion, fretting, fatique, and electrolysis; it must be skillfully sealed at the hull penetration; and it is a vital component in the torsional, longitudinal, and lateral vibration systems that may unfortunately exist in the propulsion plant. There may be clutches, couplings, brakes, and pitch-change mechanisms interacting with the shaft in addition to engine, reduction gear, and propulsor.

But let us forego consideration of shafting design and focus attention only on its interactions directly with the engine. First let us consider the thrust that the propeller produces. No type of engine is designed to absorb this thrust, so that a thrust bearing external to the engine must be provided to prevent this interaction. Often, this bearing is built into the reduction gear, though it may be a separate component. If propulsion is by a jet pump, it is likely that the bearing is integral with that unit. If electric drive is used, there is essentially no thrust developed by the generator, hence the problem does not directly concern the propulsion engine. But a thrust bearing must protect the propulsion electric motor.

The problem of alignment of the shafting to the engine, or to the reduction gear, always exists. If the shaft is so aligned that it will place an abnormal lateral load on the output coupling of the reduction gear or engine, significant distortions are likely within that unit. Since a gas-turbine engine almost certainly uses a reduction gear, it is the latter that receives the distortion, with consequent disruption of the intended mesh between output gear and its pinion. If the reduction gear is not integral with the engine, then these two components must also be aligned, most likely with the aid of a flexible coupling. The alignment problem is always complicated by the thermal movements of the couplings that take place as the components warm up in service.

Three types of vibration have been noted as afflicting the shafting, and of these three, the torsional vibration also afflicts the engine if it exists to a significant degree. A torsional vibration analysis is, therefore, a necessity, the general idea being to ensure that no natural frequencies coincide with excitation frequencies within the system.

In a less physical sense, there is always the problem of matching the characteristics of load and of engine. Basically, this involves ensuring that both torque and rev/min of the engine and load are the same at the design operating point, and this has been stated previously to be largely a matter of selecting the reduction gear ratio. So it is, but only a limited selection of gear ratios may be available. For small craft, in particular, only a few stock ratios are likely to be usable. Consequently, the designer may have to adjust propeller pitch to get the operating point at the desired torque-rev/min coordinate.

If a controllable pitch propeller is to be used, this matching process extends to a choice of the combinations of pitch, torque, and rev/min for all points of possible vessel operation.

There may be auxiliary loads in addition to the main propulsion load. A common example is a generator for ship-service electrical power driven by the propulsion engine. Its alignment and vibratory problems must also be analyzed, and so must its effect on the matching of main load and engine. For example, if such a generator must be run at constant speed—as would be expected with alternating current power—it compels use of the controllable pitch propeller, with pitch-torque-rev/min relations chosen so that rpm can be constant.

All machinery must be properly controlled. It is likely that the engine will be provided by its builder with its own individual controls, such as governor, starting sequencer, and overspeed trip, as a minimum in the case of the gas-turbine engine. But the engine control must be subservient to the system that controls the propulsion plant as a whole. When a controllable pitch propeller is used, control of pitch and control of engine fuel must be integrated. Rapid transients, such as crash reversals, must always be allowed for, and except in the smallest vessels, the response to transients of all components in the propulsion system must follow automatically from a single human command. A system using a reverse-reduction gear with shaft brake provides an example: engine fuel, air to reverse and ahead clutches, and to the brake, must be coordinated so that the engine is neither oversped nor stalled, only one clutch is engaged at the same time, the brake is engaged only when the clutches are not, and excessive torque is not applied to clutches, while overall the sequence of events occurs for these components in a way that minimizes vessel stopping time.

Control from remote stations, such as the vessel's bridge, is likely to be needed. And closely related to control is monitoring. In large installations, a complex array of indicators and alarms may be needed to assist in operation of the propulsion plant and its auxiliaries.

Firm support from the hull by a foundation seems to be an obvious and simple necessity, but simplicity must encompass several considerations. In addition to supporting mere dead weight, the foundation must resist such loads as the thrust, torque, and dynamic forces and torques from several possible sources. It must provide the basis for alignment and maintain it, yet must allow the engine to deflect due to thermal expansion without distortion. Isolation of structure-borne noise and vibration may also be an element of foundation design. In some instances, the builders of gas turbines furnish them mounted on a base in such a way that thermal deflections are accommodated by the support from this base. The base in turn is mounted rigidly on the ship-furnished foundation. If a regenerator is used, it too may require a foundation, since these components may be too large for support directly by the engine.

The engine must be supplied with certain necessities. Fuel and air are fundamental inputs, of course, to any type of combustion engine.

Some fuels require heating to reduce their viscosity; some require treatment—cleaning, usually—to enhance other properties. Cleanliness of fuel is of great importance to gas turbines, so that designing of fuel systems to meticulously avoid contamination, and to provide treatment service, if necessary, is one of the prime concerns of the marine designer in working with the gas turbine engine.

Air supply for combustion is one of the major tasks in design of a gas turbine propulsion plant, much more so than with any other type. The reason is the high air-consumption rate of gas turbines, coupled with their sensitivity to pressure loss in intake ducts and to ingestion of salts. The air must approach the engine with uniform circumferential pressure distribution, lest uneven flow into the compressor cause destructive vibration of its blades. A model test of inlet ducting appears to be an advisable design step to check for uniform inlet flow.

All engines require cooling, usually for several different purposes, so that air and water must be supplied in adequate quantities. The gas turbine requires cooling for its lubricating oil, usually by seawater in a marine installation. The space surrounding the engine must be ventilated to keep this space from becoming intolerably hot.

All engines require exhaust ducting, with provisions for thermal expansion, cooling or insulation, silencing, low pressure loss, and avoidance of hazard or nuisance outside the hull being the usual considerations. The problems are perhaps more difficult, or require greater care, in the case of the gas turbine engine, because of its comparatively large volume of exhaust gas.

Silencing is usually required with any type of internal combustion engine. With the gas turbine, sound absorbing surfaces will almost certainly be needed in inlet and exhaust paths, and the designer may have to provide for sound barriers between the engine and its crew.

Maintenance provisions often require attention from the marine designer. In the case of the aircraft-derivative gas turbine, for example, it is accepted practice to replace the gas-generator section of the engine for maintenance or repair ashore. The arrangement of machinery, structure, inlet ducting (typically, the gas generator is removed via the inlet duct), and handling equipment are, therefore, important items for the designer's attention.

These then are the several important engine-related items that in practically all cases must receive attention in the design of a successful propulsion plant. Special cases additionally have their problems. For example, a combat vessel is likely to require analysis of the deflection of the engine caused by shock impulses delivered via the shaft and foundation. The designer of the propulsion plant always faces a major engineering task, even though the engine itself may be furnished as a package. The principal theme of this text is education of the marine engineering student in this task, as it relates to gas-turbine propulsion. The topics introduced in this section, therefore, are explored in detail in the subsequent chapters.

1.5. TYPICAL VESSELS POWERED BY GAS-TURBINE ENGINES

The marine gas-turbine engine has been found to be adaptable to a wide range of vessels, from modest-sized pleasure craft requiring several hundred horsepower, to major naval vessels and the largest merchant ships, both of which may require from 20,000 to 50,000 horsepower per shaft. Given the relative newness of the gas turbine as a marine engine, many of the installations are experimental, in part or totally, but others have been chosen for regular service in full competition with possible alternative power plants. And, in addition to conventional propulsion service, the gas turbine is used for shipboard auxiliary service and for other nonpropulsion marine duty such as portable floating power plants and power generation aboard offshore oil industry equipment.

Here I do not attempt to catalog all of the many individual marine applications of the gas turbine, for doubtless the listing would quickly become obsolete; hopefully the most interesting applications still lie ahead as this is written in 1974. What follows here is a sampling, intended to give the reader an idea of what applications have been made. Brief descriptions are given, accompanied in most cases by an illustration or reference to an appropriate illustration in a later chapter.

A. Merchant Ships

Figure 1.9 is a view of the *Admiral Wm. M. Callaghan,* a roll-on/roll-off cargo ship for the U.S. Military Sealift Command, delivered by Sun Shipbuilding of Chester, Pennsylvania, in 1967. The ship is 694 ft in overall length, 7000 deadweight tons, and has a speed of 26 knots [Caldera et al. (1967)]. The propulsion plant is twin screw with an aircraft-derived gas-turbine engine on each shaft. Originally the engines were Turbo Power & Marine FT4 rated at 25,000 hp maximum continuous, 20,000 hp normal. They were subsequently replaced by General Electric LM2500 engines.

The four-ship *Euroliner* class of Seatrain Lines constitutes the first North

Figure 1.9. Roll-on/Roll-off ship *Admiral Wm. M. Callaghan.*

Atlantic container ships with gas-turbine propulsion. The ships were built by Rheinstahl Nordseewerke of Emden, Germany. Delivery of the first of the class occurred in 1971, with the others following at intervals of several months. These vessels are 799 ft in overall length, 27,984 deadweight tons, and are designed for a speed of 26.4 knots. They are twin-screw, with one Turbo Power & Marine Systems FT4A-12 engine on each shaft. Maximum continuous power rating is 30,200 hp per shaft [McAllister (1972)]. Like the *Callaghan,* the *Euroliner* vessels are intended for fast Europe-North America service. An inboard profile and the machinery arrangement are shown in Figure 10.4.

Figure 1.10 shows the *Iron Monarch,* a roll-on/roll-off cargo ship of the Broken Hill Proprietary Company of Australia, and delivered by Whyalla Shipbuilding and Engineering Company of that country in late 1973. The ship is 179.3 m long overall, 14,400 tons deadweight, and has a design speed of 20 knots. The propulsion plant is single screw, powered by a General Electric heavy-duty MS5002 engine rated at 17,500 shp normal, 19,000 shp maximum continuous. A sister ship, the *Iron Duke,* is under construction as this is being written. A unique feature of the propulsion plants is a split exhaust system that makes available space for a low vehicle deck in way of the machinery spaces. This feature is evident in Figure 1.10 in the twin stacks placed well outboard, and is shown more directly by a cross section a the machinery, Figure 1.11 [White (1973)].

Figure 1.10. Broken Hill Proprietary Steel Products carrier *Iron Monarch*.

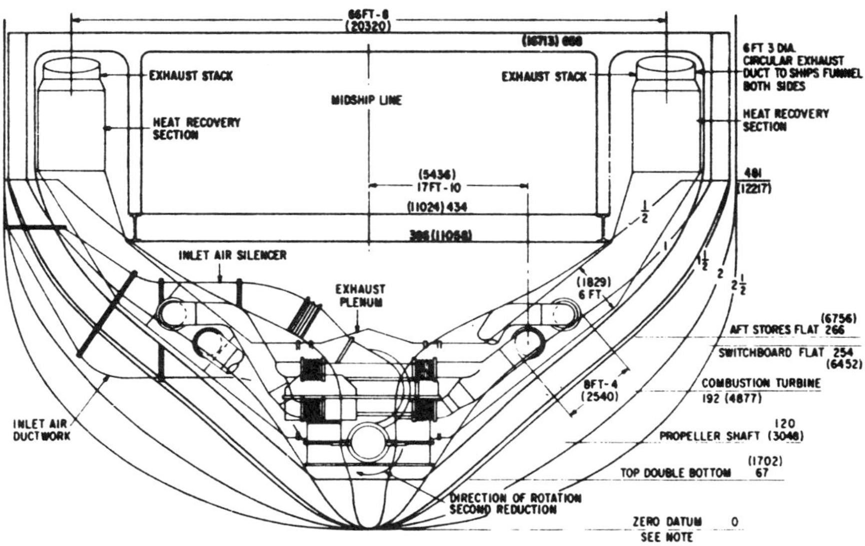

Figure 1.11. Sectional view of machinery layout in *Iron Monarch* [White (1973)].

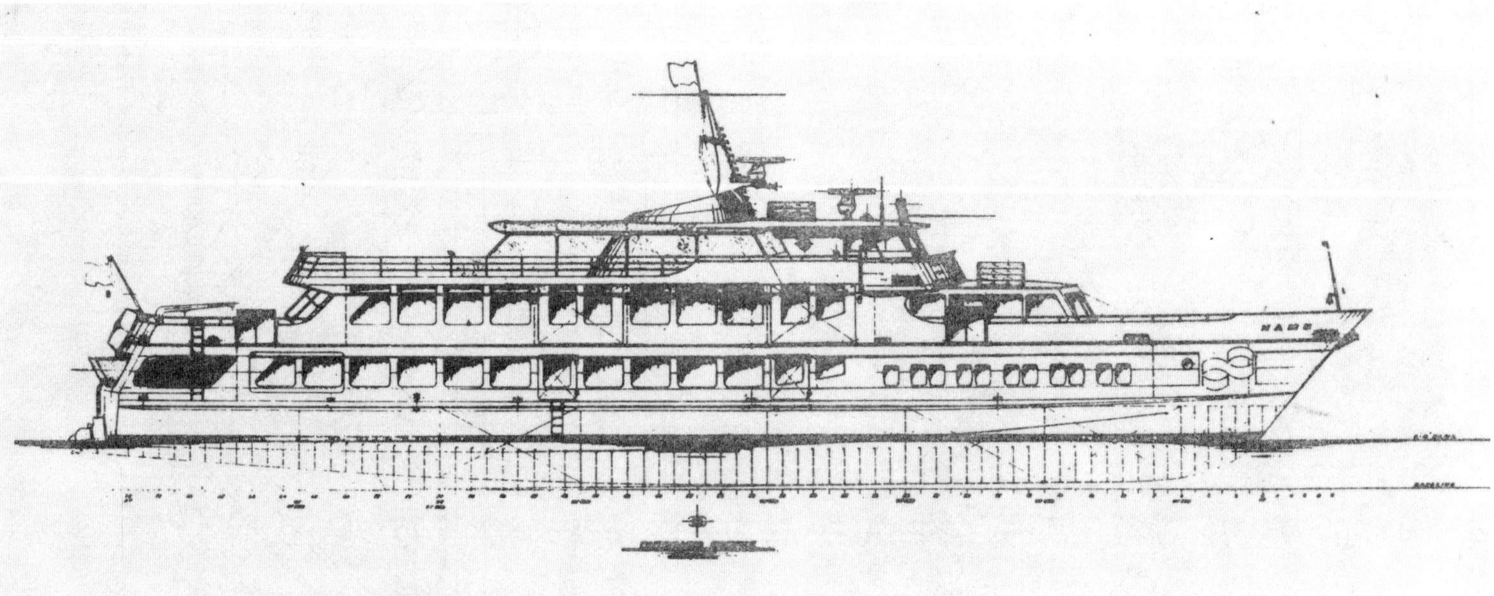

Figure 1.12. High-speed ferry for San Francisco Bay, by Campbell Industries.

Also under consideration as this is written is a class of 35,000 ton deadweight tankers for the Chevron Oil Company. They are being built at Portland, Oregon, by FMC Corporation, and are powered by General Electric heavy-duty engines, model MS3002. Propulsion is single screw, 12,500 shp, with alternate-current synchronous electric transmission. The gas-turbine propulsion engine and the generator that it drives is mounted on the main deck, with the propulsion motor directly below in what is usually considered the normal position for the driving machine. Although no photographs were available in time to be included here, this unusual machinery arrangement can be seen via an inboard profile of part of the hull, Figure 10.2.

A merchant vessel that shows major contrast to the large ocean-going ships just described appears in Figure 1.12 [*Marine Reporter and Engineering News* (1972)]. This is a passenger ferry for San Francisco Bay (Golden Gate Bridge Highway and Tunnel District), from a class of three vessels built by Campbell Industries of San Diego. They are 164 ft in overall length, displace 252 tons, and carry 750 passengers at 25 knots. Propulsion is by three Avco-Lycoming TF35 engines, each driving a Jacuzzi 36 in. water jet pump. Rated shaft horsepower is 7500 total per vessel. The three engines are arranged side-by-side in an extreme aft compartment, an arrangement indicated by the inboard profile in Figure 10.6.

B. United States Coast Guard Vessels

The United States Coast Guard uses gas-turbine propulsion in several classes of vessels, in most instances as boost engines combined with base-load diesels.

Figure 1.13. U.S.C.G. Hamilton (WPG 715).

The outstanding example of vessels in service as this is written is the nine-vessel *Hamilton* (WPG-15 et seq.) class cutters. See Figure 1.13. This type is 378 ft overall length, displace 2778 tons, cruise at 20 knots on the diesel engines, and have a top speed of 29 knots when driven by the gas-turbine engines. The propulsion plant is twin screw; each shaft is powered by a diesel engine providing 3500 shp, and by a Turbo Power & Marine FT4A-6 aircraft-derived gas turbine providing 18,000 shp. Each engine is connected to the reduction gear by a clutch so that the nonoperative engine can be disengaged [Wennberg (1966)].

Scheduled for delivery to the Coast Guard in 1975 is a large icebreaker, the *Polar Star* (from Lockheed Shipbuilding of Seattle). See Figure 1.14. The vessel is 399 ft long overall, displaces 10,863 tons, and is powered by a triple-screw propulsion plant that uses either diesel-electric base-load power or mechanical-drive gas-turbine boost power. This unique scheme is sketched in Figure 1.15. The six propulsion diesel engines are each rated at 3500 bhp; they drive alternating-current generators with this power being rectified for use by the three direct-current propulsion motors that provide 6000 shp each. The boost-power plant consists of a Turbo Power & Marine FT4A-12 gas turbine on each shaft, driving mechanically through a clutch and reduction gear. The engines are rated 20,000 shp normal, 25,000 shp maximum continuous. During diesel operation, the gas turbine is declutched; during gas turbine operation, the propulsion motors are deenergized and simply rotate as part of the propulsion shafting [Finizio (1973)].

Figure 1.14. U.S.C.G. Polar Star.

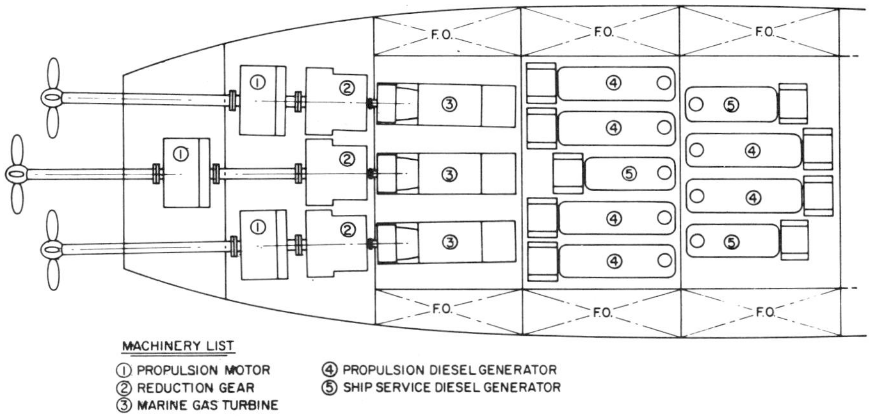

Figure 1.15. Layout of propulsion machinery, U.S.C.G. *Polar Star* [Finizio (1973)].

C. Naval Vessels

The light weight of the aircraft-derived gas turbine engine has given it great appeal to the various navies, and numerous vessels and classes of vessels are so powered. Typically, the first applications were in combined power plants where the gas turbine served only as a boost engine, with later development toward use of the gas turbine for all propulsion.

An example of the combined-plant concept is shown in Figure 1.16, this being the PG-84 (U.S.S. *Asheville*) of the U. S. Navy, the first of a class of seventeen aluminum-hull gunboats, 165 ft overall length and 240 tons displacement. The twin-screw propulsion plant is powered in the cruise mode by one diesel on each shaft of 725 bhp continuous rating, or by a single General Electric LM1500 gas turbine rated at 14,000 bhp. The arrangement is unusual since it is rare to find a single engine driving both screws of a twin-screw plant [Hauschildt and Miller (1967)].

The first U. S. Navy class of major warships with base-load gas-turbine propulsion is the DD963 (U. S. S. *Spruance*) class of destroyers, Figure 1.17. (The figure shows the first two ships of the class under construction at Ingalls Shipbuilding, Litton Industries, Pascagoula, Mississippi.) Propulsion is twin-screw with two General Electric LM2500 engines geared to each shaft, as shown in Figure 1.20. Each engine is rated 21,500 bhp, giving a total of about 80,000 shp for the ship. See also Figure 10.14 for an illustration of the propulsion plant. Although the plant is obviously all gas turbine, it has some of the features of a combined plant in that cruising power is obtained from one engine per shaft. This scheme allows cruising power to be had at reasonable

Figure 1.16. U.S. Navy gunboat PG 84 (Asheville).

Figure 1.17. DD 963 (Spruance) class destroyers of the U.S. Navy under construction at Ingalls Shipbuilding.

fuel economy, since the on-line engine can operate close to conditions of best efficiency. It is reminiscent of the once-popular naval steam propulsion concept of a "cruising turbine," a small steam turbine than ran at its rated point to provide the low power needed for cruising, and at an efficiency that the main propulsion could not provide at part load.

Figure 1.18 shows an experimental hydrofoil vessel of the U. S. Navy, the U. S. S. *Plainview,* built by Lockheed Shipbuilding in 1965. The foil-borne propulsion plant is twin screw on General Electric LM1500 (nominally 15, 000 hp) gas-turbine engine per shaft.

Several smaller U. S. Navy craft are pictured in Section D. following.

The naval use of gas-turbine propulsion was pioneered in several Western navies, especially the British. The Royal Navy experimented extensively with industrial and aircraft gas turbines throughout the 1960s, using them combined with diesel, with steam turbines, and alone. Vessels ranged from small patrol craft to frigates. By the end of the decade the gas turbine had been adopted as the sole propulsion engine for the new classes of warships where it might reasonably be applied. In late 1969, over 600,000 hp of propulsion engines had been employed by that navy [Morgan et al. (1971)]. Several papers on the British experiences have been published in the United States and contribute material to subsequent chapters.

D. Small Craft

Gas turbines have been applied to a number of small craft in naval, commercial, and pleasure services. In the United States this particular field has

Figure 1.18. U.S. Navy experimental hydrofoil craft *Plainview*.

been dominated by experimental installations, with no outstanding examples up to 1974 of regular-service classes. The experiments, though, have been extensive.

Figure 1.19 shows one of several small U. S. Navy craft that were given a trial in Viet Nam. The one pictured is jet propelled with twin Avco Lycoming engines. Figure 10.8 gives a view of the propulsion machinery arrangement.

Figure 1.20 shows the Bell Aircraft SK-5 surface-effect vehicle, which is also a military craft, used by both U. S. Navy and Army in Vietnam. Power is by General Electric LM100 engines.

Several small pleasure craft are shown in later chapters. Figure 9.18 is the inboard profile of an experimental motorboat powered by a single United Aircraft of Canada ST6 engine. Figures 9.20 and 9.21 show a Pacemaker motor cruiser, powered by twin Ford engines, that was an experimental venture, but which was also offered to regular customers for a while in the early 1970s.

The gas turbine has made essentially no penetration of the small commercial field, though several experimental craft can be noted. A small river towboat was built in the mid 1960s for NASA, using twin Solar engines (about 1000 hp) with electric-power transmission. In the early 1970s an oil industry crewboat was built for Gulf of Mexico service with twin Ford engines. Neither of

Figure 1.19. U.S. Navy jet gunboat; Avco-Lycoming power.

Figure 1.20. Bell Aircraft SK-5 surface-effect vessel.

these boats was successful enough to attract duplication. Both were powered by industrial engines. One might well suggest, however, that the most likely market for small commercial application is in high-speed craft, where the high power density of an aircraft-type engine would show more advantage.

E. Ocean Industry Uses

Chapter 13 discusses the use of the gas turbine in marine applications other than conventional ship and boat power plants. One potential, but as yet unexploited, application of this nature is for undersea power, either for propulsion of vehicles or to supply the energy needs of underwater habitations and mining installations. All such applications would require closed-cycle operation and a heat source independent of the atmosphere. Although high energy density and endurance are attractive features, actual application apparently must await a greater need for such power than has appeared up to 1974.

A nonpropulsive marine use that has arisen for the gas turbine is in meeting the needs of the offshore oil industry for compact sources of power for pumping, electric-power generation, and gas-compressor drive. Although the competing diesel engine is somewhat superior in fuel economy, the lightness and compactness of the gas turbine is an advantage, since the cost of offshore installations is sensitive to their size and to the weight that they must support. A rough and unofficial count in 1974 indicates that from several hundred up to a thousand gas turbines may be in use by the offshore oil industry worldwide.

An ocean industry power application that is unique at this time is a pipe burying barge (Brown & Root) that applies two Turbo Power & Marine FT4 engines to drive its jetting pumps. Since approximately 20,000 hp per pump is needed, the use of the gas turbines allows a significant reduction in barge size, as compared to diesel. This barge is pictured in Figure 13.8.

1.6. REFERENCES

Caldera, D. L., Hoch, C. E., and Swensson, G. C. (1967), "Gas Turbine Propulsion Machinery for the MSTS Roll-On/Roll-Off Ship," Society of Naval Architects and Marine Engineers, New York Metropolitan Section Meeting.

Finizio, A. F., (1973), "USCG *Polar Star*—Combination of a Diesel Electric Propulsion Plant with a Geared Gas Turbine Boost Plant," American Society of Mechanical Engineers Paper 73-GT-73.

Gillmer, T. (1970), *Modern Ship Design,* United States Naval Institute.

Hauschildt, Maurice R., and Miller, Charles L. (1967), "U. S. Navy PG-84 Class CODOG Propulsion Plant," Society of Automotive Engineers Paper 670952.

Kane, J. R. (1971), "General Considerations in Marine Engineering," Chapter I of *Marine Engineering,* Society of Naval Architects and Marine Engineers, New York.

Maritime Reporter and Engineering News, November 15, 1972.

McAllister, P. J., (1972), "Modularization and Installation of the Gas Turbine Propulsion System in *Euroliner,*" American Society of Mechanical Engineers, Paper 72-GT-85.

Morgan, S. G., Smith, A. J. R., and Lamport, A. W. (1971), "Gas Turbines in the Royal Navy 1965–1969," *Naval Engineers Journal,* **83,** 1.

Wennberg, P. K., 1966, "The Design of the Main Propulsion Machinery Plant Installed in the USCGC *Hamilton* (WPG-715)," *Transactions, Society of Naval Architects and Marine Engineers,* **74,** pages 29–69.

White, A. O. (1971), "Gas Turbines," Chapter VI of *Marine Engineering,* Society of Naval Architects and Marine Engineers, New York.

White, A. O., (1973), "Heavy-Duty Gas Turbine Experience on the Broken Hill Proprietary, Co., Ltd's Roll-On/Roll-Off Vessels," ISME meeting, Marine Engineering Society in Japan, Tokyo.

Chapter

TWO

CYCLE ANALYSIS

2.1. INTRODUCTION

Cycle analysis is an exercise in observing the operation of the first and second laws of thermodynamics on a substance as it is put through a closed set of processes intended to interchange thermal and mechanical energy. It is a topic thoroughly covered in textbooks on engineering thermodynamics [e.g., Van Wylen (1965)], and readers are assumed to be generally familiar with such treatments and with the accompanying terminology (e.g., enthalpy, entropy, perfect gas, etc.).

In particular terms, suited to the present purposes, cycle analysis consists of following the changes in properties of the working fluid—air—as it flows through a gas turbine engine, accounting at each step for work done and for heat transferred. After a number of processes that bring the air back to its initial conditions, a summation of work and heat transfer develops the ratio of net work to heat addition, which is the efficiency of the cycle. A look at the net work per unit of working fluid shows how much of the latter must be handled to give a specified power output. Cycle analysis thus indicates the values of two parameters of great interest to the designer of a propulsion plant, namely

1. Efficiency, or fuel rate (lb/hp-hr), which is directly calculable from efficiency, and
2. Air rate (lb/hp-hr).

The first of these is the prime information needed to calculate fuel cost and the quantity of fuel that a vessel must carry. The second is the prime information needed in designing adequate inlet and outlet ducting.

Note the use of the phrase "indicates the value" in the preceding paragraph. The implication is that exact values are not obtained because there are likely to be at least some simplifications in going from the processes that actually occur to those used for calculation purposes in the analysis. Quite typically, the analysis is performed with all irreversibilities, such as those caused by component inefficiencies and by degradation in state between components, omitted from the calculation. The cycle is then referred to as an *ideal cycle.* Often, however, the analysis is modified to incorporate losses that actually occur in compressor, combustion chamber, and turbine, and in the flow between components. Obviously the second approach provides answers that are closer to values of efficiency and air rate of the actual engine.

Even though a cycle analysis may not give an accurate prediction of actual performance values—especially if it is an ideal cycle—the results are often useful for studying trends. For example, if the effect of increasing pressure ratio were to be considered, the effect might be predicted by calculating the change in ideal cycle parameters produced by the same pressure ratio increase. However, ideal cycle predictions can sometimes be unrealistic. A case of this occurrence is noted later.

The designer of a propulsion plant usually does not find it necessary to make cycle analyses when he is dealing with the gas turbine. The reason is that he has no control over internal parameters, which are totally established by the engine builder. Nonetheless, an understanding of how these parameters affect things that the designer is vitally concerned with, such as the efficiency and the air rate, seems quite necessary for his total comprehension. Hence this chapter.

One last advice before plunging in. For accuracy in making cycle analyses, one would normally take values of the properties of air from tables, such as those of Keenan and Kaye (1948). An alternative, giving slightly less accuracy, is to calculate values by perfect gas relationships. Since this chapter aims at showing trends and generalized results, the following calculations use the perfect gas techniques unless stated otherwise at the time.

2.2. THE SIMPLE CYCLE

A *simple cycle* consists of one process each of compression, heat addition, work extraction, and cooling. It is a cycle uncomplicated by regeneration, intercooling, and reheating.

The ideal form of the simple cycle is known as the *Brayton cycle,* consisting of (1) a reversible adiabatic (isentropic) compression, (2) a constant pressure addition of heat, (3) a reversible adiabatic expansion, and (4) a constant pressure cooling to the initial state. In practice, the cooling is accomplished either by passing the air through a heat exchanger, comparable to the condenser of a steam plant, or by exhausting the hot air continuously and replacing it continuously with fresh air at ambient conditions. The latter scheme is the only one that has been used in ship propulsion. In fact, the former is rare no matter what the application. It is likely to be found only when the heat source for the cycle is a nuclear reactor, and there only because the air may be radioactive and hence unsuitable for release to the atmosphere. However, the method used to remove heat, or to add it, for that matter, has no effect on the analysis of the cycle.

The four processes of the Brayton cycle are shown by the lines 1–2, 2–3, 3–4, and 4–1 on both temperature-entropy and pressure-volume planes in Figure 2.1.

The efficiency of the cycle is its ratio of net work to heat added, where net work is the difference between work obtained from the expansion process, 3–4, and the compression process, 1–2. Work processes and heat-transfer processes are evaluated by the differences in enthalpies between their end points, for one pound of working fluid, so that the formulation is

$$W_N = (h_3 - h_4) - (h_2 - h_1)$$

$$Q_A = h_3 - h_2$$

$$\eta = \frac{(h_3 - h_4) - (h_2 - h_1)}{h_3 - h_2} \tag{2.1}$$

where

h = enthalpy, Btu/lb

W_N = net work, Btu/lb

Q_A = heat added, Btu/lb

η = efficiency

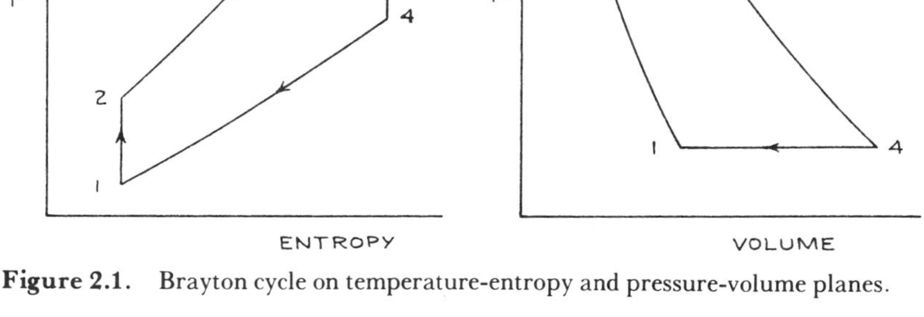

Figure 2.1. Brayton cycle on temperature-entropy and pressure-volume planes.

The air rate (air flow rate per unit of net output) is

$$w_A = \frac{2544}{(h_3 - h_4) - (h_2 - h_1)} = \frac{2544}{W_N}\,\text{lb/hp-hr} \tag{2.2}$$

where 2544 is the conversion factor in Btu/hp-hr. If a heating value of the fuel being used is known, efficiency can be converted directly into fuel rate (also called specific fuel consumption) by

$$w_F = \frac{2544}{Q_F} \tag{2.3}$$

where Q_F is the heating value in Btu/lb.

If perfect gas behavior of the air is assumed, as is done throughout this chapter, the efficiency formula is readily put into terms of temperature. It is only necessary to note that under perfect gas conditions, enthalpy is proportional to temperature. If a substitution of the temperature-enthalpy relation is made in Eq. 2.1, the constant of proportionality (constant pressure specific heat) cancels out, leaving

$$\eta = \frac{(T_3 - T_2) - (T_2 - T_1)}{T_3 - T_2} \tag{2.4}$$

The efficiency formula can also be put into terms of pressure ratio (i.e., P_2/P_1 or P_3/P_4, P being pressure, of course) by making use of the temperature-pressure relationship along an isentropic path, namely

$$T \propto R^{(k-1)/k} \tag{2.5}$$

where

R = pressure ratio

k = ratio of specific heats, C_p/C_v

C_p = constant pressure specific heat

C_v = constant volume specific heat

and the relationship along a constant pressure path, namely

$$T \propto v \tag{2.6}$$

where v = specific volume in ft^3/lb. The technique is to put all temperatures in terms of T_2, using Eqs. 2.5 and 2.6, as follows:

$$T_1 = T_2 R^{(1-k)/k}$$

$$T_3 = T_2 \frac{v_3}{v_2}$$

$$T_4 = T_3 R^{(1-k)/k} = T_2 (v_3/v_2) R^{(1-k)/k}$$

When these are substituted into Eq. 2.4, the result is

$$\eta = 1 - R^{(1-k)/k} \tag{2.7}$$

Temperatures drop out completely, showing that efficiency of the Brayton cycle is a function of pressure ratio only, and that the higher this ratio, the higher is the efficiency.

Here is the case of a misleading result of an ideal cycle analysis mentioned in the preceding section, for in actuality, efficiency of a simple-cycle gas turbine engine is very much a function of T_3. The explanation for the discrepancy lies in the influence of component efficiencies: even though these efficiencies may be quite high, they affect cycle efficiency significantly, and their neglect in the ideal cycle analysis makes a major difference in the result. This point is illustrated by a numerical example, comparing two cycles having $T_3 =$ 900 F and $T_3 = 1600$ F, respectively, both having turbine efficiency of 85 %, and compressor efficiency of 85%. The effect on the cycle of these efficiencies is shown by Figure 2.2, which is Figure 2.1 suitably modified by the increases in entropy that take place along the paths representing compression and expansion. The changes are most evident, of course, on the temperature-entropy plane. Equations 2.1 through 2.4 are still applicable if T_2, h_2, T_4, and h_4, are replaced by T_2', h_2', T_4', and h_4'. The calculations are tabulated, and further details may be gleaned from the table.

	Units	*Cycle A*	*Cycle B*	*Source*
T_3	°F	900	1600	Specified
T_3	°R	1360	2060	$T_3 + 460$
R	—	6	6	Specified
T_1	°R	535	535	Specified
T_2	°R	894	894	$535 \times R^{(k-1)/k}$
T_2'	°R	957	957	$T_1 + (T_2 - T_1)/0.85$
T_4	°R	814	1234	$T_3/R^{(k-1)/k}$
T_4'	°R	896	1358	$T_3 - 0.85\,(T_3 - T_4)$
W_T	BTU/lb	111	168	$C_P\,(T_3 - T_4')$
W_C	BTU/lb	101	101	$C_P\,(T_2' - T_1)$
W_N	BTU/lb	10	67	$W_T - W_C$
Q_A	BTU/lb	97	265	$C_P\,(T_3 - T_2')$
η	—	0.103	0.253	W_N/Q_A
w_A	lb/hp-hr	252	38	Eq. (2.2)
w_F	lb/hp-hr	1.35	0.55	Eq. (2.3)

Values of $C_p = 0.24$ Btu/lb-°F and $Q_A = 18{,}000$ Btu/lb are used.

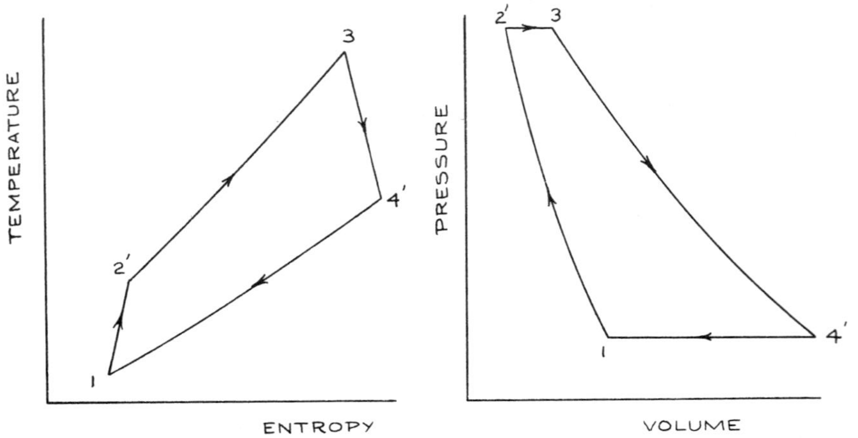

Figure 2.2. Figure 2.1 modified to show effect on cycle of turbine and compressor efficiencies.

The strong dependence on T_3, the highest temperature in the cycle, is clearly shown. If the calculations are repeated for a number of combinations of T_3 and R, the pattern depicted by Figure 2.3 appears. It is evident that efficiency is increased by higher temperature at any pressure ratio, and that at a given temperature, there is an optimum pressure ratio.

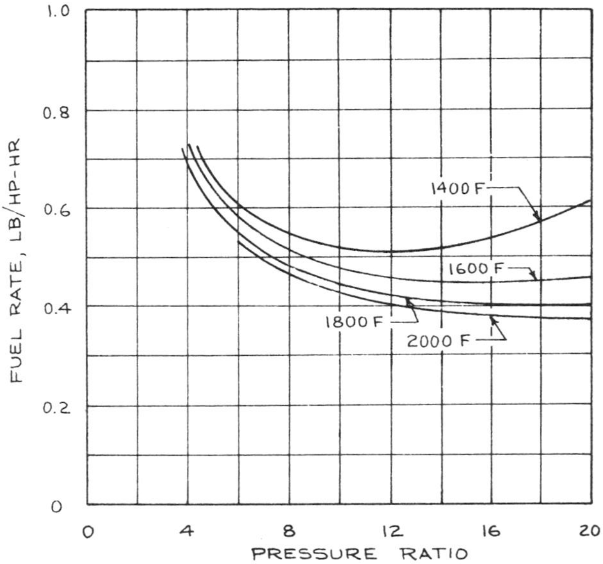

Figure 2.3. Fuel rate as a function of pressure ratio.

The analysis of the simple cycle also illuminates the high air throughput that is a feature of gas turbine engines. The example preceding has shown an air rate of 38 lb/hp-hr for the cycle with the realistic value of T_3 (cycle B). Among other types of engines, the highest air rate is likely to be found among the slow-speed diesels. These big engines typically have a full-load air/fuel ratio of about 40/1 and a fuel rate of 0.40 lb/hp-hr (or better). Combining these two figures gives an air rate of 16 lb/hp-hr. It is thus obvious that the gas turbine must be supplied with considerably more air. Note, though, that the example has also shown that air rate declines with increasing efficiency, so that increases in efficiency that have characterized the early years of marine gas turbines have this beneficial side effect.

A further lesson from the cycle analysis is the demonstration of sensitivity to pressure losses, this demonstration being accomplished by minor manipulations of the figures already set down. Suppose, for example, that there is a 5% loss in pressure as the air flows through the combustion chambers and exhaust duct, so that the pressure ratio across the turbine in the preceding example is 5.7 rather than 6. The turbine work must obviously be less, and if the calculations for the example cycle B are adjusted by using the new turbine pressure ratio, it is seen that cycle efficiency is reduced by about 5½%. The loss in gross output, that is, turbine work, is only about 2½%. Now, if this were a steam plant (Rankine cycle), the pump work would be such a small fraction of the gross work that loss of gross work and loss of cycle efficiency would be essentially the same. With the gas turbine, the pump work is such a large fraction of the gross (about 0.60 in the example here) that a small loss in gross shows up as a much larger loss in net. It is the "small difference between large numbers" phenomenon. This sensitivity to pressure loss is a further reason for the large duct areas necessary for inlet and exhaust to gas turbine engines.

The relatively small difference between turbine work and compressor work also makes it apparent that high component efficiencies are essential for a successful gas turbine. If, for example, the 2½% loss in turbine output occurred because of a degradation in its efficiency, the 5½% loss in cycle efficiency would likewise occur. And as a further example, if turbine and compressor efficiencies were both 0.75 rather than the 0.85 used in the example, the cycle B efficiency would be only about 0.125, or roughly one half of the value found previously.

2.3. THE REGENERATIVE CYCLE

The exhaust from a gas turbine is usually at a higher temperature than the compressed air entering the combustion chamber. The opportunity thus exists for a reduction in the fuel input required, since the heat being exhausted can be used (at the cost of a heat exchanger and connecting ducting) to provide part of the heat input, with consequent increase in efficiency. Looking at the simple cycle diagram, either Figure 2.1 or 2.2, the opportunity is evident, as is illustrated more directly in Figure 2.4. This figure illustrates the Brayton cycle modified into the *regenerative cycle.*

The efficiency formulas, Eqs. 2.1 and 2.4, are modified for this cycle by changing the heat added (denominator) to $h_3 - h_x$ or $T_3 - T_x$. Since an ideal counterflow heat exchanger should be assumed with this ideal cycle, $h_x = h_4$ or $T_x = T_4$, so that the efficiency formulas become

$$\eta = 1 - \frac{h_2 - h_1}{h_3 - h_4} \tag{2.8}$$

$$\eta = 1 - \frac{T_2 - T_1}{T_3 - T_4} \tag{2.9}$$

If the process used in deriving Eq. 2.5 is applied to Eq. 2.9 in an attempt to show efficiency in terms of pressure ratio, the result is

$$\eta = 1 - \frac{T_1}{T_3} R^{(k-1)/k} \tag{2.10}$$

In this case the effect of temperature is shown, and it is seen that high temperature benefits efficiency, while high pressure ratio is predicted to be detrimental to efficiency. These predictions are approximately in accord with reality, but are not to be depended on explicitly. For example, note that $R = 0$ would produce an efficiency of 1.00 (if you believe the formula without a bit of skepticism), a trivial result, since R must be finite to cause the working fluid to flow through the machine.

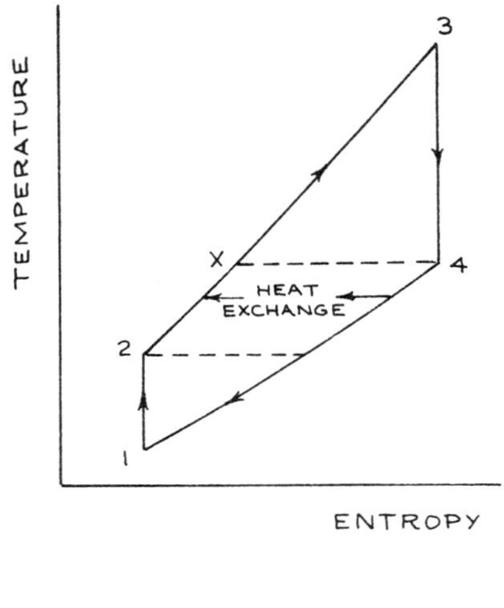

Figure 2.4. Regeneration added to Brayton cycle.

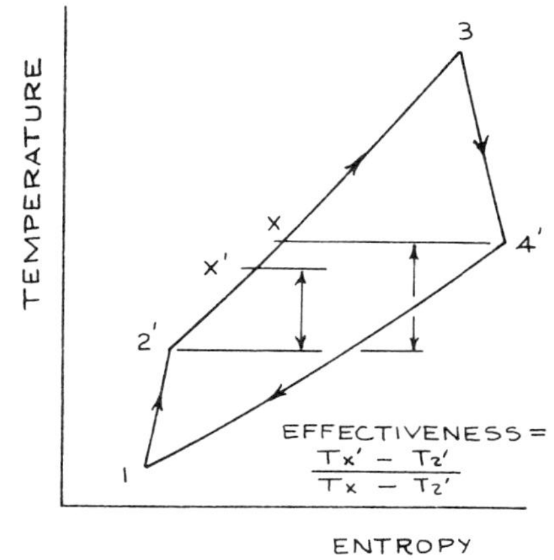

Figure 2.5. Defining effectiveness.

The efficiency analysis can be rendered more realistic if component efficiencies and heat exchanger (the regenerator) effectiveness are incorporated into the analysis. The effectiveness is the fraction of heat available for transfer that is actually transferred. Figure 2.5 demonstrates the definition in terms of a formula. Efficiency formulas 2.8 and 2.9 cannot be used, but are easily modified to suit. Equation 2.8 becomes

$$\eta = \frac{(T_3 - T_4') - (T_2' - T_1)}{T_3 - T_x'} \tag{2.11}$$

A numerical example follows. R, T_1, T_2, T_2', and compressor and turbine efficiencies are the same as in the previous example. T_3 is 1250 F (1710 R), and a regenerator of 0.80 effectiveness has been added. The essential calculational steps are summarized in the table following.

	Units	*Value*	*Source*
T_3	°R	1710	Specified
T_2'	°R	957	Earlier example
T_4	°R	1023	$T_3/R^{(k-1)/k}$
T_4'	°R	1126	$T_3 - 0.85\ (T_3 - T_4)$
T_x'	°R	1092	$T_x' - T_2'/T_4' - T_2'$
ΔT_A	°R	618	$T_3 - T_x'$
ΔT_W	°R	162	$(T_3 - T_4') - (T_2' - T_1)$
η		0.262	$\Delta T_W/\Delta T_A$

Efficiency is roughly the same as before (cycle B), although the highest temperature is 350° less, thus illustrating the ability of regeneration to increase efficiency.

2.4. INTERCOOLING AND REHEAT

Figure 2.6 pictures a Brayton cycle that has been modified by addition of intercooling to the compression process and reheat to the expansion process.

The intercooling step consists of a constant-pressure cooling process interposed between two compression processes. Its purpose is to reduce the work required to compress through a given pressure ratio. This it does, both in the ideal process and in a realistic compression process with losses. A negative result, however, is that the air is cooler at the end of the compression process and consequently requires more addition of heat to reach a required turbine-inlet temperature. The net result is an efficiency less than that of the simple cycle.

The reheat step consists of a constant-pressure heating interposed between two expansion processes. Its purpose is to increase the work done by expansion through a given pressure ratio. The consequences are parallel to those of intercooling: the work done is indeed increased, both in the ideal process and in a realistic one, but enough extra heat input is required that cycle efficiency is worsened.

Intercooling produces a cooler air leaving the compressor; reheating produces hotter air leaving the turbine. If regeneration is used with these addi-

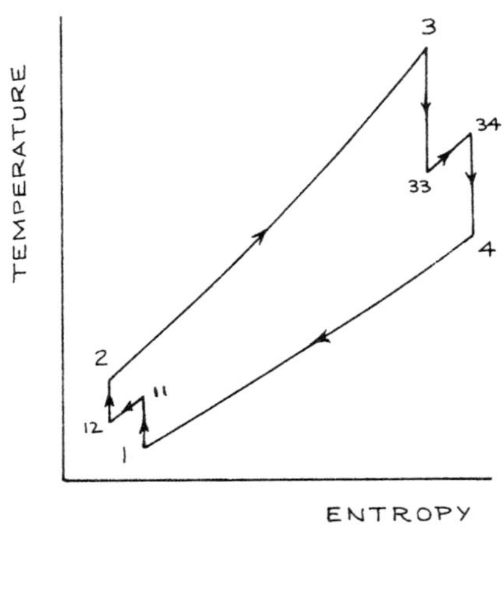

Figure 2.6. Brayton cycle with addition of intercooling and reheat.

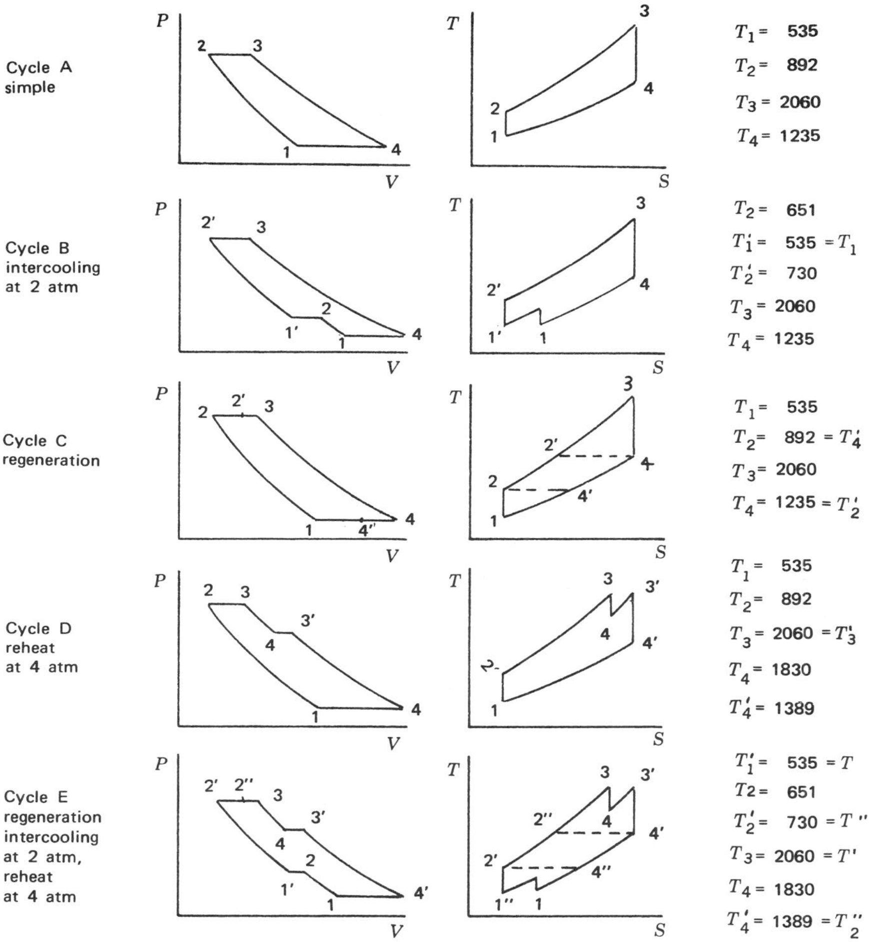

Table 2.1.

tions, a greater temperature difference is therefore available, and more heat can be transferred in the regenerative heat exchanger. The net result may be an increase in cycle efficiency; that is, the efficiency of a regenerative cycle may be increased by either intercooling or reheating, or by both.

Ideal cycle efficiency calculations with typical values are summarized in Table 2.1. A simple cycle is shown to be degraded in efficiency by addition of intercooling and reheat; a regenerative cycle is shown to be improved in efficiency by these additions. [Table 2.1 is reproduced from Parsons (1972)].

2.5. A NOTE ON COMBINED CYCLES

Gas turbines have been combined in several ways with other engines to drive a common shaft. The scheme most frequently seen in practice uses the gas turbine to provide boost power to a base-load diesel propulsion plant (the CODAG or CODOG schemes). It involves, however, no thermodynamic linking of the two engines; hence no cycle analysis is appropriate or possible.

A combination that has attracted interest, but has yet to be put to practice as this is written, has the hot exhaust from a gas turbine produce steam for use in a steam-propulsion plant.* The gas turbine and steam turbine outputs would be geared to the same propeller shaft. Since the heat rejected by one engine is the heat added to the other, the two are most definitely linked in the thermodynamic sense. Analyses of a cycle such as those discussed in the preceding sections, and analyses of a steam cycle (i.e., Rankine cycle with or without regeneration), can be combined with due regard for the steam being produced, and perhaps superheated, in the heat-addition process of the steam cycle. Presumably, an analyst familiar with both gas-turbine and steam cycles could carry this out readily. In other words, good luck. I'm not going to proceed any further down this path, since it would take this text too deeply into steam work and too far away from its expositions on gas-turbine propulsion.

2.6. REFERENCES

Keenan, J. H., and Kaye, J. (1948), *Gas Tables,* John Wiley, New York.

Parsons, M. G. (1972), unpublished lecture notes, University of Michigan.

Van Wylen, G. J., and Sonntag, R. (1965), *Fundamentals of Classical Thermodynamics,* John Wiley, New York.

* Waste-heat steam for ship-service purposes has been used, for example, in the John Sergeant plant.

Chapter

THREE

COMPONENT FUNDAMENTALS

3.1. INTRODUCTION

This chapter presents the fundamentals of compressors, turbines, combustors, and heat exchangers—the essential components of a gas turbine engine—in an elementary fashion. The purpose is to give the student a background from which to understand the characteristics of gas-turbine engines as they are presented in the following chapter. The term "characteristic" as here used, and generally throughout this text, refers to a relationship linking two or more operational parameters of the engine, such as its torque/(rev/min) characteristic. This chapter outlines the principles underlying operation of the several components and shows how the significant characteristics evolve from them.

It must be explained that the treatment following is deliberately kept at an elementary level, since it intends only to assist the beginning student of marine engineering. Because the topic of this book is marine *application* of the gas-turbine engine, it is inappropriate to include here material of sufficient depth and scope for the purposes of a reader who wishes to analyze or design a compressor, say, or other component. Such a reader should consult one of the numerous texts devoted to turbomachinery principles. Other readers who may be interested only in the marine application, and who feel themselves to be reasonably well acquainted with the principles of compressors, turbines, combustors, and heat exchangers, should skip this chapter.

3.2. ELEMENTARY PRINCIPLES OF TURBOMACHINERY

A rotating wheel interacting with a fluid stream can serve as a model for either a turbine or a compressor. Figure 3.1 shows such an interaction in simple

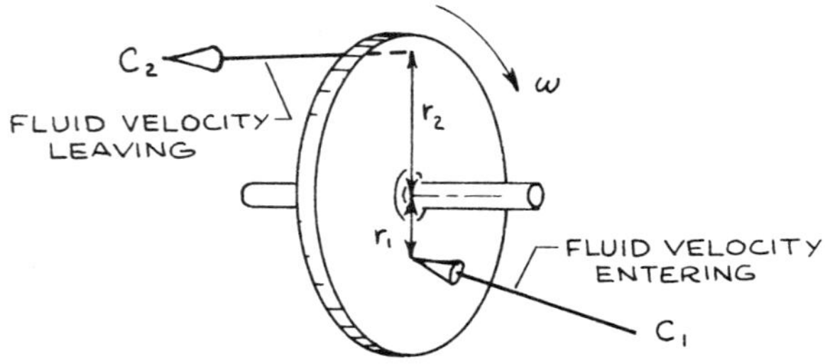

Figure 3.1. Elementary wheel-fluid interaction.

fashion. Of course, an actual machine accomplishes significant interaction only by means of blades (also called vanes or buckets) that impell the fluid, or are impelled by it (several of the illustrations in Chapter 1 show the blades of both turbines and of compressors), but they need not be considered in the beginning. The only important points here are that the wheel rotates with a steady angular velocity and that the fluid undergoes a change in momentum due to its passage over or through the wheel. Newton's second law provides the fundamental information needed, based on change of momentum, namely that the total moment of external forces equals the time rate of change of moment of momentum. Our only interest is in the moments that produce rotation of the wheel. With this limitation, the Newton's law statement can be symbolized by

$$Q = \dot{m}(r_1 C_{u_1} - r_2 C_{u_2}) \tag{3.1}$$

where Q is torque, $\dot{m}$ is mass flow rate, r is radius, and C_u is the tangential component of the fluid velocity, C.

Since the rate of energy transfer between wheel and fluid is the product of torque and angular velocity, ω, it is given by

$$E = \dot{m}\omega(r_1 C_{u_1} - r_2 C_{u_2}) \tag{3.2}$$

This can be put in terms of linear wheel speed, U, which is the product of r and ω, hence,

$$E = \dot{m}(U_1 C_{u_1} - U_2 C_{u_2}) \tag{3.3}$$

If $U_1 C_{u_1} > U_2 C_{u_2}$, E is positive, and work is being done by the fluid on the wheel—the wheel is thus a *turbine*. If the magnitudes are reversed, work is being done on the fluid by the wheel—the wheel is thus a *compressor, pump, fan,* or *blower*.

Three new terms are suddenly (and temporarily) injected here; let's explain. A pump is a device that imparts energy to an incompressible fluid, that is, a liquid. A fan or blower (essentially synonymous terms) is a device that imparts energy to a compressible fluid, that is, a gas, but does not raise the pressure enough to affect the fluid density significantly. It is thus, in essence, a pump working on a gas. A compressor pumps gas also, but radically increases its pressure and density in so doing. (A fan or blower is a mover of gas; a compressor is a squeezer.)

On the elementary level of the present discussion, it is not necessary to explore the differences in principles underlying these several related energy-adding devices. Let's pursue the question of a characteristic for all of them, this characteristic being most appropriately a relationship between flow rate and pressure increase. Factors unique to the compressor, those caused by compression of the fluid, will be noted in passing.

A. Centrifugal Compressors

Consider first a centrifugal device, that is, one in which the fluid leaves the wheel (or impeller) in a radial direction. Figure 3.2 is a picture of the centrifugal compressor impeller used in the Ford gas turbine (Figure 1.8). As can be seen by comparison with axial flow compressors that appear in Figures 1.5 and 1.6, the centrifugal compressor is a relatively simple device. It is, therefore, most typically found in the smaller engines where the complexity of the axial-flow machine is not justified.

Figure 3.2. Centrifugal compressor impeller from Ford engine. Reprinted with permission, Copyright © Society of Automative Engineers, Inc., 1972, all rights reserved.

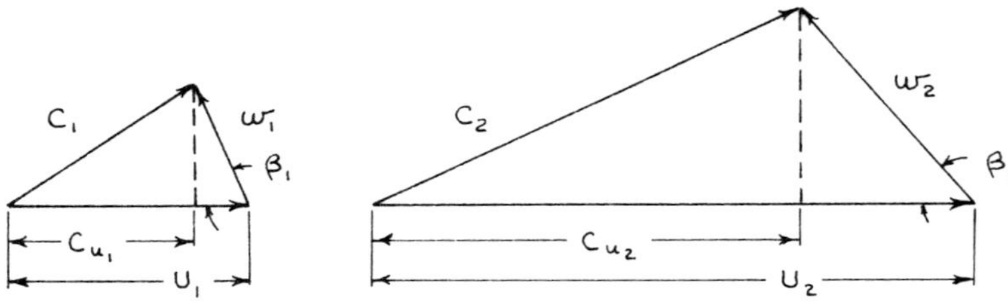

Figure 3.3. Vector diagrams of flows and wheel velocities at inlet and outlet.

The analysis represented by Eq. 3.3 can be continued with the aid of a vector diagram of fluid and wheel velocities in the plane of the wheel, Figure 3.3. Here two vector triangles represent, for inlet and outlet, respectively, the fluid velocities, C (as projected on the radial plane, if necessary), its tangential component, C_u, the linear velocity of the wheel, U, and the velocity, w, of the fluid relative to the wheel (vector sum of U and C). Using these sketches, several steps are now carried out on Eq. 3.3, as follows:

1. The $U_1 C_{u_1}$ term is dropped. This is justified because $U_2 > U_1$, and $C_{u_2} > C_{u_1}$, making the product, $U_1 C_{u_1}$, insignificant compared to $U_2 C_{u_2}$. The sign of the $U_2 C_{u_2}$ term is thereupon changed to positive.
2. The substitution, $C_u = U - w \cos \beta$, is made, this relationship being evident in Figure 3.3.
3. The flow rate is put in terms of volume flow rate by
$$V = w(\pi DB \sin \beta) \tag{3.4}$$
where V is the volume flow rate, D is the wheel diameter, and B is depth of the flow passage at the wheel rim (i.e., perpendicular to D)
4. The energy rate, E, in Eq. 3.3 is converted to an equivalent pressure rise* by multiplying the equation by ρ/m, where ρ is the density of the fluid being handled.

The result of these steps is

$$\Delta P = \rho U^2 - \frac{\dot{m} U}{\pi DB \tan \beta} \tag{3.5}$$

This equation relates mass flow rate to pressure rise for a given speed, U, and given geometric parameters, D and B. The angle, β, is also usually regarded as a geometric parameter, it being assumed equal to the angle between vane tips (Figure 3.2) and a tangent to the rim of the wheel (e.g., straight radial vanes have $\beta = 90°$). Note that the equation is linear in ΔP and $\dot{m}$.

* The conversion to a static pressure actually occurs in the *diffuser*, a section of the casing surrounding the impeller in which C_2 is slowed by an increase in flow area.

The equation is known as *Euler's head equation,* apparently from an association with Leonhard Euler, Swiss mathematician of the eighteenth century. Being based on the most elementary concept of flow through the machine, it is inadequate for many analyses. It is nonetheless adequate for predicting the effects of changes (such as changes in speed) on performance and for predicting qualitative trends. It is for these uses that it has been introduced here.

The value of the Euler equation is enhanced, for the second use in particular, by including modifications to account for losses in ΔP that occur as the fluid flows through the wheel. The losses that occur within a compressor (or pump, blower, fan) may be looked upon in two ways. First, they may be regarded as a requirement for a greater-than-theoretical amount of energy to reach a given pressure. Second, they may be regarded as pressure losses that decrease the pressure level reached with a given expenditure of energy. It is this second view that is appropriate here in explaining the pressure rise curve, since the explanation is conveniently based on the simple expendient of subtracting losses from the curve predicted by Eq. 3.5.

First, though, one should consider a few points regarding the losses. There are several components of loss, some of which can be calculated from theoretical considerations, others that cannot. In general, the losses are found by comparing the actual performance of the machine to its theoretically attainable performance; hence it is not feasible to state losses by formula. The discussion here is perforce qualitative and only sketches the behavior, as a function of flow, of the sum of the components of loss. This sum is conveniently looked upon as being composed of two principal categories. One of these categories includes those losses that are zero at zero flow, but increase proportional (approximately, at least) to the square of the flow (flow would be varied at constant rev/min by changing the external resistance). The other category contains those losses that are finite at all flows, including zero flow, but are a minimum near the design point, that is, that flow for which the machine is designed. Figure 3.4 shows these two categories plotted as functions of flow, then summed to a total loss.

The actual pressure rise characteristic may be represented as an ideal straight line, less the sum of losses. Such a representation is made in Figure 3.5. Note that the ideal line is shown as a horizontal straight line, consistent with radial vanes such as those in Figure 3.2 ($\beta = 90°$).

A second element in the construction of the centrifugal characteristic curves is the effect of speed changes. It is perhaps intuitively obvious that if the wheel slows, the flow will decrease, and vice versa. Given that a characteristic curve such as that of Figure 3.5 has been obtained for a stated speed, the problem is to find from it a curve that represents the characteristic for another speed. More specifically, the problem is to find how any point in the original curve is

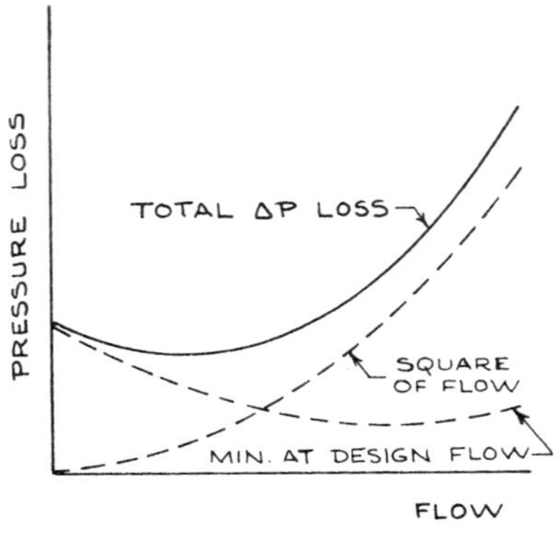

Figure 3.4. Categories of loss and their summation.

moved when the speed changes. When this is known, whatever process is required can be repeated at sufficient number of other points to establish a new curve.

Take, for example, point A in Figure 3.6. Suppose the vector triangle at the wheel exit from Figure 3.3 is represented by the lower triangle in this figure. (Subscript F is for "fast," subscript S for "slow.") If the wheel is slowed to a lower speed, the vector, U, shrinks proportionally. Now, if the flow rate, which is proportional to the magnitude of w, decreases in the same ratio (i.e., if $U_F/U_S = w_F/w_S$ = speed ratio), the upper triangle representing exit conditions at slow speed, will be similar to the lower triangle. Presumably the same similarity prevails for triangles drawn anywhere in the flow path through the machine. The effect on head can be shown by the following manipulation applied to Eq. 3.5:

$$\Delta P_F = U_F^2 - \frac{\dot{m}_F U_F}{\pi D B \tan\beta}$$

Pressure rise thus varies as the square of the speed ratio if flow varies in the

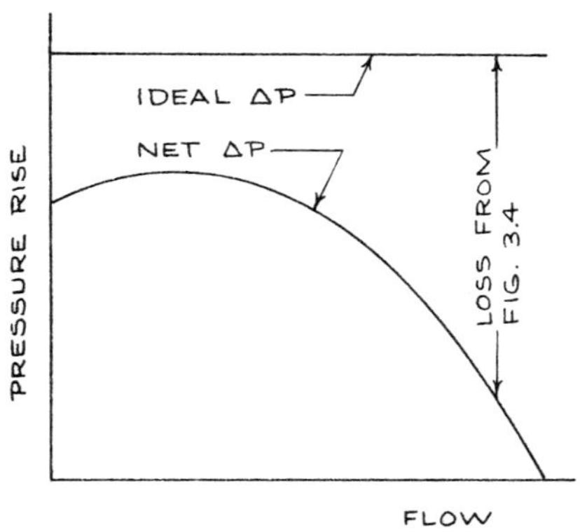

Figure 3.5. Construction of pressure rise curve from ideal curve minus losses.

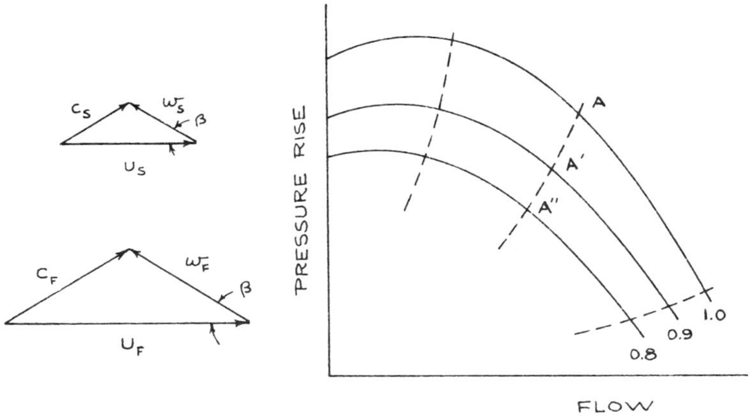

Figure 3.6. Illustrating the derivation of the speed-change rule.

same ratio as speed. What about the losses that cause such a discrepancy between Eq. 3.5 and the actual characteristic? If the velocity triangles remain similar, as has been specified here, then dynamic similarity is maintained during the speed change (geometrically similar flow patterns, resulting in the same ratio among all individual forces in fast and slow conditions). Under dynamic similarity, losses remain in the same ratio; for example, pressure losses also vary as the square of the speed. The efficiency of the machine therefore stays the same, and the use of Eq. 3.5 to predict a change in ΔP turns out to be quite acceptable.

Curves for three speeds, labeled 1.0, 0.9, and 0.8 are shown. Point A′ on the 0.9 speed curve is found from point A on the 1.0 curve by multiplying flow by 0.9, ΔP by 0.81. Likewise, point A′′ on the 0.8 speed curve is found using the multipliers, 0.8 and 0.64. Since, as has been noted, efficiency stays the same, the dashed line through A, A′, and A′′ is a contour of constant efficiency on the ΔP-flow plane. Two other such contours are shown at arbitrary (and unlabeled) points. Points placed along these contours by the 0.9/0.81 and 0.8/0.64 ratios demonstrate how enough points can be found to draw the curves. Each efficiency curve represents a different value of efficiency, although the present discussions do not allow any quantitative relationship among them to be established.

Figure 3.7 shows characteristics that would be typical of an actual centrifugal compressor, as established by a performance test. The curves of pressure rise and the efficiency contours are much like those of Figure 3.6. Several new features are introduced, however. The coordinates are slightly changed from Figure 3.6 in order to show them in more conventional form: pressure rise is plotted as a ratio of outlet pressure to inlet pressure, and flow

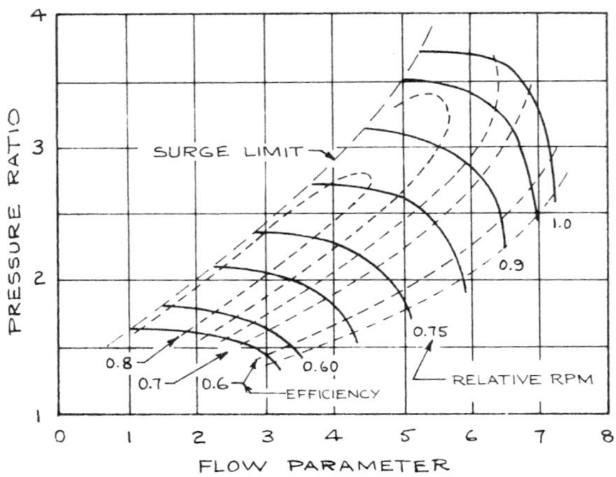

Figure 3.7. Typical characteristic curves for a centrifugal compressor.

rate is designated by the flow parameter, $WT^{1/2}/P$, where W is the weight rate of flow, P is the inlet pressure, and T is the inlet absolute temperature. Alternatively, P may be the ratio of inlet temperature to a standard temperature.

A factor of great importance introduced in Figure 3.7 is the *surge limit,* delineated by a dashed line that bounds the left edge of the area covered by the plots. This line is the locus of peak pressure ratios for several curves. Surge is a pulsing flow that is likely to occur if the compressor is operated to the left of the limiting line, and is to be avoided because of its effect on the subsequent combustion and because the surges may be severe enough to damage the machine. If the compressor is operating to the left of the highest point on its characteristic curve (Figure 3.6), some random increase in flow resistance can cause surging through the following sequence of events:

1. Because the drop in flow rate due to the resistance increase causes less pressure to be developed by the compressor, the rate decreases still farther, perhaps to zero flow.
2. Since, however, resistance to flow is a function of that flow, it falls to zero as compressor delivery ceases.
3. When the resistance has fallen to zero, the compressor is capable of delivering a flow much greater than the original flow. (Note from Figure 3.6 that zero flow does not imply zero pressure rise).
4. The abnormally high flow rate occurs momentarily, but it in turn produces excessive resistance, which causes the rate to decrease toward the original operating point.
5. The cycle repeats.

B. Axial Flow Compressors

The axial flow compressor is almost universally used in marine gas turbines, excepting only the smallest engines (several hundred horsepower) that may use the centrifugal. The name arises from the nearly axial direction of the air flow from inlet to outlet, through the several stages of the machine. (A *stage* is in fact a compressor itself; a multistage compressor thus is a set of individual compressors stacked on the same shaft.)

An elementary analysis sufficient to explain the major qualitative differences between centrifugal and axial compressors can also be based on Eq. 3.5, although the equation can be put in more convenient form if the derivation is revised slightly. This can be done from Eq. 3.3 as a starting point and using Figure 3.8. Note in this figure that C_{u_1} is zero, since C_1 is axial, that $C_{u_2} = U_2 - C_{A_2}/\tan \beta_2$, and that $\rho C_A = \dot{m}/A$. These produce, analogous to Eq. 3.5,

$$\Delta P = \rho U^2 - \frac{\dot{m}U}{A \tan \beta} \tag{3.6}$$

where C_A is the axial component of velocity, A is flow area, and other symbols are as used in Eq. 3.5. Again, numerical subscripts are dropped, and the sign of ΔP is arbitrarily made positive. The equation is linear in ΔP and $\dot{m}$, with slope determined by the value of β. Since β is typically close to 90°, the slope is steeply negative.

Figure 3.9 shows a ΔP curve for a typical axial flow compressor operating

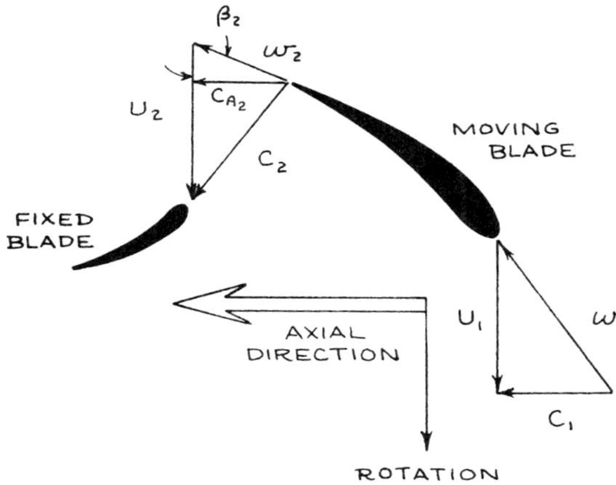

Figure 3.8. Vector diagram of flows and wheel velocities for an axial compressor stage.

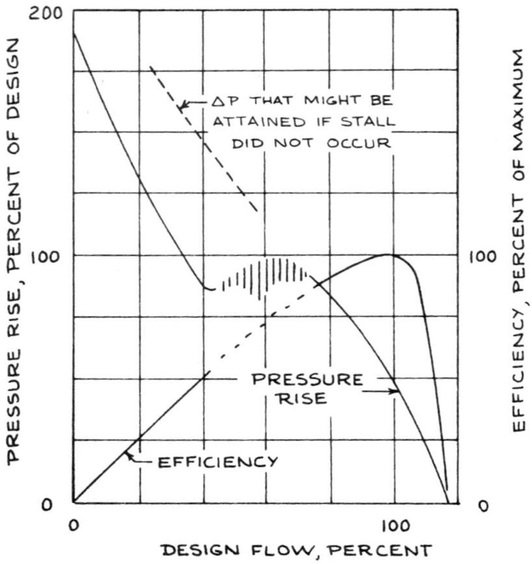

Figure 3.9. Typical pressure rise and efficiency curves for an axial flow compressor.

at a single speed. As with the centrifugal compressor, there are losses that make the pressure rise less that than the simple theory predicts, but the steep negative slope predicted by Eq. 3.6 is quite apparent in the figure. Another significant phenomenon, quite evident in the figure, is flow separation that occurs at flow rates somewhat less than the design flow. The result is *stall* (or stalling) of the blade, a word borrowed from aircraft terminology, wherein it refers to the loss of lift that occurs when a severe angle of attack causes air flow over a wing surface to separate from that surface. In the compressor, this loss of lift manifests itself externally as a loss in pressure rise. At the flow rates where it begins, flow is unstable; no continuous or repeatable relationship between ΔP and flow rate can be ascertained, as indicated by the cross-hatched area replacing the smooth curve in Figure 3.9.

Figure 3.9 also includes an efficiency curve. The intended, or design, flow would be near the flow producing highest efficiency. The curve is, therefore, shown here to indicate where the flow separation is likely to begin, relative to the design flow.

The phenomenon of flow separation cannot be fully explored here, but a basic idea can be gained from Figure 3.8. Look at the inlet velocity triangle (U_1,w_1, C_1). It is presumably drawn here for the design flow rate; note that w_1, which represents the air flow relative to the moving blade, is aimed in what might intuitively be called the "right" direction, that is, so that the air will flow smoothly past both surfaces of the blade. Now, if the flow is reduced while speed (U) stays constant, the vector, C_1, representing flow relative to a fixed reference, shrinks proportionally. The vector, w_1, being the sum of U_1 and C_1, is shifted, therefore, in a direction to impinge more on the underside of the blade—its angle of attack increases. As this angle is increased, a value is reached at which the air can no longer turn with short enough radius to follow the upper blade surface. It separates from that surface, producing the operational phenomenon known as stall.

A set of typical characteristic curves, including the effect of speed change and showing efficiency contours, is given by Figure 3.10. (Although this figure represents an actual compressor, the figure is simplified here for ease of presentation; cf. Figure 17, Chapter VI, of *Marine Engineering* [White (1971)].) Note that the ΔP curves are generally steeper than the centrifugal

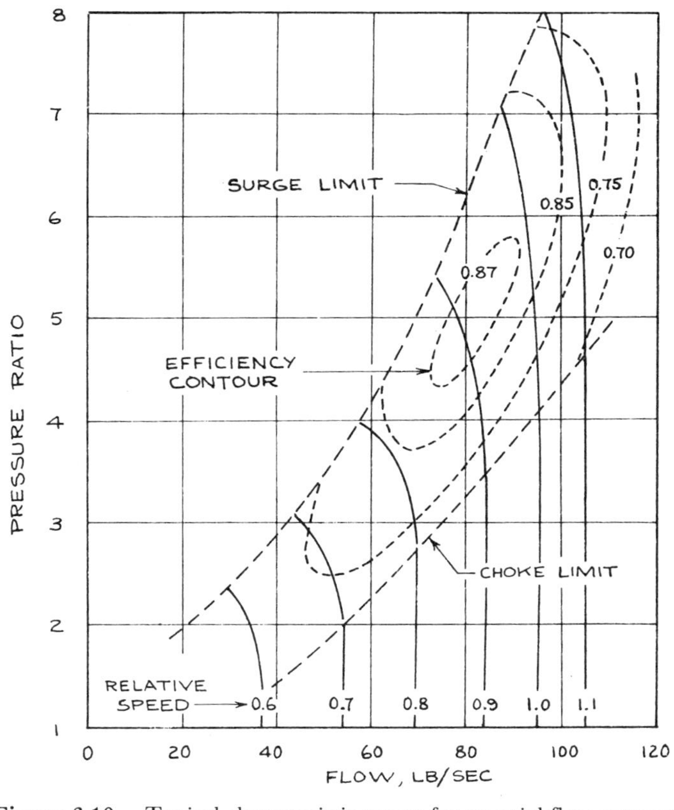

Figure 3.10. Typical characteristic curves for an axial flow compressor.

ΔP curves of Figure 3.7. The surge limit that was introduced in Figure 3.7 appears here, established in this case by the stall phenomenon, rather than by the peak pressure rise. A line labeled "choke limit" also appears in Figure 3.10. This is the locus of maximum flow rates, these maxima being established by the phenomenon of choking (inability to pass additional fluid) that is encountered when a fluid reaches its sonic velocity. Choking, however, is a phenomenon of compressible fluid flow, not just of axial compressors; it also is possible in the centrifugal machine.

C. Turbines

Turbines are also built in both axial and centrifugal configurations, but this discussion is confined to the axial flow turbine, since it is the only type likely to be encountered in marine propulsion work.

An analysis based on the concept that the fluid passing over the wheel follows the path delineated by the blades is sufficient for the present purpose of explaining characteristics. It begins with Eq. 3.3, aided by Figure 3.11. This figure shows a vector, C_1, representing a jet of fluid developed by a nozzle (not shown in the figure). The jet impinges on the blades (two shown) of the rotating wheel, which has linear speed, U_1, at the mean height of the blade. The tangential (horizontal in the figure) component of C_1 is C_{u_1}, as before, and the velocity of fluid relative to the moving wheel is represented by w_1. Corresponding quantities are shown at the exit from the blade passages, with subscript 2.

In Figure 3.11 it can be seen that

$$C_A^2 = C^2 - C_u^2$$

$$C_A^2 = w^2 - (U - C_u)^2$$

where C_A is the axial component of C. From these, it is readily seen that $U\,C_u$

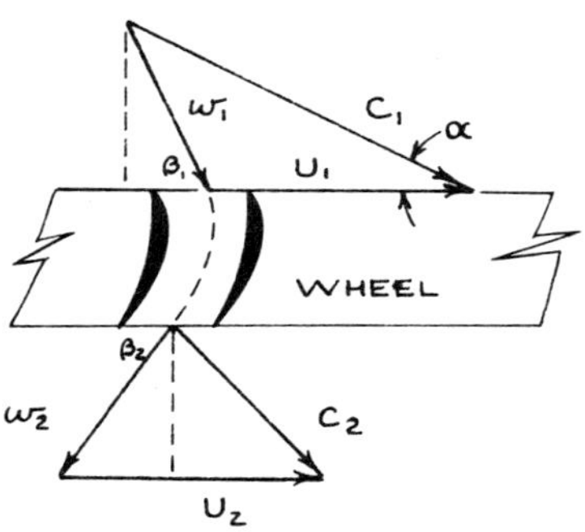

Figure 3.11. Vector diagram for discussion of turbines.

$= \frac{1}{2}(C^2 + U^2 - w^2)$, and when this is applied to each term of Eq. 3.3, the result is

$$E = \dot{m}(C_1^2 - C_2^2) + (w_2^2 - w_1^2) \tag{3.7}$$

(A $U_1{}^2 - U_2{}^2$ term also is formed, but it is zero for axial flow.) The $C_1{}^2 - C_2{}^2$ term in Eq. 3.7 represents the energy exchange due to impulse, or the change in kinetic energy of the fluid stream. The $w_2{}^2 - w_1{}^2$ term represents the energy exchange due to the reaction that results from an increase in relative velocity as the fluid traverses the wheel. If $w_2 = w_1$, the turbine is known as an *impulse turbine*; if $w_2 \neq w_1$, it may be called a *reaction turbine,* although most of the energy transfer may still be due to impulse. Figure 3.11 is drawn for zero reaction.

Whether or not reaction accompanies impulse, absorption of entering kinetic energy is a major part of the turbine wheel's function. But the considerable lengths of C_2 and w_2 indicate that kinetic energy leaves the wheel, as well. Unless a following stage (i.e., a second turbine in series with the first) uses this leaving energy, it is wasted and may constitute a large part of the losses that prevent the turbine from being 100% efficient. The turbine designer, therefore, considers means of minimizing the leaving energy, or in terms of Figure 3.11, making C_2 as short as possible. Or, he attempts to maximize the utilization [Shepherd (1956)] of energy, especially by choosing the best ratio of wheel speed, U, to jet speed, C_1.

The utilization for an impulse turbine can be written

$$\epsilon = \frac{U(C_{u_1} - C_{u_2})}{\frac{1}{2}C_1^2} \tag{3.8}$$

where the numerator is E from Eq. 3.3 and the denominator is the kinetic energy of the entering jet. Note that if $\beta_1 = \beta_2$ which is often the case in practice) in Figure 3.11, $C_u = C_1 \cos \alpha$, and $C_{u_2} = -(C_1 \cos \alpha - 2U)$. Substitution of these into Eq. 3.8 gives

$$\epsilon = \frac{4U}{C_1}\left(\cos \alpha - \frac{U}{C_1}\right) \tag{3.9}$$

This relationship is plotted in Figure 3.12. From it, or from a simple exercise in finding a maximum by differential calculus, the maximum value of ϵ is $\cos^2 \alpha$, occurring at a U/C_1 value of $\frac{1}{2} \cos \alpha$. In the figure, a value for α of 24° is used, placing the maximum at a speed ratio of 0.46. Since α is usually less than 24°, causing $\cos \alpha$ to be nearer to 1.0, a reasonable rule of thumb is to say that speed ratio should be approximately 0.5.

If some reaction is present ($w_1 \neq w_2$), the derivation doesn't work out so neatly, though the concept is the same. In one special case, that in which half of the kinetic energy is developed in the fixed nozzle and half in the moving

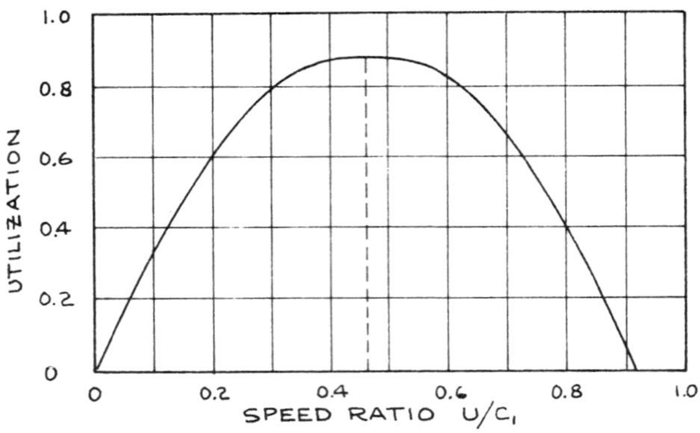

Figure 3.12. Utilization for an impulse turbine with $\alpha = 24°$.

blades ($C_1^2 = w_2^2 - w_1^2$), the derivation does give a simple result, namely,

$$\epsilon = \frac{U}{C_1}\left(2\cos\alpha - \frac{U}{C_1}\right) \tag{3.10}$$

from which maximum ϵ is $\cos^2 \alpha$ at a speed ratio of $\cos \alpha$.

The discussion above explains why the turbine is basically a high-speed machine and also provides the key for understanding why a reduction gear is always necessary between engine and propeller. Cycle analysis (Chapter 2) shows that a large pressure and temperature difference across the turbine is necessary for good efficiency, that is, $h_3 - h_4$ in Figure 2.1 must be large. Now, jet velocity, C_1, is a function of $h_3 - h_4$ [$C_1 \propto (\overline{\Delta h})^{1/2}$] and may therefore be several thousand feet per second in magnitude. As just seen, linear wheel speed should be roughly half of this for an impulse turbine, still quite a high speed. Of course, a large value of U can be had either by large diameter or by large rotational velocity. Lightness, compactness, and economical use of material dictate a small diameter, the minimum size being established by the need for sufficient area to pass the requisite flow. Consequently, wheel rotational speed is high.

The concept of a best ratio of jet speed to wheel speed also applies to the marine propeller. (Indeed, it applies to any turbomachinery. The stall problem confronting axial compressors could be explained in terms of a poor speed ratio.) Its "jet" is the water flowing into it due to the speed of the vessel, a speed that is usually measured in tens of feet per second rather than in thousands or hundreds. Again, there is a choice between use of diameter or rotational speed in attaining the desired U. The usual approach in propeller design is to make the diameter as large as possible in order to minimize the thrust loading. Since U is comparatively low, and diameter should be large for reasons other than speed ratio, rev/min is usually low compared to that of the turbine that drives the propeller.

If the propulsor is a jet pump, the same principles apply. The "jet" in the present context is the water approaching the pump via the inlet duct, again at a speed not much different from the forward speed of the vessel. The best speed for the pump impeller is established by a best speed ratio to this jet. For both the pump impeller and the propeller, this speed ratio is a function of an angle corresponding to the α of the turbine discussion above, but the choice of a value for this angle is not unlimited, because, for example, a large α produces a large w_1. The magnitude of w_1 is the velocity of the fluid relative to the turbine blades, and, consequently, fluid losses within the blade passages increase as α does.

This loss of energy, or its degradation to heat, should be distinguished from the nonuse of energy, $E(1 - \epsilon)$. The latter represents energy that can be used by a second wheel, and, indeed, it is common practice to have a second row of moving blades into which C_2 is directed by a set of stationary guide vanes. Also, the available energy ($h_3 - h_4$ in Figure 2.1) may be split among several stages. Each stage is, in effect, a turbine, since it has its own nozzles to develop its C_1 from the hot, high-pressure gas exiting from the preceding stage. In this case, the kinetic energy from each stage also is energy available to the following one. The jet velocity for each turbine is reduced by the split into stages, making it more practicable to attain a good value of U/C_1 without going to excessive rotational speeds. The concept of multistaging is often used in turbine technology, both steam and gas, to obtain high efficiency in the ways mentioned here. It is stated in Chapter 1 that two mechanically independent turbines are nearly always used in marine gas turbine engines—one to drive the compressor and the other to drive the output shaft. These two are in series with respect to gas flow and so effectively constitute two stages of the same turbine. Each of these may also consist of more than one stage. Usually, though, the compressor turbine is single stage (unless the engine is a twin-spool machine), with the multistaging, if any, being used in the power turbine. For example, the General Electric industrial-type gas turbine shown by Figure 1.5 has a single stage for each of these two turbines; the General Electric aircraft-type engine shown by Figure 1.1 has a single-stage compressor-drive turbine with two rows of moving blades and a power turbine of six stages.

Some engines employ adjustable nozzles for the power turbine, with the objective of improving efficiency at part loads, that is, when slowing down of the turbine moves the speed ratio away from its design value. By rotating the blades that form the nozzles, both direction and magnitude of the jet to the power turbine can be adjusted, thus changing U/C_1 toward a better value.

Utilization of a turbine is always less than 1.0, since the last stage—no matter how many there may be—must have a finite C_2. Efficiency must always be somewhat less than utilization, for in addition to the energy thrown away in the leaving fluid, there are degradations in energy caused by fluid friction and turbulence, leakages, and bearing friction. These are usually small, being minimized by what may be classified as "good design." Generally speaking, the key problem in obtaining good efficiency in design, and maintaining it in operation, is obtaining the best possible speed ratio.

The engine characteristic of greatest significance to the propulsion plant designer is the relationship between torque and rev/min. For the impulse turbine of a single stage, it can be derived readily from the foregoing material on utilization. Recall that in deriving Eq. 3.2 it was noted that the energy exchanged between fluid and wheel is proportional to the product of torque and rev/min. Therefore, if rev/min is factored out of the Eq. 3.9, the result is what might be called the "torque utilization," or

$$\epsilon_q = \frac{8\pi r}{C_1}\left(\cos\alpha - \frac{U}{C_1}\right) \tag{3.11}$$

If C_1 is a constant, then torque is shown by this equation to be a linear function of U and hence of rev/min.

When $U/C_1 = \frac{1}{2}\cos\alpha$ (the best value), then

$$\epsilon_q = \frac{4\pi r \cos\alpha}{C_1} \tag{3.12}$$

and when $U/C_1 = 0$ (zero rev/min).

$$\epsilon_q = \frac{8\pi r \cos\alpha}{C_1} \tag{3.13}$$

thus showing that for a constant supply of gas to the turbine, torque at zero rev/min is twice the value at the speed of highest utilization, which would normally be the designed operating speed.

The torque-rev/min characteristic for the ideal single-stage impulse turbine is, therefore, a straight line with steep negative slope. The characteristics for actual turbines closely follow the prediction of this simple analysis, so that there is no need to expound upon the differences between ideal and actual, as was done for compressors. Actual torque curves appear in the next chapter, Figure 4.10.

3.3. THE COMBUSTION PROCESS

This section sketches the determination of the air required for combustion and the resulting temperature, and briefly outlines the exhaust emission problem.

In most marine applications the gas-turbine engine burns a liquid hydrocarbon fuel. These oils are not simple compounds, but are rather a mixture of many compounds of carbon and hydrogen, plus (often) compounds of several minor constituents. For purposes of this discussion, however, they can be represented by a single chemical formula that gives the correct overall proportions of the two major constituents. For example, light diesel oil can be represented by the formula, $C_{12}H_{26}$ (dodecane), since the weight ratios of carbon and hydrogen stated by this formula are close to those found in a typical sample of the actual oil.

The amount of air required for complete combustion of a fuel depends upon the oxygen requirement, and this can be calculated from the elementary chemical formulas for oxidation of carbon and hydrogen, namely,

$$C + O_2 \rightarrow CO_2$$
$$H_2 + \tfrac{1}{2}O_2 \rightarrow H_2O$$

These indicate that 1 mole (mole = a weight numerically equal in pounds or grams to the molecular weight) of oxygen combines with 1 mole of carbon to produce 1 mole of carbon dioxide, and that ½ mole of oxygen combines with 1 mole of hydrogen to produce 1 mole of water. Molecular weights of carbon and oxygen are 12.01 and 32.06, respectively, indicating that 2.665 (= 32.06/12.01)lb oxygen per lb carbon is required; the molecular weight of hydrogen is 2.02, indicating that 7.94 (= 32.06 × ½/2.02)lb oxygen per lb hydrogen is required. The formula, $C_{12}H_{26}$, which is used here as an example, shows that its weight fraction of carbon is $12 \times 12.01/(12 \times 12.01 + 26 \times 2.02/2) = 0.846$, leaving 0.154 as hydrogen. The amount of oxygen required to completely oxidize $C_{12}H_{26}$ is consequently

$$(0.846)(2.665) + (0.154)(7.94) = 3.48 \text{ lb/lb}$$

The air that must supply this oxygen is, of course, composed of nitrogen and oxygen, plus several minor constituents. In most practical situations, the air is also mixed with a significant amount of water vapor. A standard composition, usually used in marine work, [*Recommended Practices for Preparing Marine Steam Power Plant Heat Balances* (1961)], is the following:

Oxygen	0.2315 weight fraction of dry air
Nitrogen	0.7685 weight fraction of dry air
Moisture	0.0165 lb per lb of dry air

Argon, the principal minor constituent, is here lumped with the nitrogen; other minor constituents are neglected.

The gas turbine enjoys the advantage of a continuous combustion process, but, nonetheless, satisfactory combustion is not easily accomplished, since the compactness of the engine requires that combustion take place within a small volume. The fact that the air is highly compressed is a great help, but careful design is also essential. Figure 3.13 shows a diagramatic view of the combustor used in the Ford gas turbine and well illustrates a technique of obtaining thorough mixing within a small volume. The fuel nozzle is on the centerline at the top of the figure; air for combustion enters through swirler vanes and the primary holes. A strong toroidal vortex is created (dashed lines) in which most of the combustion takes place.

The temperature reached in combustion is a vital consideration in the gas turbine. As seen in Chapter 2, efficiency of the engine is a function of the highest temperature reached in the cycle, this being the temperature of the air as it leaves the combustion chamber. Conversely, the integrity of the compressor turbine nozzles—the first component encountered by the hot fluid—and to a lesser extent, of the parts downstream of these nozzles, depends on avoidance of excessively high temperatures.

The heat released by the combustion of a fuel is usually designated by its *heating value,* this being the heat released during complete combustion at constant temperature. ("Constant temperature" implies cooling the product of combustion to the temperature at which the process began.) For fuels containing hydrogen, the heating value depends on whether the resulting water vapor is condensed during the measurement of heating value. If it is, the heat released during condensation is included, and the result is known as the

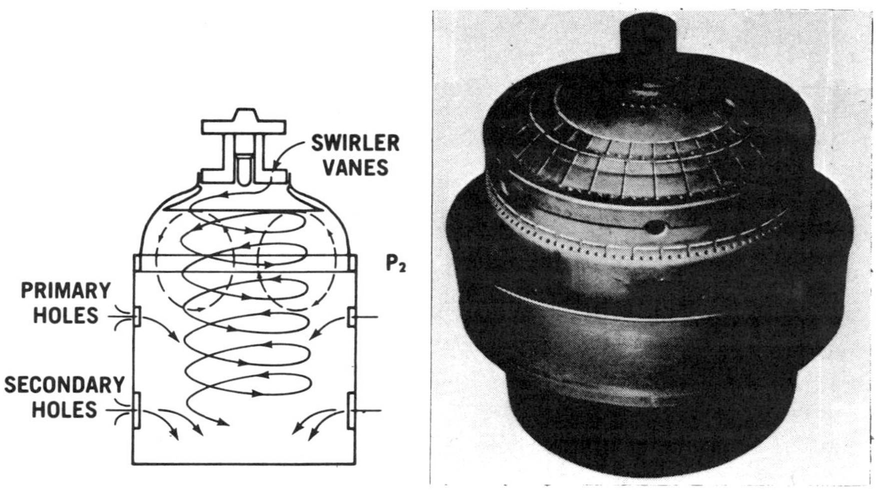

Figure 3.13. Combustor for Ford Motor Company gas turbine [Cadwell, Chapman, and Walch (1972)]. Reprinted with permission, Copyright © Society of Automative Engineers, Inc., 1972, all rights reserved.

higher heating value; otherwise, it is the *lower heating value*. For $C_{12}H_{26}$, tested at 25 C (77 F), the two values are 20,140 and 18,968 Btu/lb, respectively [Van Wylen and Sonntag (1965)].

The temperature resulting from combustion of a fuel-air mixture can be calculated by means of the first law of thermodynamics. For the constant-pressure, steady-flow process, such as that followed in the gas turbine combustor, the appropriate expression of this law is

$$Q + H_R = H_P \tag{3.14}$$

where Q is the lower heating value of the fuel, H_R is the enthalpy of the reactants (air + fuel), and H_P is the enthalpy of the products of the reaction. Units are typically Btu per lb of fuel. For practical use, the equation must be expanded so that enthalpies of fuel, air, and exhaust gas can be evaluated separately, perhaps by perfect gas equations for the gaseous constituents. More accurate calculations can be accomplished with the aid of charts of properties of fuels and products, such as those prepared by Hottel et al. (1949). These charts also include the conversion from enthalpy to temperature.

If approximate results are satisfactory, a quick answer can be found by treating the air and products of combustion as perfect gases, with some appropriate constant values of specific heat. For example, the temperature reached by the burning of $C_{12}H_{26}$ stoichiometrically and adiabatically (without heat leakage) can be estimated by Eq. 3.14, numerically evaluated as follows:

$$18{,}968 + 0 = (16.27)(0.25)(T_P - 77)$$

where a base temperature of 77 F and 0.25 Btu/lb F is used for the sum (16.27) of air and product weight per pound of fuel. The result is $T_P = 4740$ F.

The temperature found in this example is impractically hot, and this is the reason for the high air consumption associated with gas turbines: a considerable excess of air, far beyond that required for good combustion, must be used to reduce this temperature by dilution. For example, if 300% stoichiometric air is used, the first factor on the right becomes $46.8 (= 3 \times 15.27 + 1)$, and the resulting temperature is 1680 F, a temperature in the practical range for the hot parts of an engine.

The diluting air is often supplied after combustion as secondary air. The secondary-air supply path for a typical combustor is evident in Figure 3.13.

All combustion equipment burning the common fuels produces combustion products ("emissions") that are biologically harmful. These may be impurities from the fuel or oxides of these impurities; sulfur dioxide is the most common example. They may be the results of incomplete combustion; carbon monoxide and hydrocarbon vapors are two undesirable and legally restricted constituents of automobile exhaust that are due to incomplete combustion. Or they may be the end product of unwanted chemical reactions that accompany combustion; the oxides of nitrogen are the significant items in this class.

The continuous combustion process as used in the gas turbine can readily accomplish complete combustion. The federal standards on unburned hydrocarbons and carbon monoxide, being implemented during the 1970s by the automotive industry, for example, are generally regarded as being within the capability of the existing gas turbines without auxiliary equipment. The oxides of nitrogen are another matter, since it is not completeness of combustion that governs their appearance. The principal parameter is temperature: the higher the temperature reached at some point in combustion, the greater will be the concentration of these oxides in the exhausts. For automotive use, the gas turbine appears to face essentially the same problem here as the piston engine, but solution via modifications to the combustor may be feasible. However, as this is written, no legal standards have been imposed on marine engines.

Sulfur dioxide has been a significant pollutant only when large quantities of high-sulfur fuel is being used, as in large coal-burning electric-power plants. Although some fuel oils used in marine propulsion may have sulfur concentrations approaching those of high-sulfur coals, deleterious effects on the gas turbine itself usually restrict the use of these fuels.

3.4. HEAT EXCHANGERS

Heat exchangers—devices in which heat is transferred between one fluid and another—may be part of the gas turbine engine in three distinct services, as listed following:

1. Lube oil cooling. In marine service, this is typically accomplished in a small (i.e., of small size compared to the engine itself) water cooler. However, if the engine is mounted high in the vessel, an air cooler may be used.
2. Compressor intercooler. As noted in Chapter 2, intercooling is a refinement that improves efficiency slightly when used with regeneration. Because of the additional complexity of arrangement, and because of the bulkiness of the exchanger, it is rarely used in marine service, there being no active examples in American practice.
3. Regeneration. Also discussed in Chapter 2, where it is observed that regeneration can contribute a large increase in efficiency. Because the necessary heat exchanger may be quite large, regeneration is never used with the aircraft engine. It is commonly used with the industrial type, but even there the size of the regenerator may be a handicap in marine service.

The size of the regenerator is perhaps the heat-exchanger problem most likely to interest the marine engineer concerned with the design of a gas-turbine power plant. This section, therefore, takes a brief look at heat-exchanger principles with an emphasis on the types of exchangers likely to be used for regenerators, their size, and their effectiveness (a term introduced in Chapter 2).

The simplest type of heat exchanger is shown by the pictorial sketch in Figure 3.14. Here, two tubes are concentric (the outer one is labeled "shell" in the figure), with each carrying one of the two fluids that are to exchange heat. The performance of this device, and indeed any heat exchanger, is expressed by the fundamental heat exchanger formula

$$Q = UA \, \Delta T_{m} \tag{3.15}$$

where Q is the heat transfer rate (Btu/hr), U is the overall heat transfer coefficient (Btu/hr ft^2 F), A is the heat-transfer surface area (ft^2), and ΔT_m is the mean temperature difference between the two fluids (F).

The area is the surface area of the tube separating the two fluids; it is usually taken to be the area of the outer surface of this tube.

The temperature difference between the fluids varies along the length of the exchanger as one heats and the other cools. This is illustrated by two graphs in Figure 3.14. One shows the temperature variation if the fluids flow in the

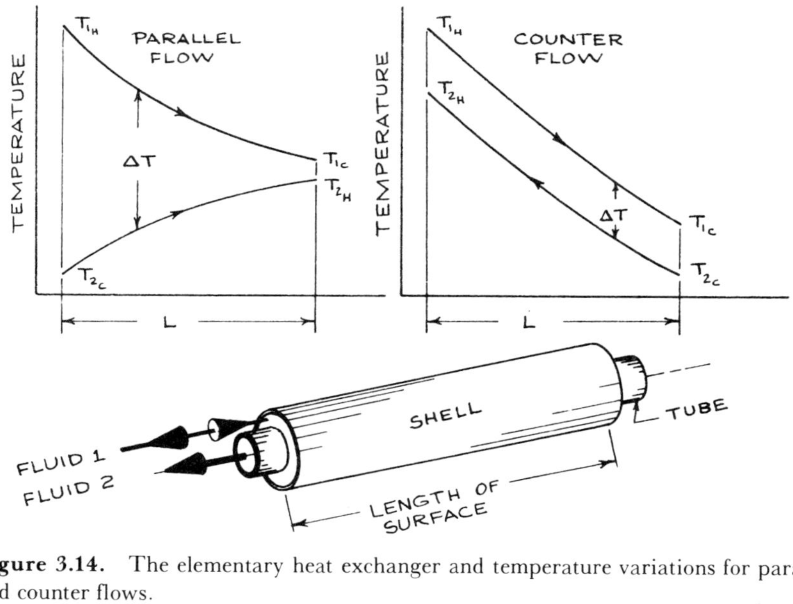

Figure 3.14. The elementary heat exchanger and temperature variations for parallel and counter flows.

same direction (parallel flow); the other shows this variation if they flow in opposite directions (counter flow). However, the heat-exchanger equation requires a mean value of ΔT. This, usually referred to as the *log mean temperature* difference, is given by

$$\Delta T_{\mathrm{m}} = \frac{\Delta T_{\mathrm{L}} - \Delta T_{\mathrm{S}}}{\ln\,(\Delta T_{\mathrm{L}}/\Delta T_{\mathrm{S}})} \tag{3.16}$$

where ΔT_{L} and ΔT_{S} are the temperature differences at the ends (e.g., $T_{1_{\mathrm{H}}} - T_{2_{\mathrm{H}}}$ in the left graph), with L designating the larger difference, S the smaller.

The overall heat-transfer coefficient is expressed by the inverse of the sum of three terms giving, respectively, the thermal resistivities of the interface between the outer fluid and the tube wall. The expression is

$$U = \left(\frac{1}{h_0} + \frac{x}{ka'} + \frac{1}{h_{\mathrm{i}}a''}\right)^{-1} \tag{3.17}$$

where h_{o} is the surface heat-transfer coefficient at the outer surface of the tubes separating the fluids (Btu/hr ft^2 F), h_{i} is the surface heat-transfer coefficient at the inner surface of this tube, x is the thickness of the tube wall (ft), k is the thermal conductivity of the tube material (Btu/hr ft^2 F/ft), and a is the factor to correct for the increase in area between inside and outside of a cylindrical tube (if A is outside area, as suggested above, a'' is the ratio of inner area to outer area, and a' is the ratio of mean area to outer area).

The fluid-tube interface usually provides the major resistance to heat flow, especially when the fluid is a gas. Attention should thus focus on the two h terms. These are functions of fluid properties, flow conditions, and surface geometry. Generally, any of these that tend to reduce the stagnant film of fluid that forms on the surfaces increases the value of h, and hence increases U. For example, higher fluid velocity usually increases h. Because of the complexity of the relationships, and because there are many different ones needed to suit different situations, we shall not attempt here to give a formula for h.

The designer of a regenerator for a gas turbine is faced with the task of contriving a compact device in face of the low values of h that are inherent with gas-to-solid (i.e., to tube wall) heat transfer. His principal efforts toward compactness are consequently directed toward fitting the greatest heat transfer area into the smallest volume. Let's sketch how this is done.

The first step away from the elementary exchanger of Figure 3.14 is to replace the inner tube with several smaller tubes. For example, suppose that the shell in this figure is a 2-in pipe (2.375 in. o.d., 2.067 in. i.d.), and that the inner tube is a 1¼-in. pipe (1.660 in. o.d., 1.380 in. i.d.), a pair of sizes that provide approximately equal flow areas for the two fluids. The area, A, is 0.435 ft^2 per foot of length. A brief exercise in drawing circles shows that eight tubes of 0.5-in. o.d. could be fitted into the 2-in. pipe, and that slightly less cross-sectional area would be occupied than by the single pipe. The outer surface area of these tubes would be 1.255 ft^2 per foot of length, or an increase of A by a factor of 2.9. And if the tubes were made yet smaller, an even greater increase in A could be obtained within the original 2-in. shell. This exercise points the way to the common method of obtaining compactness in heat exchangers: subdivision of flow passages.

The limiting feature in subdivision is the increase in pressure drop and pumping power required for the fluids, especially the inner one in this case. With the eight 0.5-in. o.d. diameter tubes, the cross-sectional area for flow is reduced from 1.495 in^2 to 1.016 in^2 (assuming tube wall thickness of 0.049 in.). The consequent increase in velocity, in addition to the increase in area contacting the fluid, would raise the pressure drop of the inner fluid by a factor of 7.5. Since the gas turbine is especially sensitive to pressure losses, this purchase of compactness at the cost of greater pressure losses presents a dilemma to the designer of a regenerator and limits the compactness he might otherwise hope to attain by subdividing the flow passages.

The designer, of course, exercises his ingenuity in attempting to maximize heat transfer area within a given volume while minimizing the increase in pressure drop. Figure 3.15 depicts four common types of heat exchangers from among the many that are found in services of all kinds. The top two illustrate

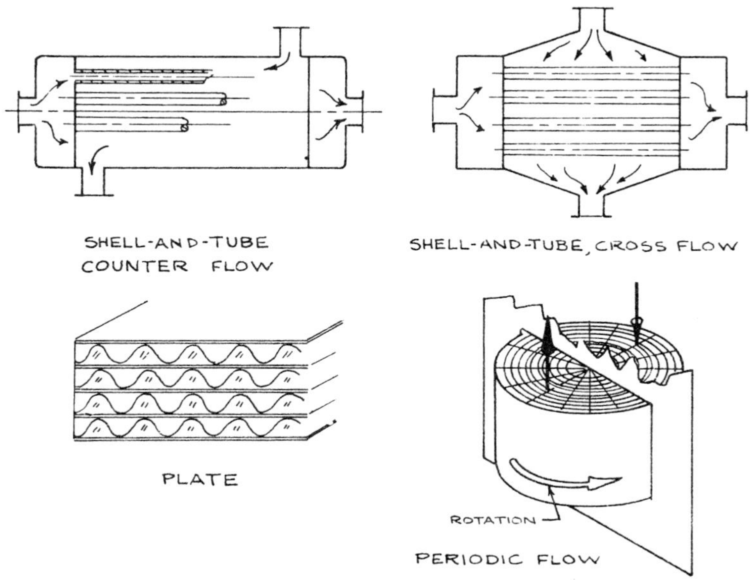

Figure 3.15. Heat-exchanger configurations.

the multitube concept introduced in the preceding two paragraphs, while the lower two are examples of schemes adopted by the designers of gas turbines in their quest for more compact regenerators. Top left shows the simplest form of the multitube heat exchanger, and top right is essentially the same thing, but arranged for cross flow of the two fluids. Bottom left shows the essential features of a plate-type exchanger, constructed of alternating flat plates and corrugated plates. The spaces designated by ticks (′′) contains one of the fluids, while the other flows in the unmarked spaces. The plate-type exchanger is favored for gas-turbine regenerators, being used, for example, by the General Electric machine of Figure 1.5. Reference to that figure shows that the regenerator is bulky, in spite of the high density of heat transfer surface area that this type of construction allows. The exchanger at bottom right is the periodic type (sometimes called rotary regenerative). The cylindrical drum rotates, driven by an auxiliary shaft from the engine output. It contains a matrix of metal or ceramic surfaces that alternately absorb and release heat as the rotation passes them through the fluids that flow on opposite sides of the mid-plane barrier. For a given pressure loss, this type of exchanger allows a much greater compactness than the stationary types previously noted and is, therefore, used in engines that require both high efficiency and compactness. It does have its unique handicaps, such as the necessity for a mechanical drive system and the leakage between the two fluids that must inevitably occur as its internal volume is rotated from one side to the other.

Effectiveness of a regenerator has been defined from a power-cycle standpoint in Chapter 2; it is the fraction of a possible temperature change that is actually accomplished by a heat exchanger. From the standpoint of the exchanger, its definition is expressed by the following equation [Kays and London (1955)]:

$$\epsilon = \frac{C_h(T_{hi} - T_{ho})}{C_{min}(T_{hi} - T_{co})} = \frac{C_c(T_{co} - T_{ci})}{C_{min}(T_{hi} - T_{ci})} \tag{3.18}$$

where C_h is the product, $w \times c$, for the hotter fluid, w being the weight rate of flow, and c the specific heat (constant pressure specific heat if the fluid is a gas); C_c is the corresponding product for the cold fluid; C_{min} is either C_h or C_c, whichever is smaller, and T is the temperature, with the subscripts, hi, ho, ci, and co, designating hot fluid in, hot fluid out, cold fluid in, and cold fluid out, respectively. Since the amount of heat transferred is a function of the area, A, effectiveness is related to A, and hence to the size of the heat exchanger. This should be readily apparent in Figure 3.14, where it can be seen that the temperature change accomplished depends upon the length, and hence the area, of the exchanger pictured. This figure also illustrates the effectiveness advantage of counterflow. While 100% effectiveness is approachable with counter flow, 50% is the limit (assuming the same $w \times c$ product for both fluids) with parallel flow.

The relationship between area and effectiveness can be reduced to a formula for each type of exchanger. For convenience, this relationship can be expressed in terms of heat transfer units, a nondimensional parameter defined as follows [Kays and London (1955)]:

$$H = \frac{AU}{C_{\min}} \tag{3.19}$$

where all terms are are as defined previously. This parameter, in effect, is the number of units of heat transfer capability per unit of flow. Its relationship to effectiveness for a counterflow heat exchanger is [Kays and London (1955)]:

$$\epsilon = \frac{1 - \exp[-H(1 - C_{\min}/C_{\max})]}{1 - (C_{\min}/C_{\max}) \exp[-H(1 - C_{\min}/C_{\max})]} \tag{3.20}$$

where $C_{\max}$ is either C_h or C_c, whichever is larger; exp is the exponential of the base of natural logarithms; and other symbols as defined earlier. This relationship is plotted in the left-hand graph of Figure 3.16. The "diminishing returns" effect, implying very large areas for high effectiveness, is apparent. ($C_{\min}/C_{\max} = 1$ is the case of interest for gas-turbine regenerators.) Figure 3.16 also shows the effectiveness plot for the periodic-flow exchanger. A note in the figure indicates that H has a slightly different definition for this type of unit, this difference stemming from the absence of area for the heat to flow through. Also, the parameter, C_r, is introduced in the figure; C_r is the product of rotational speed, rotating mass exposed to the fluids, and the specific heat of this mass. Since the figure indicates that a high value of C_r increases effectiveness, the designer has incentive to cram as much surface as possible into the rotating element and to turn it at high speed. But, as with heat exchangers in general, more surface means more resistance to flow. The penalties associated with

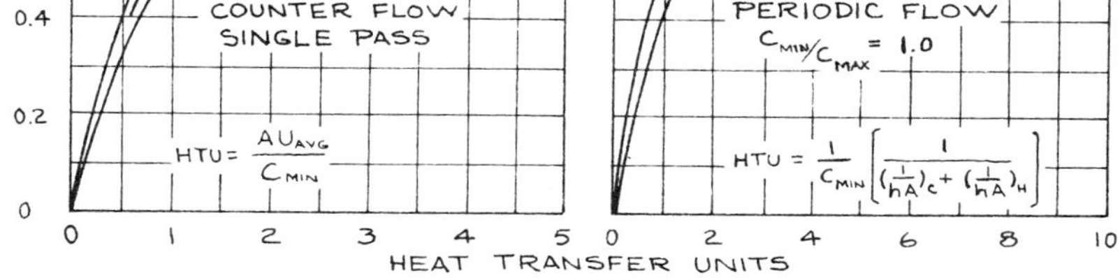

Figure 3.16. Effectiveness for two types of heat exchangers.

higher speed are unique to the periodic-flow type: they are the increased mixing of the two fluids by the rotating element and the increased difficulty of sealing against ordinary leakage from the high-pressure side to the low-pressure side.

Comparison of the two parts of Figure 3.16 shows the counter-flow unit to be slightly more effective for any particular value of H. What the figure does not show is the relationship between H and the volume in which the area, A, is enclosed, and this is a relationship that cannot be readily demonstrated by formula or by graph. It is found in practice, however, that for given pressure loss and volume, the periodic-flow unit can provide significantly more surface. One contributing factor is the absence of any need for the surfaces to be pressure barriers. Consequently, the surfaces can be composed of thinner material than in exchangers that must resist the large pressure differences typical in gas-turbine regenerators.

3.5. REFERENCES

Cadwell, R. G., Chapman, W. I., and Walch, H. C. (1972), "The Ford Turbine—An Engine Designed to Compete with the Diesel," Society of Automotive Engineers paper 720168.

Hottel, H. C., Williams, G. C., and Satterfield, C. N. (1949), *Thermodynamic Charts for Combustion Processes,* John Wiley, New York.

Kays, W. M., and London, A. L. (1955), *Compact Heat Exchangers,* The National Press, Palo Alto.

Recommended Practices for Preparing Marine Steam Power Plant Heat Balances (1961), Technical & Research Bulletin 3-11, Society of Naval Architects and Marine Engineers, New York.

Shepherd, D. G. (1956), *Principles of Turbomachinery,* MacMillan, London.

Van Wylen, G. J., and Sonntag, R. E. (1965), *Fundamentals of Classical Thermodynamics,* John Wiley, New York.

White, A. O. (1971), "Gas Turbines," Chapter VI of *Marine Engineering,* Society of Naval Architects and Marine Engineers, New York.

Chapter

FOUR

GAS-TURBINE CHARACTERISTICS

4.1. INTRODUCTION

The fundamental characteristic of a propulsion engine is the combination of torque and rev/min (or power and rev/min) produced at its output coupling. Other external manifestations of its performance, such as fuel rate, exhaust and intake flows, and exhaust temperature, are also of interest to the marine designer.

An important consideration in all of these things is how they vary as functions of operational parameters, since a propulsion engine must operate over a range of conditions. For example, the variation of torque with rev/min must be such that at part-load operation an excess of torque is available to accelerate the shaft to full speed. And the variation in fuel rate with load—it can be quite large—is often a prime factor in determining the fuel economy of the propulsion plant. In fact, the word *characteristic* is most often used here to mean the functional relationship between an important parameter and the independent variables that determine its values.

A further point is the effect of ambient conditions on the engine characteristics. The density of the atmosphere is a factor in determining the air mass flow rate through the engine, hence ambient temperature and pressure, as well as changes in them that may be imposed on them by the inlet system, must be accounted for. Gas-turbine performance is sensitive to pressure losses from any source, so that the effect of the exhaust duct also has a bearing on characteristics. The ambient temperature affects temperatures throughout the engine cycle. In general, then, the designer must make corrections to characteristics before applying them in any analysis of interest.

Characteristics of an engine, and any correction factors that accompany them, usually appear as graphical data supplied by the manufacturers of the engine. This chapter is largely devoted to illustrating the several characteristics with typical manufacturers' curves of this sort. The source of the data must be acknowledged, of course, but the reader is admonished that this text is not intended to be a handbook for direct use in design; if he is engaged in an actual design using perchance one of the engines whose characteristics appear here, he should obtain the current information directly from the engine builder.

4.2. TORQUE-REV/MIN AND POWER-REV/MIN CHARACTERISTICS

An elementary analysis derives the relationship between torque and rev/min for the ideal single-stage impulse turbine. This has been done in Chapter 3 and demonstrates that torque is a linear function of rev/min, with torque at zero rev/min being twice the value at the rpm of best utilization. Expressed as a formula, this is

$$\frac{Q}{Q_r} = 2 - \frac{N}{N_r} \tag{4.1}$$

where Q is torque; N is rev/min; and r is a subscript denoting conditions at the rated point, implied to be at the rev/min of highest utilization. If Eq. 4.1 is multiplied by N/N_r, the result is the corresponding power-rev/min relation

$$\frac{P}{P_r} = \frac{N}{N_r}\left[2 - \frac{N}{N_r}\right] \tag{4.2}$$

where P is power. This is the equation of a parabola as plotted in Figure 4.1.

Figure 4.2 is a reproduction of power-rev/min curves for a Turbo Power & Marine (United Aircraft) engine, for three different ratings or levels of output.* These curves resemble the ideal curve of Figure 4.1, being parabolic (most easily checked on the normal curve), although the peaks lie off the curves to the right. It seems apparent that these peaks lie farthest to the right for the highest curve (max intermittent). It should be recalled that the ideal torque relationship is derived in Chapter 3 on the basis of a constant jet velocity, C_1, and that the rev/min for best utilization of that jet's energy was established by a speed ratio, U/C_1 (U = linear wheel speed). Or said another way, each torque curve or each power curve represents the behavior of the power turbine (or free turbine) with constant input from the gasifier: the compressor,

* The implications of *rating* are discussed in Section 4.3.

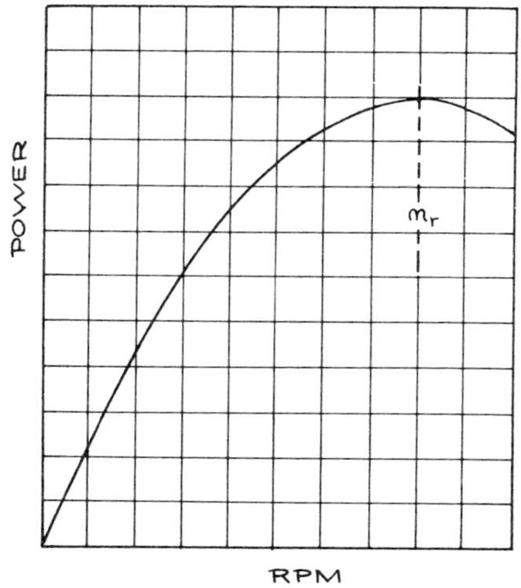

Figure 4.1. Power-rev/min curve for ideal impulse turbine.

combustor, and compressor-drive turbine. In going from one curve to another that shows higher output, it is thus implied that the output of the gasifier is correspondingly increased; the jet velocity is thus increased, and rev/min must be increased, therefore, in the same ratio if utilization is to be kept the same. If, instead, rev/min is kept constant, utilization must decline, and a given rev/min will represent a lower point on the utilization curve. Such is the case in Figure 4.2, where 4000 rev/min is apparently the maximum rev/min allowed for the free turbine.

Torque curves can be directly calculated from power curves, since

$$Q = 5252 \frac{P}{N} \tag{4.3}$$

for torque in lb-ft, and P expressed in horsepower. The curves of Figure 4.3 are calculated in just this way from the three power curves of Figure 4.2. The torque curves are indeed essentially straight and rise with decreasing rev/min as predicted for the ideal turbine. It is difficult to say if 0-rev/min torque is exactly twice best-rev/min torque, since the rev/min for highest utilization is not stated. However, an extrapolation of the normal curve in Figure 4.2 suggests that its peak should be at about 4200 rev/min. If so, then the corresponding (lowest) torque curve of Figure 4.3 can be seen to conform closely to the ideal.

Generally speaking, actual power-rev/min and torque-rev/min characteristics do conform quite closely to the shapes predicted by elementary turbine theory, even if the actual turbine is multistage and may involve some reaction.

The power-rev/min curves of Figure 4.2 stop well short of 0 rev/min, while those of Figures 4.1 and 4.3 are drawn all the way to 0, prompting a question concerning the ability of a gas turbine engine to operated at low rev/min or to produce torque at 0 rev/min. Like internal combustion engines in general, there is a minimum speed below which the engine is not self-sustaining. But it is the compressor-combustor combination to which this applies in gas turbine

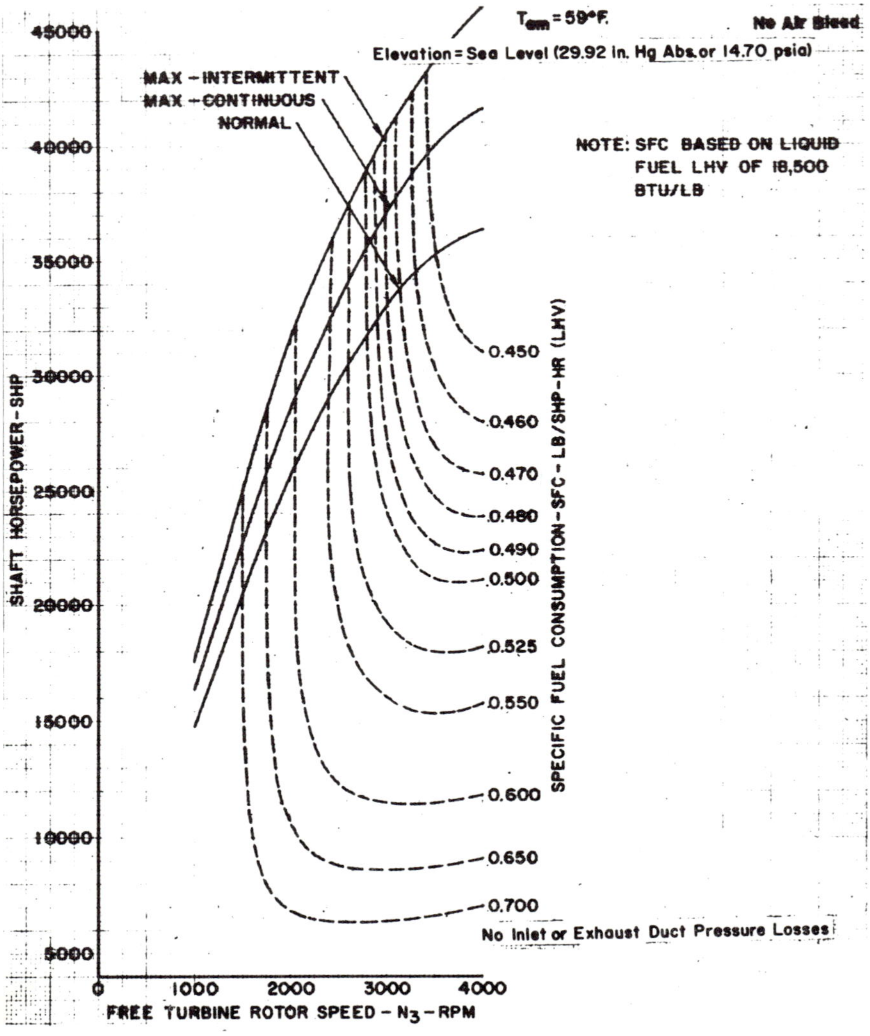

Figure 4.2. Power-rev/min curves for FT4C-2 engine (Turbo Power & Marine Systems).

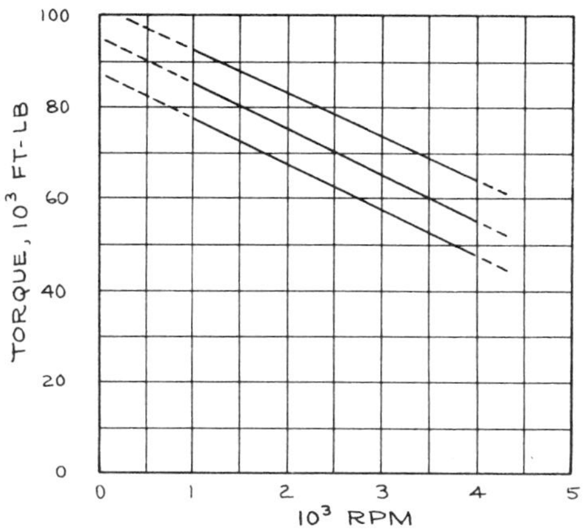

Figure 4.3. Torque curves corresponding to the power curves of Figure 4.2.

technology, and the power turbine may in fact run at low rev/min—or even stop—while the compressor is run at a self-sustaining speed. On the other hand, the marine propulsion load is actually so small at low rev/min that the compressor speed and fuel input must be cut back below the point at which its self-sustaining operation is possible. Thus as a practical matter in marine propulsion, very low rev/min operation under nontransient conditions is usually not feasible.

A word should be said about the torque-rev/min characteristic of a single-shaft gas turbine, even though its use for marine propulsion is rare. Since any change in rev/min of its output shaft is a change of compressor rev/min, no constant-input torque characteristic can be constructed. However, comparable information is conveyed by a curve that shows maximum torque that can be produced as a function of rev/min at a given turbine inlet temperature. The typical form of such a curve is shown by Figure 4.4, as constructed from data given in a paper that discusses propulsion by a particular single-shaft engine [McCoy (1968)].

4.3. ENGINE RATING

The *rating* of an engine is a term that usually designates the power that it produces in a particular application. A certain rev/min is always implied or stated also, since the development of the intended power depends on the requisite rev/min being reached. Thus the engine whose power-rev/min curves are pictured in Figure 4.2 is properly said to have a rating, 35,500 hp at 3600

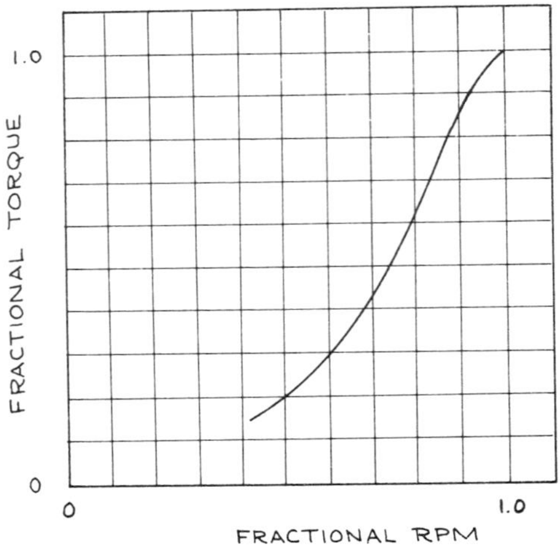

Figure 4.4. Maximum torque available from a single-shaft gas-turbine engine (typical form).

rev/min (using the normal curve). Values of other parameters under this condition are also described as being rated values, for example, rated torque and rated exhaust temperature.

The rather simple concept of a rating is complicated by several factors that are evident in Figure 4.2. The first is the presence of three power curves, hence three ratings, rather than one. The highest of these curves shows the power of which the engine is capable; the other two stem from the relationship between rating and service deterioration of the engine: the lower the rating—and hence temperature of the hottest parts—the longer is this time. It is sometimes said that this relationship is exponential, that is, time between overhauls increases exponentially as rating is decreased [*LM2500 Marine Gas Turbine Installation Design Manual* (1973)]. The maximum continuous curve and the rating it represents (40,400 hp at 3600 rev/min) is, therefore, the manufacturer's recommendation for continuous use with resulting reasonable time between overhauls. The "reasonable time" cannot be quantified in general, since its actual value depends on things other than rating, for example, quality of the fuel.

The normal curve represents a recommended level of operation for an extended life, or in effect, a "conservative" rating. Again, the life cannot be generally stated. The designer should expect to develop an estimate of life for a particular design by direct consultation with the engine builder.

The use of three ratings is arbitrary; many manufacturers publish only two.

The other complicating factor in ratings is the effect of external conditions.

Note in Figure 4.2 that the curves are specified to be for a definite ambient pressure and temperature, and for zero inlet and exhaust duct losses. Since duct losses are nearly always greater than zero, and ambient temperature is nearly always greater than the stated 59 F over some of the vessel's operating experience, the designer must carefully correct a rating for actual conditions. Procedures for this, and sample data, are discussed in Section 4.6.

4.4. FUEL-RATE CHARACTERISTICS

Fuel-rate characteristics are usually presented as contours on the power-rev/min plane. Figure 4.2 is an example, and the characteristics shown there are of typical form.

Two major factors are responsible for the worsening of fuel rates at low outputs. One of these is the changing speed ratio of the turbines and compressor; the other is the decrease in cycle efficiency that accompanies lower turbine-inlet temperature.

Recall that the torque and power curves of Figures 4.1, 4.2, and 4.3 are characteristics of the power turbine for a constant input, constant input implying constant fuel flow, and compressor rev/min. Along the power-rev/min curves in Figure 4.2, therefore, fuel rate increases toward lower powers are caused mostly by worsening utilization of the power turbine. A calculation shows that fuel consumption (= rate × power) along these curves is almost constant, indicating a constant output and rev/min for the gasifier portion of the engine.

If the power were to be changed along a vertical line, for example, in Figure 4.2 (this would occur in practice if propulsion power were being changed at constant rev/min by propeller pitch change), it would be accomplished by fuel-flow reduction. The consequences to the gasifier would be a reduction in its rev/min, a decrease in the fuel/air ratio, and a resultant lowering of gas temperature. The change in fuel rate seen along any constant-rev/min line in the righthand portion of Figure 4.2 is thus largely due to the reduction in efficiency caused by lower temperature and pressure at the turbine nozzle pressure: temperature lowers the nozzle velocity, thus lowering the optimal speed of the power turbine. Note that a minimum is evident in each of the fuel rate contours, a minimum that moves to the left as power is reduced; this is a consequence of the decrease in optimal speed as nozzle velocity decreases.

On left side of Figure 4.2, the fuel rate contours steepen and become essentially vertical, so that a change of power at constant rev/min is accomplished with no change in fuel rate. In this region of the figure, changes due to cycle efficiency and to utilization are offsetting along constant-rev/min lines.

If a propeller power-rev/min curve is superimposed on Figure 4.2 (as is done in Chapter 6; see Figure 6.2), we observe that the line of actual operation in marine service passes through the right-hand region of the figure, except at the lowest powers and rev/min. The increase in fuel rate at part loads experienced in marine service, therefore, is due mainly to the decrease in turbine inlet temperature. Since the same is true of most other loads likely to be served by gas turbines, features are sometimes incorporated in the engine to lessen the change in temperature with changes in load.

About the only practicable method of lessening temperature changes is to adjust the area of the power turbine nozzles. Since these nozzles are formed by the passages between blades (similar to the blades on the turbine wheel), rotating the blades simultaneously about their axes changes the flow area between each adjacent pair. It also changes the nozzle angle, α (see Eq. 3.9), but the major consequence comes from the change in area. Intuitively, it seems that an area reduction decreases the air flow rate for a given compressor speed, thus raising the fuel/air ratio for a given fuel flow and, consequently, raising the temperature leaving the combustor; such is generally the case in practice. The process is depicted graphically in Figure 4.5. It is illustrated there how shifting from one nozzle characteristic to another as a consequence of reducing

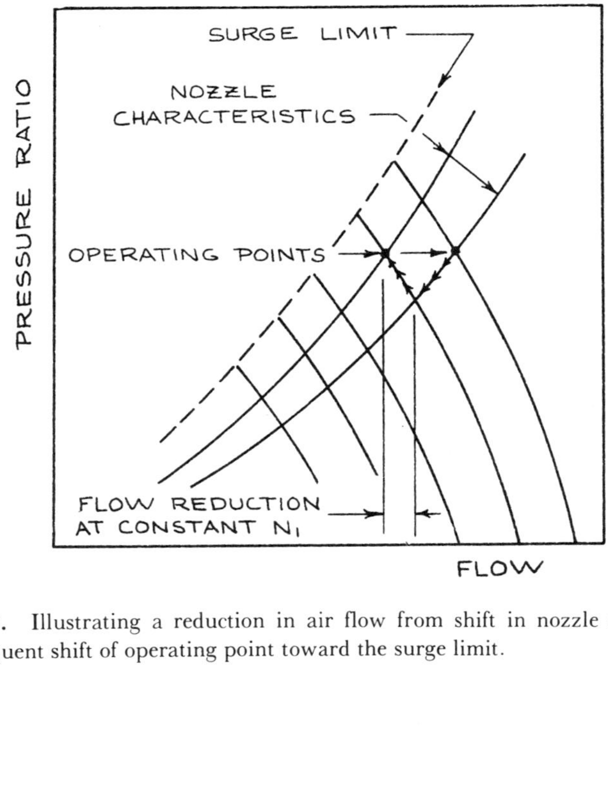

Figure 4.5. Illustrating a reduction in air flow from shift in nozzle characteristic, with consequent shift of operating point toward the surge limit.

area decreases air flow for a given compressor speed. Note, though, that such a shift moves the compressor operating point toward the surge limit. The process is, therefore, limited in range.

The Ford truck gas turbine is an example of those that use the nozzle-change technique to improve part-load efficiency. In this engine, the primary power control sets gasifier speed to a value appropriate to the power specified. A control that senses turbine-inlet temperature adjusts the nozzle angle to maintain design temperature. However, the process cannot be carried below 86% compressor design speed because the surge limit interferes in the manner just noted [Cadwell, Chapman, and Walch (1972)].

Because of the paucity of comparative data, it is difficult to say with assurance how much the nozzle adjustments improve part-load fuel rate. It is possible to compare fuel rate data of different engines, one having this feature, the other not, and one such comparison is tabulated here:

Percent power at constant rev/min	100	80	60	40
Percent fuel rate, Engine A	100	103	107	115
Percent fuel rate, Engine B	100	103	111	125

But this comparison should be viewed as a rough indication only, since there are many other differences between the engines. For example, one is a regenerative engine, the other is not.

4.5. EXHAUST FLOW AND TEMPERATURE

The weight rate of flow of air and exhaust gas are very nearly the same, the obvious difference being the weight of combustion products in the exhaust. Both flows are nearly proportional to compressor speed, unless modified by the nozzle control scheme discussed in the preceding section.

Exhaust temperature is a function of turbine-inlet temperature and of the energy extracted by the turbine. The former predominates, and, in fact, exhaust temperature is often monitored by control and safety systems as a good indicator of turbine inlet temperature. The latter is a function of load, as previously discussed, but is also significantly affected by ambient temperature. A rise, for example, in ambient temperature must appear as a rise in all temperatures throughout the engine cycle, but turbine inlet and exhaust temperature will change an even greater amount if power is kept the same.

The lesser density of the ambient air increases the fuel/air ratio, hence the additional temperature increase.

Figure 4.6 illustrates these points with graphical data on exhaust flows and temperatures for the General Electric LM2500 engine [*LM2500 Marine Gas Turbine Installation Design Manual* (1973)]. Note that an increase of ambient temperature at constant power causes a decrease in exhaust weight flow and a change of approximately twice the magnitude in the exhaust temperature. This particular engine does not employ nozzle control.

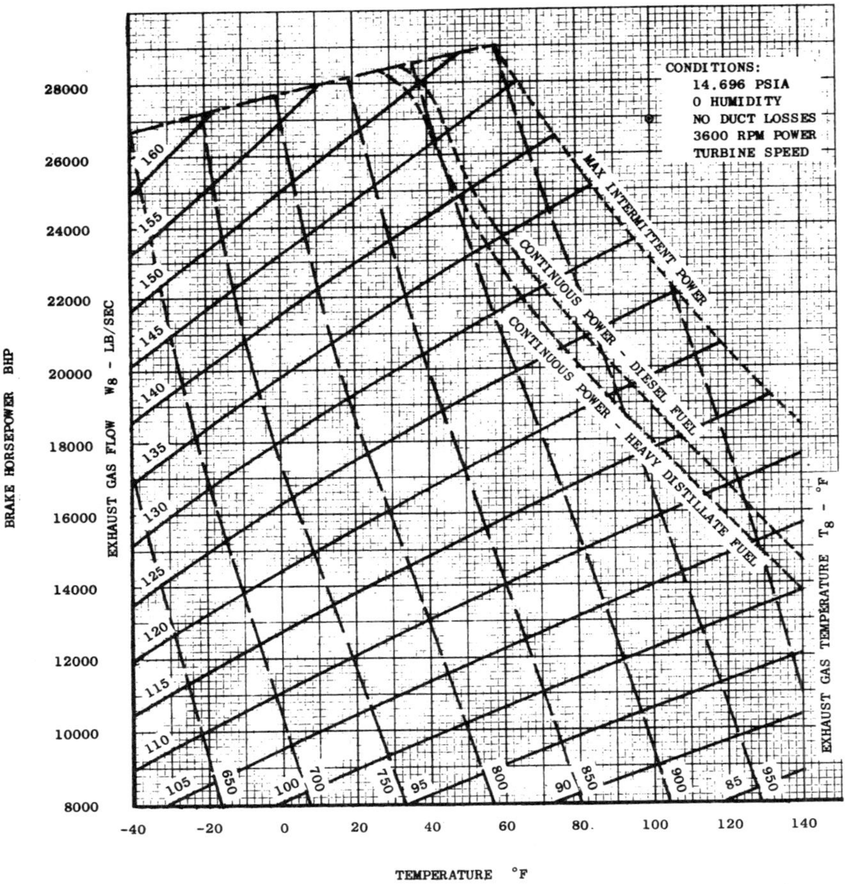

Figure 4.6. Ambient temperature corrections to rating—illustrated by curves for GE LM2500 engine [LM2500 Marine Gas Turbine Installation Design Manual (1973)].

4.6. CORRECTIONS

The output of a gas-turbine engine is affected by pressure and temperature of the air at its compressor inlet, by its exhaust pressure, and by the type of fuel. A rating, or curves such as those of Figure 4.2, must consequently always be accompanied by information on these factors. But, in nearly all cases, actual service conditions are different from those stated, and the ratings must be changed to suit. Fuel rate is likely to be different also.

Figure 4.6, used earlier for other illustrative purposes, serves as the rating correction curve for the General Electric LM2500 engine [*LM2500 Marine Gas Turbine Installation Design Manual* (1973)]. For instance, this figure shows a continuous rating of 24,000 hp (diesel fuel) at the standard temperature of 59 F; at 100 F the continuous power curve shows the rating to be 19,000 hp, and maximum intermittent power to be 23,000 hp (all at the rated 3600 rev/min). In other words, this engine cannot produce more than 23,000 hp if operated with 100 F inlet air, and if its time between overhauls is not to be shorter than that obtained at 24,000 hp at 59 F, it must be derated to 19,000 hp.

Ambient-pressure corrections are usually not provided for marine gas turbines, since sea-level pressure is not likely to vary enough to be significant. Of course, inland marine service at high elevation is possible: Lake Superior is about 600 ft above sea level. When such cases are faced, the engine builder must supply the additional correction data. Figure 4.8 is an example, being the correction curves of both pressure and temperature for the Ford engine.

Pressure at inlet and exhaust are often significantly affected by losses that occur in the ducts between engine and atmosphere. Power losses due to inlet and exhaust duct pressure drops are typically presented by curves such as those of Figure 4.7. Ratings are usually given for zero duct loss, since it is the ship designer, not the engine designer, who determines what these losses are to be. The former must, therefore, adjust the rating with the aid of curves such as those of Figure 4.7 after establishing the magnitude of duct losses.

Fuel rate is also affected by ambient conditions and duct losses, and correction data is usually supplied by the engine builder. Figures 4.9 and 4.10 are examples, again curves for the General Electric LM2500 engine [*LM2500 Marine Gas Turbine Installation Design Manual* (1973)] Figure 4.9 has correction curves for brake horsepower and rev/min also, a circumstance that requires a bit of explanation. These two curves are needed to establish the power and rev/min to use in finding fuel rate *before* correction. For example, if the rate were wanted for 20,000 hp at 3000 rev/min, with ambient temperature of 100 F, power and rev/min factors of 1.026 and 1.038, respectively, would be applied, giving 19,500 hp (= 20,000/1.026) at 2890 rev/min as an entry point

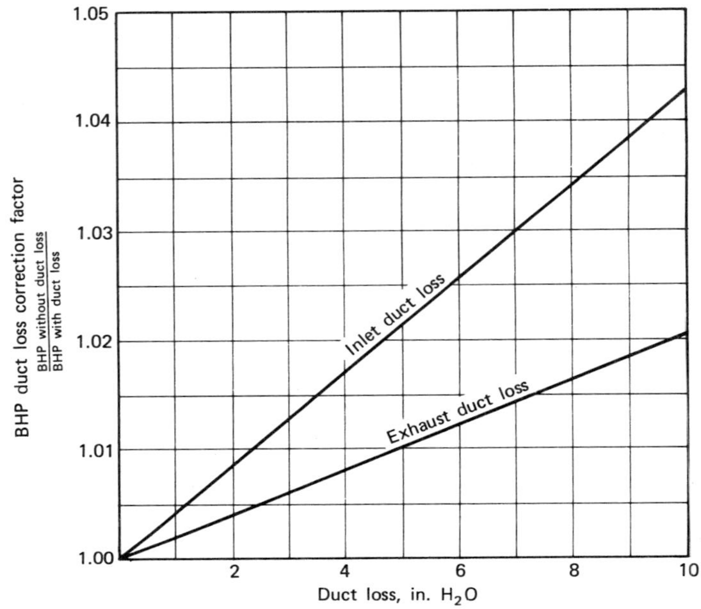

Figure 4.7. Corrections to rating for duct losses—illustrated by curves for GE LM2500 engine [LM2500 Marine Gas Turbine Installation Design Manual (1973)].

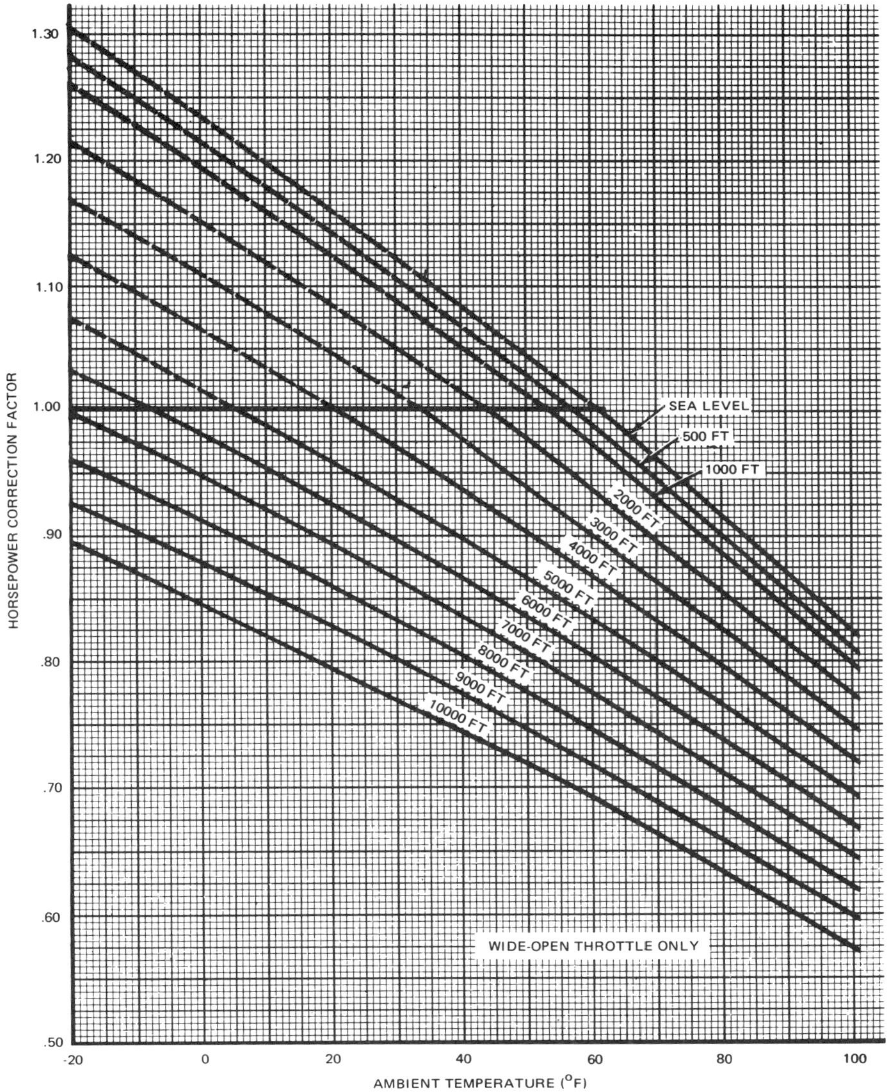

Figure 4.8. Temperature and altitude corrections to rating—illustrated by curves for Ford engine [Turbine Application and Installation Manual (1972)].

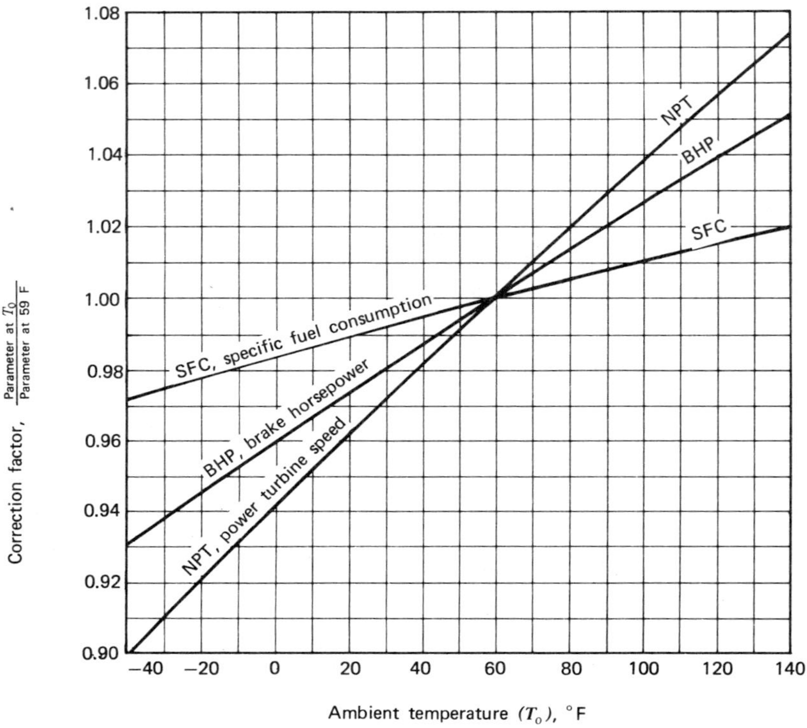

Figure 4.9. Corrections for ambient temperature—illustrated by curves for GE LM2500 engine [LM2500 Marine Gas Turbine Installation Design Manual (1973)].

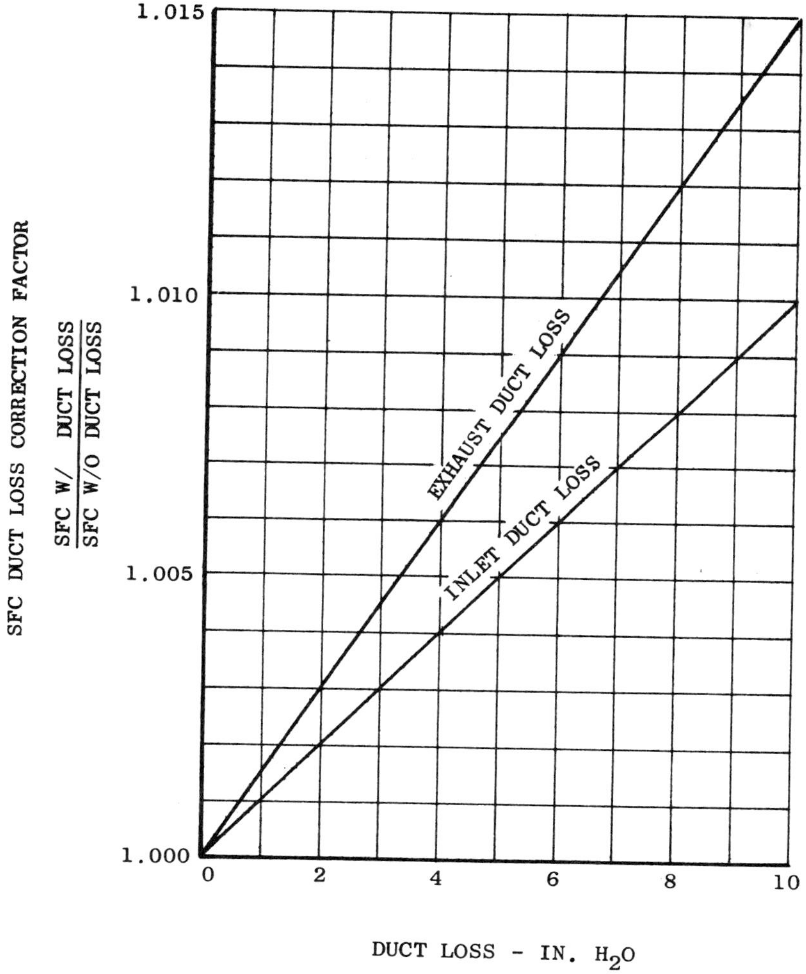

Figure 4.10. Corrections to specific fuel consumption for duct losses—illustrated by curves for GE LM2500 engine [LM2500 Marine Gas Turbine Installation Design Manual (1973)].

(on a chart such as Figure 4.2, usually) for finding the uncorrected fuel rate. The correction factor for fuel rate, 1.010, would then be applied (multiplied) to the value read to get the corrected rate. Correction factors from Figure 4.10 would subsequently be applied to correct fuel rate for the effect of duct losses.

4.7. REFERENCES

Cadwell, R. G., Chapman, W. I., and Walch, H. C. (1972), "The Ford Turbine—An Engine Designed to Compete with the Diesel," Society of Automotive Engineers Paper 720168, Automotive Engineering Congress.

LM2500 Marine Gas Turbine Installation Design Manual (1973), General Electric Company, MID-IDM-2500-2.

McCoy, A. W. (1968), "Gas Turbine Propulsion for High-Speed Small Craft," *Marine Technology,* **5,** 3.

Turbine Application and Installation Manual (1972), Ford Motor Company, Industrial Power Products Division.

Chapter

FIVE

PROPELLER CHARACTERISTICS

5.1. INTRODUCTION

Marine propulsion is effected by a device that converts the thermal energy obtained from consumption of a fuel into a thrust. Although this conversion can be accomplished by a single component (e.g., if a rocket were used for propulsion), it conventionally requires the coordinated working of three distinct units, namely,

1. The *engine,* which converts the thermal energy into mechanical energy. The mechanical energy is characterized by the torque and the rotational speed (revolutions per minute, or rev/min) of an output shaft. In essence: fuel in, torque out.
2. The *transmission,* which transmits the mechanical energy to its point of use. In its simplest form, the transmission is merely a rotating shaft; it often, however, includes a speed-reducing gear set and may include clutches, brakes, and couplings of various types; if the transmission is electric, motors, generators, cabling, and switchgear are used. But, whatever its form, it receives energy from a rotating shaft and delivers it to a rotating shaft. In essence: torque in, torque out.
3. The *propulsor,* which converts mechanical energy into the hydraulic energy of an accelerating fluid, or, speaking otherwise, into thrust horsepower. Most often, this propulsor is the familiar screw propeller, though the jet pump may be used for high-speed propulsion. Principles are much the same as either propulsor. In essence: torque in, thrust out.

A fundamental task in the design of a marine propulsion plant is that of matching these units, a process of ensuring that engine and propeller combine to produce an optimal fuel-to-thrust conversion under design operating conditions, while ensuring that all operating conditions are acceptable to each. In the preceding paragraph, it is noted that the engine delivers torque at a certain rev/min and that the transmission delivers torque at a certain rev/min to the propeller. The linking characteristic of engines and propellers is thus that which expresses the relationship between torque and rev/min.

A knowledge of torque-rev/min behavior of engine and propeller is, therefore, essential to propulsion plant design. As an alternative, power and rev/min serve the same purpose, because mechanical power transmitted to a rotating shaft is given by

$$P = \frac{2\pi QN}{33000} \tag{5.1}$$

where P = horsepower, Q = torque in ft-lbf, and N = rev/min, so that if you know power and rev/min, you also know torque. Thus the power-rev/min characteristic is equally useful. Sometimes it is more convenient, as when an engine builder furnishes this characteristic, rather than torque-rev/min. You will see in subsequent discussions that these two alternative ways of presenting the same information are used interchangeably.

The purpose of this chapter is to outline the torque-rev/min or power-rev/min characteristics of propulsors, with emphasis on screw propellers. The reader is assumed to be generally familiar with the principles of propellers, and with their terminology, to a level and scope implied by knowledge of Chapter VII in *Principles of Naval Architecture* [Todd (1967)]. The present chapter is not, therefore, intended to be a complete treatise on propellers, but treats only their characteristics that are important relative to the engine, which—to repeat—are torque-rev/min or power-rev/min. A section on the torque-rev/min characteristics of jet pumps is included.

A brief review of definitions of terms that may be encountered in engine-propeller work is given in the section on notation at the front of this book.

5.2. BASIC CHARACTERISTICS OF FIXED-PITCH PROPELLERS

Propeller behavior is typically presented by plots of nondimensional coefficients, derived from systematic model tests of a family of propellers. The most familiar coefficients are perhaps those of Taylor (B_P, B_U, and δ), and the most familiar propeller family is perhaps the Wageningen Series B [Todd (1967)]. Examples of the B_P–δ plots for Wageningen B propellers are shown in *Principles of Naval Architecture,* page 413 and following.

The B_P–δ and B_U–δ charts are useful in selecting a propeller for a particular application. More convenient for use in deducing torque-rev/min or power-rev/min behavior, or in analyzing the behavior of an individual propeller, are plots of the torque coefficient (K_Q) and the thrust coefficient (K_T) against the advance coefficient (J). They are defined by

$$K_Q = \frac{Q}{\rho D^5 n^2} \tag{5.2}$$

$$K_T = \frac{T}{\rho D^4 n^2} \tag{5.3}$$

$$J = \frac{V_A}{nD} \tag{5.4}$$

where Q is torque, T is thrust, ρ is density of water, D is propeller diameter, n is rotational speed (rev/sec), and V_A is propeller speed of advance. Units are chosen to make these nondimensional.

Open water efficiency (η_O) is usually presented with these parameters and is related to them by

$$\eta_o = \frac{JK_T}{2\pi K_Q} \tag{5.5}$$

K_Q, K_T, and J are related to B_P, B_U, and δ by the following relations:

$$K_Q = \frac{1.89 \times 10^7}{\frac{\rho}{g_0}} \frac{B_P^2}{\delta^5} \tag{5.6}$$

$$K_T = \frac{1.173 \times 10^6}{\frac{\rho}{g_0}} \frac{B_U^2}{\delta^4} \tag{5.7}$$

$$J = \frac{101.3}{\delta} \tag{5.8}$$

where ρ is in lbm/ft^3 and g_0 is 32.17 lbm-ft/lbf sec^2, for B_P, B_U, and δ in English units.

A typical plot of K_Q, K_T, and η_O against J is given in Figure 5.1. It was calculated for the Wageningen B4.55 propeller from the corresponding B_P–δ and B_U–δ charts by means of the formulas just stated.

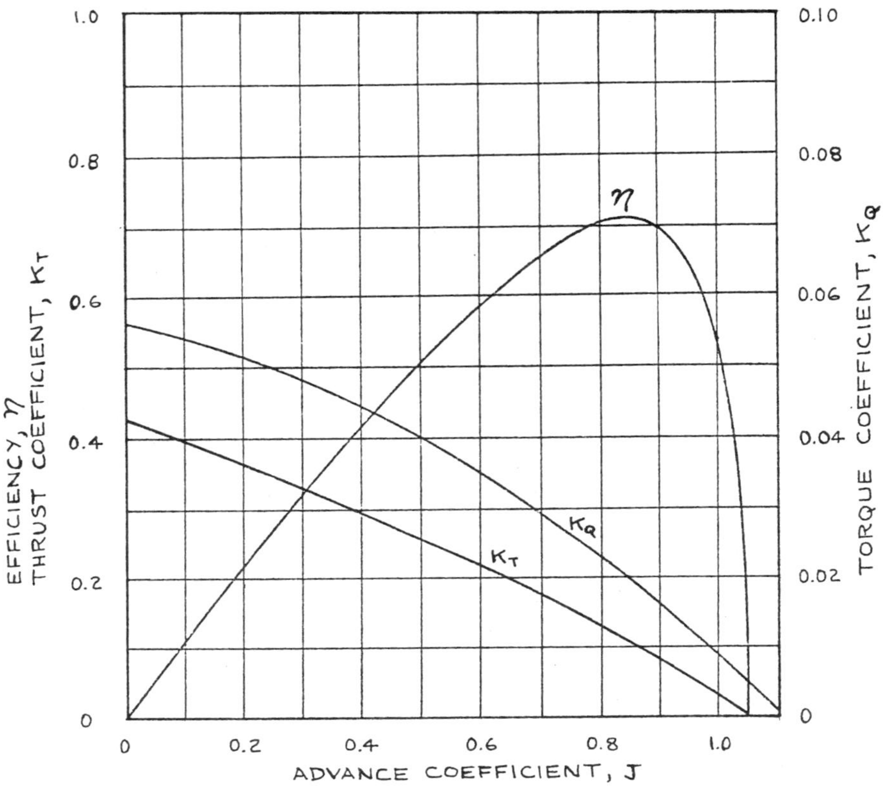

Figure 5.1. Thrust coefficient, torque coefficient, and open-water efficiency for a typical propeller.

5.3. TORQUE AND POWER CHARACTERISTICS OF FIXED-PITCH PROPELLERS

A. The Propeller Law

The torque-rev/min or power-rev/min characteristics of propellers can be found from the basic characteristics just discussed *if* the effects of interaction between hull and propeller are also known. For general discussions, an idealized concept of this interaction is often used, leading to the so-called "propeller law." The essential assumptions of this idealization are

1. Resistance is proportional to the square of vessel speed.
2. Wake fraction and thrust deduction are constant over vessel speed range, so that thrust is also proportional to speed squared.

A consequence of these assumptions is that J, K_Q, and K_T do not change as speed of the vessel varies. To show this, T in Eq. 5.3 is replaced by V^2 (using the assumptions above). Comparison with Eq. 5.4 shows that now $K_T \propto J^2$. The K_T line in Figure 5.1 obviously doesn't obey this parabolic relation, so the latter can be satisfied only at a point, that is, only if J doesn't change as V does. If J doesn't change, then V and N must be proportional over the speed range, η_O doesn't change, and because K_Q doesn't change, torque is proportional to (rev/min)^2 and shaft power is proportional to (rev/min)^3. These statements constitute the "propeller law," which may be summarized as follows:

1. Propeller rev/min is proportional to vessel speed.
2. Propeller torque is proportional to square of rev/min.
3. Propeller power absorbed and delivered is proportional to the cube of rev/min.

B. Deviations from the Propeller Law

A displacement hull operating at low speed-length ratios usually obeys the propeller law reasonably well. As the speed increases beyond the neighborhood of $V/L^{1/2} = 1$, however, resistance increases at rates greater than the square of the speed. The torque-rev/min and power-rev/min curves consequently become steeper, and the shaft rev/min likewise increases at a greater rate than the hull speed.

More radical departures occur with planing craft. In the planing regime, the resistance typically varies much less steeply than second power over a wide speed range. Figure 5.2, borrowed from *Principles of Naval Architecture* [Todd (1967)], shows a few examples, including a displacement hull for comparison. Figure 5.3 gives a further example for a planing boat. The right-hand

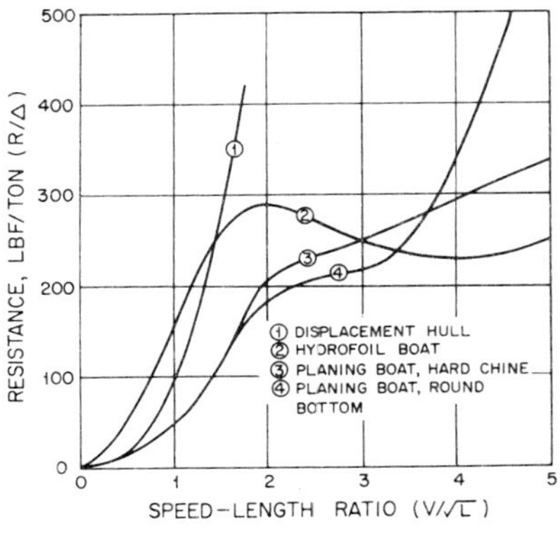

Figure 5.2. Typical resistance curves for several hull types.

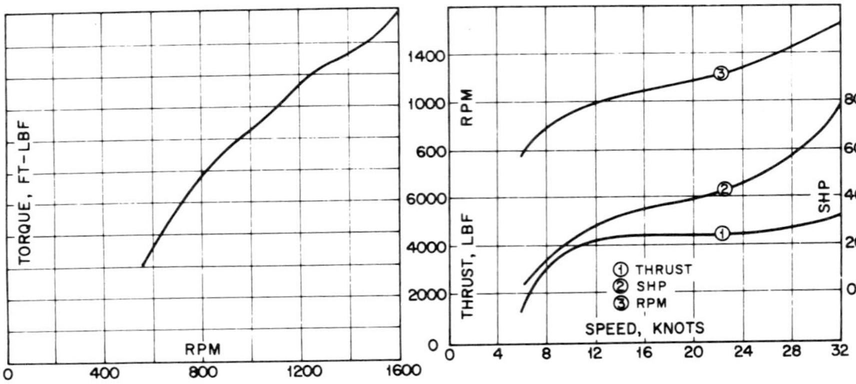

Figure 5.3. Typical thrust, power, and rev/min curves for a planing boat [Blount (1966)].

curves show thrust, power, and rev/min for a particular boat [Blount (1966)], and on the left is the torque-rev/min curve calculated from the power and rev/min curves.

Another important deviation from the propeller law occurs because of the many factors that can change the resistance characteristics of a hull during operation. Fouling of the underwater surfaces, in effect, changes the constant of proportionality between resistance and speed. Variations in weather and sea state have the same effect. Vessels that perform towing duties suffer the same phenomenon; their resistance at a given speed may be increased many times over its free-running value at that speed.

If resistance at a particular speed increases, so must power and rev/min required to produce that speed. Figure 5.4 gives an idea of the expected increases in rev/min and power, and Figure 5.5 shows the corresponding torque-rev/min characteristics. Note that some arbitrary increases in the resistance-speed constant of proportionality are chosen for illustration, but after each increase, the propeller law is presumed still in force (i.e., the new resistance curves are parabolic, new power curves cubic, etc.).

Calculation of the power, torque, and rev/min curves in Figures 5.4 and 5.5 requires use of the K_Q–K_T–J characteristics for a particular propeller. The method of calculation is outlined in Section 6.4, where it is used to find engine and propeller behavior in the presence of resistance increases such as those of Figure 5.4.

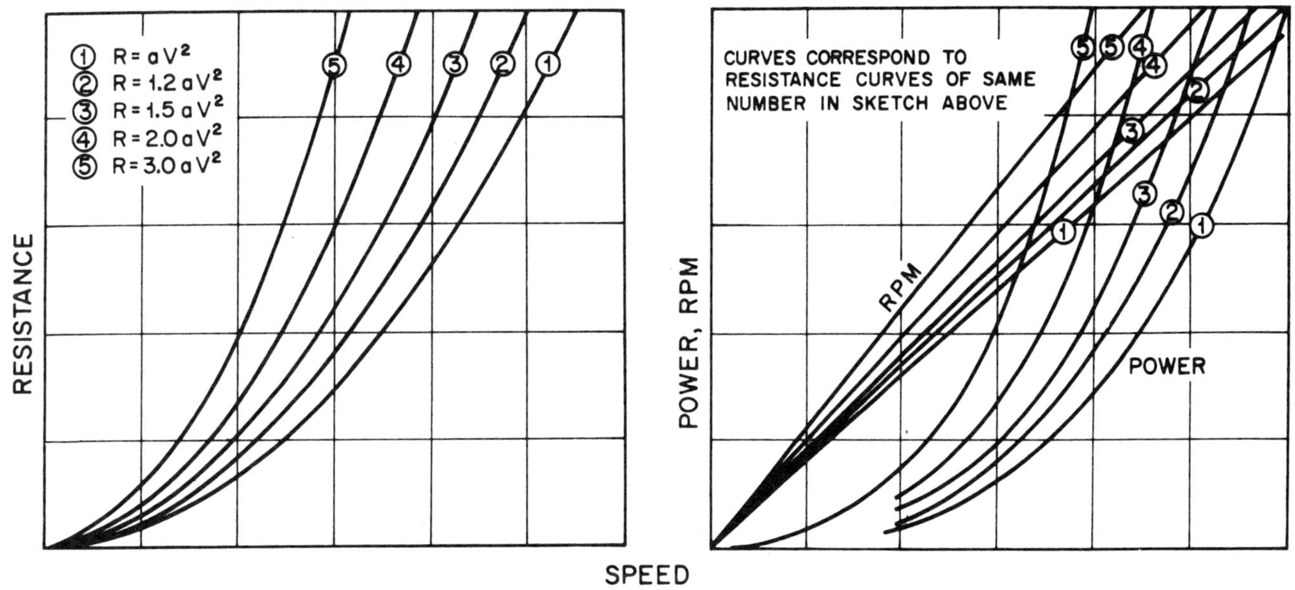

Figure 5.4. Resistance, power, and rev/min curves according to the propeller law.

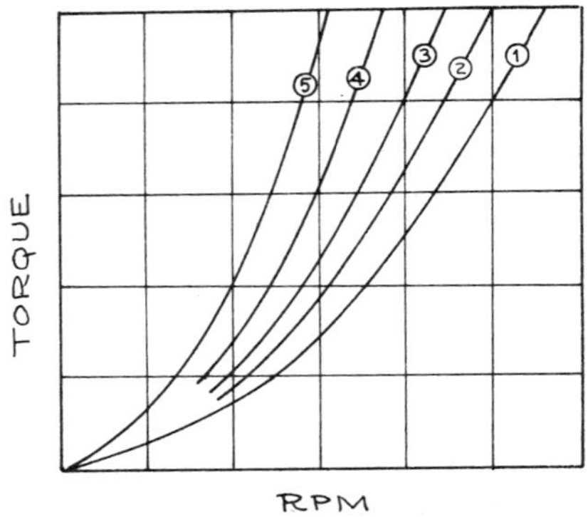

Figure 5.5. Torque curves corresponding to the power and rev/min curves of Figure 2.4.

5.4. CONTROLLABLE-PITCH PROPELLER CHARACTERISTICS

If the pitch of a propeller can be changed, the K_Q, K_T, and η curves must be repeated for each pitch ratio of interest. As an example, Figure 5.6 shows these parameters for pitch ratios of 0.6, 0.8, 1.0, and 1.2, B4.55 propeller, as in Figure 5.1. The values plotted are calculated from the B_P–δ and B_U–δ charts in the same manner as the curves of Figure 5.1.

Figure 5.7 shows a set of power-rev/min curves derived from the curves of Figure 5.6. These are cubic curves, calculated with the aid of propeller law assumptions and of parameter definitions, as follows:

1. The design point is assumed to be at $J = 0.8$, $H/D = 1.0$. On the power-rev/min plane, this point is plotted at (1.0, 1.0).
2. Points having the same rev/min, but different pitch ratios, are observed to bear the following relationships to the point just plotted:

$$\frac{J}{J_0} = \frac{V}{V_0} \tag{1}$$

$$\frac{K_T}{K_{T_0}} = \frac{T}{T_0} \tag{2}$$

$$\frac{K_Q}{K_{Q_0}} = \frac{Q}{Q_0} = \frac{P_S}{P_{S_0}} \tag{3}$$

$$\frac{T}{T_0} = \left(\frac{V}{V_0}\right)^2 \tag{4}$$

In all cases, the zero subscript implies the condition at the design point;

absence of such subscript implies the condition at some other pitch ratio. The first three equations are just the defining relations for J, K_T, and K_Q, respectively, all with $n/n_0 = 1$; the fourth is a statement of propeller-law conditions. All are written in ratio form to eliminate factors that do not change as pitch ratio changes.

3. The first, second, and fourth relations are combined to produce $K_T/K_{T_0} = (J/J_0)^2$, and this relation is plotted on the K_T–J plane, passing through the design point. (See Figure 5.6.)
4. The J value at each intersection of $K_T/K_T = (J/J_0)^2$ with a K_{T_0} characteristic is read.
5. The K_Q value for the same J (and at the same pitch ratio, of course) is read and observed to give the power ratio, P_S/P_{S_0}, directly by the second relation above.

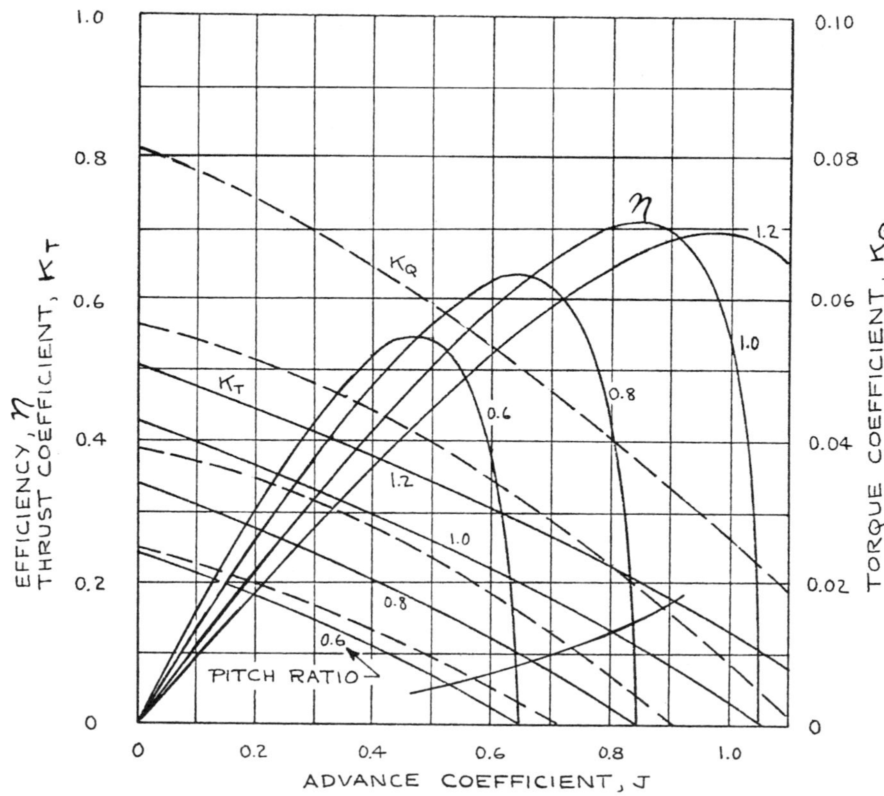

Figure 5.6. Curves of thrust coefficient, torque coefficient, and open-water efficiency for a controllable-pitch propeller.

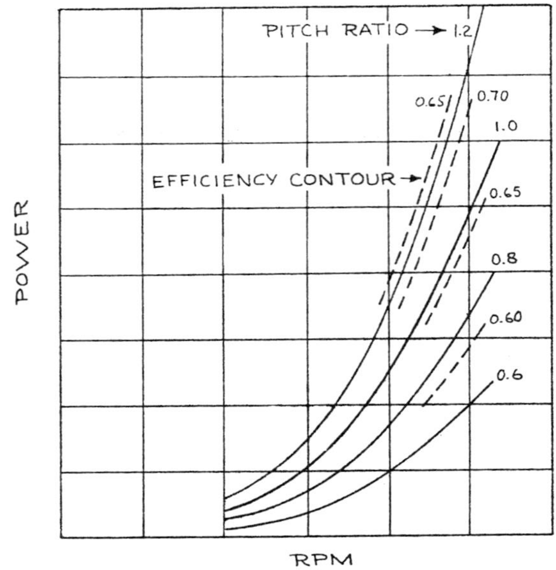

Figure 5.7. Power-rev/min curves calculated from Figure 5.6.

6. These values of P_S/P_{S_0} are plotted along the N/N_0 line on the power-rev/min plane.
7. Through each point now plotted, a cubic curve is passed to the origin.

A particular value of efficiency is associated with each power curve in Figure 5.7. The values can be read from the efficiency plots in Figure 5.6. These values are interpolated in Figure 5.7 to locate the efficiency contours at values of 0.60, 0.65, and 0.70.

A set of actual power-rev/min characteristics for a displacement vessel is given in Figure 5.8, principally to show the slight difference in efficiency contours from the propeller-law version just calculated from the propeller charts and plotted in Figure 5.7.

5.5. TRANSIENT CHARACTERISTICS

All vessels, needless to say, must accelerate and decelerate. During these transients, propeller thrust and torque differ widely from steady-state design values, since the ratio of its rev/min and speed of advance (i.e., the advance coefficient, J) is at times radically different from the design point. For example, during a crash reversal, the propeller is turning in reverse, developing negative thrust, with the hull decelerating but still going ahead.

If a designer wishes to analyze a transient, he will need to know both thrust and torque for all possible combinations of rev/min and speed of advance, that is, for either or both being negative. Very little propeller data of the necessary sort has been published; the principal source is a paper by Miniovich (1960). Figure 5.9 is a sample of his data, showing on separate plots the torque coefficient and thrust coefficient for $V_A > 0, n > 0$; $V_A > 0, n < 0$; $V_A < 0, n > 0$; and $V_A < 0, n < 0$. Note that both coefficients are presented in two different forms: K_Q and K_T versus J as herein previously defined, and K_Q' and K'_T versus J'. The latter set is defined by

$$J' = \frac{1}{J} \tag{5.9}$$

$$K'_Q = \frac{Q}{\rho D^5 V_A^2} \tag{5.10}$$

$$K'_T = \frac{T}{\rho D^4 V_A^2} \tag{5.11}$$

and is made necessary by the undefined state of J at $n = 0$. The J' coefficient suffers the same defect at $V_A = 0$, so that both of these sets are necessary.

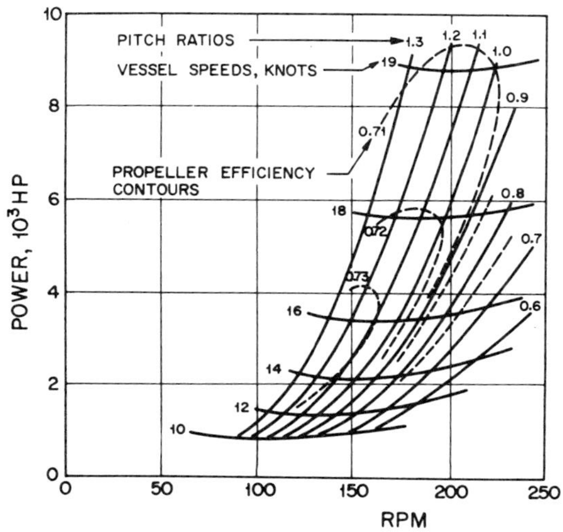

Figure 5.8. Power-rev/min curves for an actual controllable-pitch propeller.

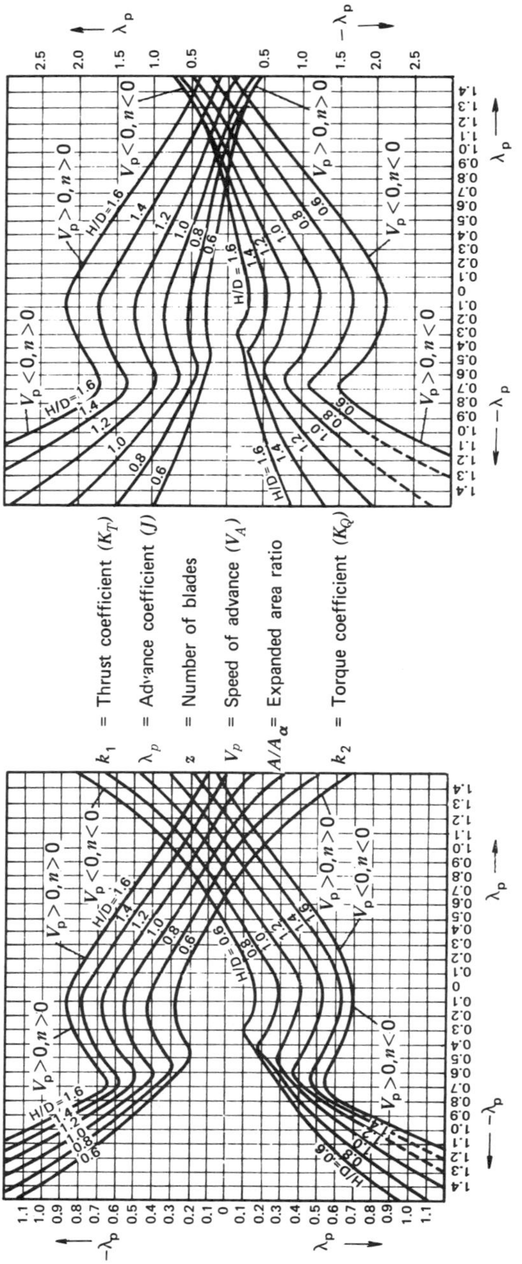

k_1 = Thrust coefficient (K_T)
λ_p = Advance coefficient (J)
z = Number of blades
V_p = Speed of advance (V_A)
A/A_α = Expanded area ratio
k_2 = Torque coefficient (K_Q)
$V_p > 0, n > 0$
$V_p < 0, n < 0$
$V_p < 0, n > 0$
$V_p > 0, n < 0$
H/D = 1.6
1.4
1.2
1.0
0.8
0.6
λ_p
$-\lambda_p$

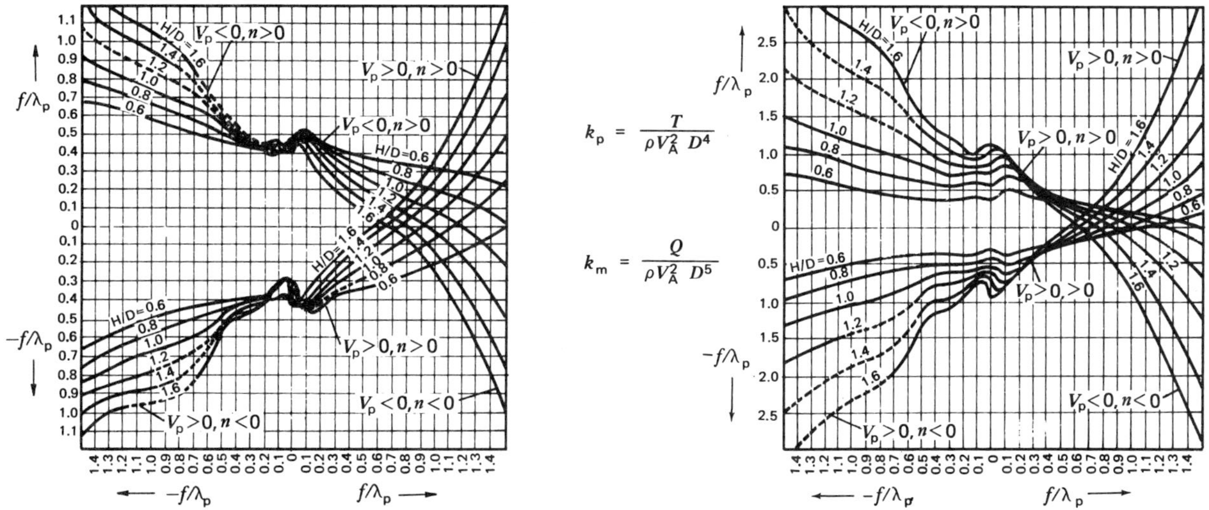

Figure 5.9. A sample of propeller transient data from Miniovich (1960).

A conventional way of representing the torque-rev/min characteristic during a transient is the Robinson diagram (named after an early user). Figure 5.10 is an example. It shows torque versus rev/min for several constant values of ahead ship speed. Its data can be calculated directly from the K_Q and K'_Q curves, although accuracy is improved if information is available on how wake fraction may change during the transient.

If a reversing transient with fixed-pitch propeller is being analyzed with the aid of a computer—digital, analog, or hybrid—the fact that both n and V_A become zero makes the storage and look-up of data awkward because of the multiplicity of curves with overlapping regions of applicability. The task is considerably eased by the angle-of-advance coefficient methods developed by Baker and Patterson (1972). One of their several similar techniques makes use of the following coefficients:

$$C_Q = K_Q \cos^2 \theta_P = K'_Q \sin^2 \theta_P \tag{5.12}$$

$$C_T = K_T \cos^2 \theta_P = K'_T \sin^2 \theta_P \tag{5.13}$$

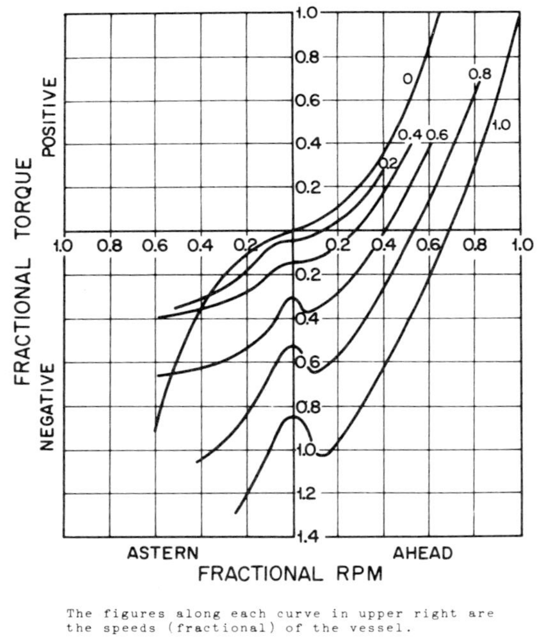

Figure 5.10. Torque-rev/min characteristics during stopping and backing (Robinson diagram).

where

$$\theta_P = \tan^{-1} J = \cot^{-1} J' \tag{5.14}$$

A big advantage is that thrust data and torque data for all combinations of V_A and n are now each shown by one single-valued curve. To illustrate, Figure 5.11 shows C_Q calculated from the K_Q and K'_Q curves of Figure 2.9 ($H/D = 1.2$) by means of the formulas just cited.

A controllable-pitch propeller should not approach 0 rev/min during a transient, hence J should remain well within bounds. The K_T and K_Q curves for $n > 0$, such as shown in Figure 5.10, therefore, should be adequate for a transient analysis.

5.6. CHARACTERISTICS OF JET PROPULSORS

The conventional propeller and the jet pump share many features. Both are members of the turbomachine family, and if the pump is of the axial-flow type, it resembles a propeller mounted in a duct. On the other hand, the pump may be of the centrifugal or mixed-flow type, and, in any case, the duct has an influence on propulsion behavior. The duct inevitably causes energy losses stemming from fluid friction. It also tends to isolate the pump from flow caused by forward motion of the hull, so that flow through the pump is determined principally by pump rev/min.

Two equations suffice as a basis for a discussion of characteristics. They are

$$T = \frac{\dot{m}}{g_0} (V_J - V) \tag{5.15}$$

$$H_P = \frac{V_J^2}{2g_0} + H_E + H_L - \frac{V^2}{2g_0} \tag{5.16}$$

where T is thrust, $\dot{m}$ is mass flow rate through the pump, V is speed of the boat, V_J is speed of the jet, H_P is head developed by the jet pump, H_E is elevation head (head to lift water to jet centerline), H_L is head lost in flow through the jet system, and g_0 is the unit conversion constant (lbm-ft/lbf-sec^2 for $\dot{m}$ in lbm; V and V_J in ft/sec; H_P, H_E, and H_L in ft-lbf/lbm).

Equation 5.15 is a momentum equation (see the discussion of the momentum theory of propulsion in *Principles of Naval Architecture* [Todd (1967)]. Equation 5.16 is an energy balance (*head* being mechanical energy per unit mass of water flowing) expressing pump head as the summation of kinetic energy supplied to the jet, the energy to lift the water to the jet cen-

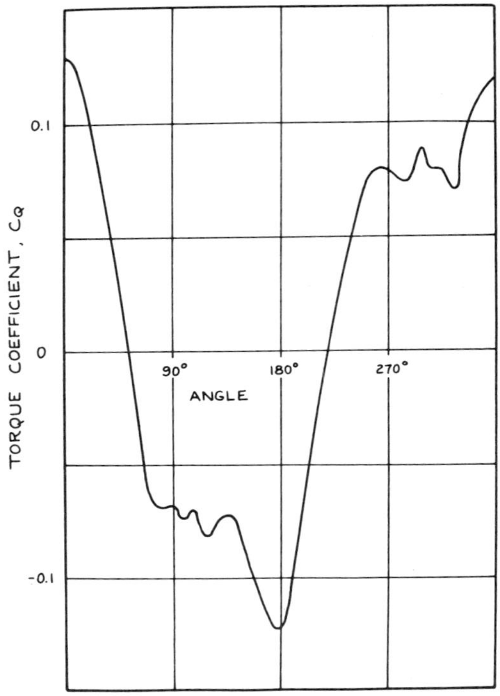

Figure 5.11. Torque coefficient, C_Q, calculated from data of Figure 5.9.

terline, the energy to overcome losses within the ducting, less the kinetic energy contributed by the forward motion of the boat.

Figure 5.12 represents Eq. 5.16 graphically, using example parameter values. H_P is represented by typical pump head-flow characteristic curves for several different rev/min (these curves are discussed in Chapter 3). These curves are intersected by curves representing the right-hand side of Eq. 5.16 for values of V ranging from 0 to 50 ft/sec. H_E is 10 ft-lbf/lbm (i.e., a 10-ft lift), and H_L is proportional to the square of V_J with a value of 10 ft-lbm/lbm when $\dot{m}$ (which is proportional to V_J) is 1420 lbm/sec. The relative influences of boat speed and of pump speed on flow through the pump can readily be seen. For example, stopping the boat from 50 ft/sec reduces flow about 12% if pump speed is constant at 2000 rev/min. But if pump speed changes at constant boat speed, flow changes almost proportionally to rev/min. These results do depend somewhat on the type of pump: some axial-flow pumps may have steeper head-flow curves than those shown, making flow more nearly

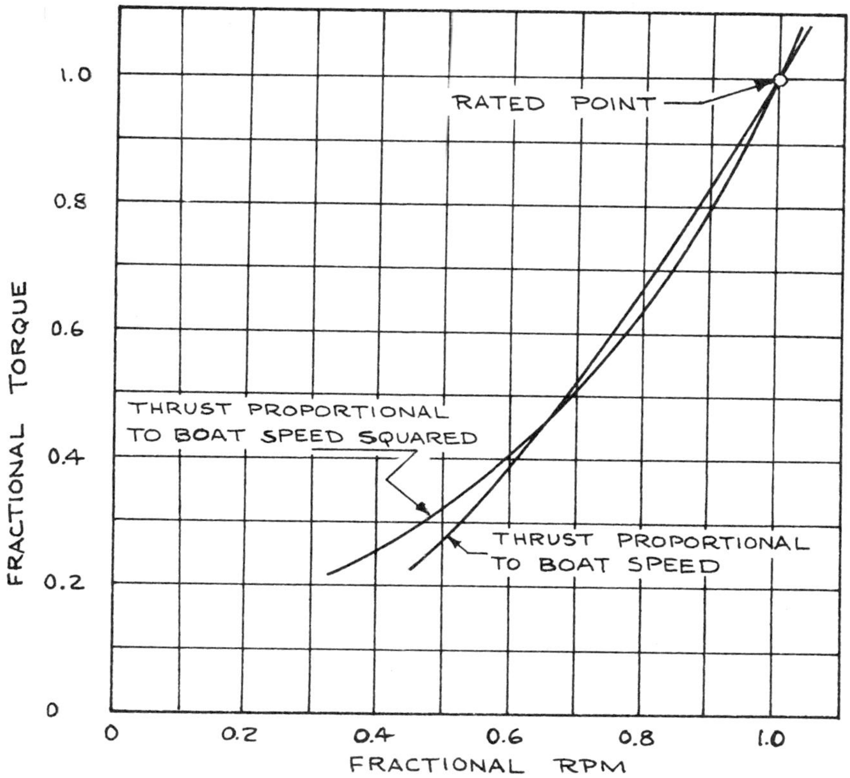

Figure 5.13. Example of jet pump torque-rev/min curves, calculated from Figure 5.12 data.

The *Taylor coefficients,* B_P, B_U, and δ, are

$$B_P = \frac{NP_D^{0.5}}{V_A^{2.5}} \tag{5.17}$$

$$B_U = 0.05541 \frac{NS^{0.5}}{V_A^2} \tag{5.18}$$

$$\delta = \frac{ND}{V_A} \tag{5.19}$$

where V_A is speed of advance in knots, N is rotational speed in rev/min, P_D is delivered horsepower, S is thrust in lb, and D is diameter in ft.

Pitch ratio is the ratio of effective pitch (H) to tip diameter (D).

Effective horsepower (ehp or P_E) is the power required to move the vessel at speed, V, against resistance, R:

$$P_E = \frac{RV}{550} \tag{5.20}$$

for R in lbf and V in ft/sec. It is sometimes referred to as *towrope horsepower,* since it is the power required to tow and thus is the power predicted by a model resistance test.

Thrust horsepower (thp or P_T) is the power delivered to the water by a propeller-producing thrust, T, and moving at speed of advance, V_A:

$$P_T = \frac{TV_A}{550} \tag{5.21}$$

for T in lbf and V_A in ft/sec.

The relationship between P_T and P_E is conventionally represented as follows:

$$R = (1 - t)T \tag{5.22}$$

$$V_A = V(1 - w) \tag{5.23}$$

where t is the *thrust deduction* and w is the *wake fraction.*
Therefore,

$$P_T = \frac{RV}{550}\frac{(1 - w)}{(1 - t)} = P_E\frac{(1 - w)}{(1 - t)} = \frac{P_E}{\eta_H} \tag{5.24}$$

The definition of *hull efficiency* (η_H) is implicit in Eq. 5.24.

Delivered horsepower (P_D or dhp) is the power delivered to the propeller by its shaft. It is, of course, greater than P_T because of propeller inefficiency. Thus

$$\eta_P = \frac{P_T}{P_D} \tag{5.25}$$

Propeller efficiency (η_P) is usually regarded as being the product of two terms, the *open-water efficiency* (η_O) and the *relative rotative efficiency* (η_R), the former being efficiency predicted for the propeller operating far from the distrubing effects of a hull and the latter being a correction factor to account the hull influence.

Now relating P_E to P_D via Eqs. 5.24 and 5.25, and the preceding discussion:

$$P_E = \eta_H\eta_O\eta_R P_D \tag{5.26}$$

constant at constant rev/min, while centrifugal pumps may have curves less steep in their best operating region.

But what of torque? First note that boat speed and jet speed (hence flow through the pump) are almost certainly going to vary together in some way, that is, the speed of the boat is changed by changing the speed of the pump. If the propeller law assumption of resistance and thrust proportional to vessel speed squared is applicable, then Eq. 5.15 is satisfied over a range of speeds only if V_J is proportional to V. The dashed line in Figure 5.12 indicates the right-hand side of Eq. 5.16 in the case of such a proportionality. This curve is a parabola passing through the elevation head, H_E, at zero flow. The contours (not shown in this figure) along which flow is proportional to rev/min, pump head is proportional to (rev/min)^2, and pump efficiency is a constant, are parabolas passing through the origin. Along such a contour, also, the torque required by the pump is proportional to (rev/min)^2. Except at low flows and speeds, the dashed curve almost coincides with the contour passing through the

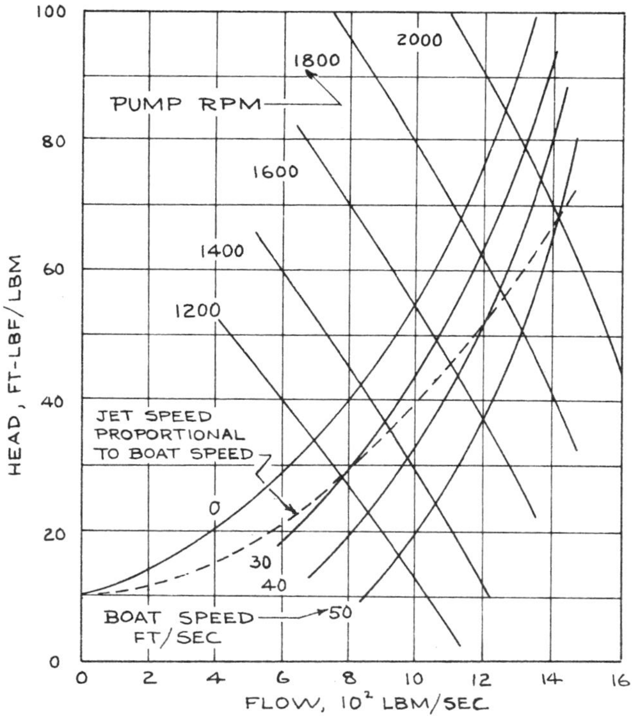

Figure 5.12. Equation 5.16 represented graphically, using typical parameter values.

rated point ($V = 50$ ft/sec, flow = 1420 lbm/sec), so that torque is nearly proportional to (rev/min)^2. The propeller law thus produces the same torque-rev/min characteristic for both propeller and jet pump, save for the distortion in the latter case produced by the elevation head, H_E. As in the example here, H_E is usually small compared to the other components of pump head.

Since jet propulsion is most commonly used in high-speed vessels that are expected to operate in the planing regime, resistance-proportional-to-speed-squared is usually an unrealistic assumption. If the relationship between resistance and speed is known for a case of interest, it can be used with Eqs. 5.15 and 5.16, and pump characteristic curves, to find the appropriate torque-rev/min curve. The procedure for finding a point on this curve is sketched by the following steps:

1. Since T and V are presumably known, Eq. 5.15 gives $\dot{m}$ and V_J ($\dot{m}$ and V_J are proportional).
2. H_P can be calculated from Eq. 5.16, assuming that $H_L \propto V_J^2$.
3. Spotting the H_P, $\dot{m}$ point on the pump-characteristic curves allows rev/min to be read.
4. Pump power can be calculated from H_P, $\dot{m}$ and pump efficiency (efficiency would be shown on the pump curves, though omitted here in Figure 5.12).
5. Pump torque can be calculated from its power and rev/min.

Figure 5.13 shows two torque-rev/min curves calculated in the manner just outlined, using the data from Figure 5.12 and the 2000 rev/min, 50 ft/sec intersection as the starting point for both. One curve is based on thrust (or resistance) proportional to boat speed, and the other on thrust proportional to boat speed squared. Pump efficiency is assumed to be essentially constant over the range shown and was taken thus in the calculation.

5.7. DEFINITIONS

As stated earlier, the reader is expected to be generally familiar with propellers, hence some terms are used in this chapter without being defined. However, just in case a reminder is needed, those terms not previously defined are given in this section, along with several others that may be encountered elsewhere in engine-propeller work.

matching to the propeller needs to be considered, as well as the problem of matching the auxiliary itself.

If a controllable-pitch propeller is used, some aspects of the matching problem are eased, since pitch can be adjusted in service. But this second degree of freedom also allows operation of the propulsion plant at unfavorable points, so that the designer usually removes the second degree of freedom by linking pitch control and throttle control in a definite program. Construction of this program is considered here as a major adjunct of the matching problem.

6.2. POWER-SYSTEM PRINCIPLES

The matching of engine and propeller is an application of the principle of conservation of energy. In the present context, the principle is this: power produced by the engine must equal the power absorbed by the load. In a simple application, this statement comes close to being trivial and may be intuitively obvious. Things are a bit less obvious in complex cases, as when several engines drive the same load, or when an engine simultaneously drives several diverse loads. It is, therefore, worth the effort to pay heed to some generalities before looking specifically at engines and propellers.

Consider first that however power is transmitted, it is characterized by two factors. In the case of transmission by a rotating shaft, these factors are speed (rev/min) and torque. Electrical transmission involves voltage and current; hydraulic involves pressure, flow rate; and so forth. Always, one factor must be the same at both ends of the transmission link (barring leakage). In the mechanical case, rev/min is the factor that must be the same at both ends. Torque may be less at the output end than at the input end if bearings or other devices absorb energy along the shaft. If there are no such absorbers, then torque also must be the same at both ends.

Consider then a simple case: a driver (engine) linked to a load (propeller) by a shaft without bearings or other intermediate absorbers of energy. Shaft rev/min must be the same for both, and since conservation of energy requires that power (i.e., energy rate) be the same for both, torque must be the same for both also. Each of the two units will have a torque-rev/min characteristic, that is, a relationship between its rev/min and the torque that it can produce, or absorb, at that speed. If these characteristics are plotted on the same graph, their intersection shows the torque-rev/min coordinates at which both factors are equal for both units and, therefore, shows the operating point for the load-driver combination. (See Figure 6.1.) This is the point of equilibrium between load and driver, and is their only possible steady-state operating point.

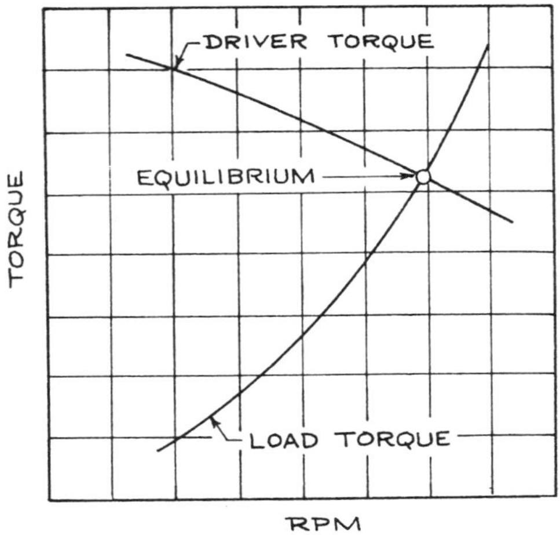

Figure 6.1. Driver-load equilibrium.

Usually this pair of curves intersects in only one point, and usually the intersection is such that the equilibrium is a stable one. Stability can be inferred from the characteristic curves. Note in Figure 6.1 that if a momentary aberration of some kind caused speed to increase above the point of intersection, torque demanded by the load to maintain that rev/min would exceed that available from the driver, hence the machine would decelerate toward the intersection. Likewise, if speed slowed, the driver torque would exceed load torque, and the machine would accelerate toward equilibrium. Since deviations are automatically reduced to zero, the equilibrium is stable.

Changes in torque due to bearing losses, and in rev/min due to reduction gearing, occur in practice, and complicate this discussion—but only slightly. If a reduction gear is used to change rev/min between driver and load, one should lump the gear with the driver, that is, take driver rev/min (and hence load rpm) to be engine rev/min divided by gear ratio and take driver torque to be engine torque multiplied by the gear ratio, less any torque losses in the gear set. Torque losses in the shaft bearings can be subtracted from the driver torque, or added to the load torque, before the plot of Figure 6.1 is made. In this fashion, the problem is reduced easily to the simple one of driver and load.

The preceding paragraph gives a hint of the technique to follow in analysis of yet more complicated cases: in a word, lump. This means to combine (lump) all load characteristics into a single characteristic, combine all driver characteristics into a single characteristic, and then find the intersection as in Figure 6.1. The following example illustrates: a diesel engine and a gas turbine drive a propeller through a common reduction gear, with an electric generator being attached to the same gear set; the problem is to find propeller

The product, $\eta_H \eta_O \eta_R P_D$, is called the *quasi-propulsive coefficient* (QPC) (sometimes *quasi-propulsive efficiency*).

Shaft horsepower (P_S or shp) is frequently used almost synonymously with P_D. It is, however, power delivered to the propeller as measured inboard of the stern tube and so is larger than P_D by the amount of stern tube losses. The engine builders' definition of P_S is the power as measured at the reduction gear output coupling and thereby agrees with the standard definition only if there are no intermediate bearings. When the propeller is driven through an inboard-outboard drive, P_S is defined as the power delivered to the propeller and in this case is, therefore, the same as P_D.

The ratio of P_E to P_S is known as the *propulsive coefficient* (PC) (or *propulsive efficiency*).

Brake horsepower (P_B or bhp) is the power delivered by the engine, usually thought of as being measured at the engine output coupling.

A word about notation: although P_B, P_E, and so forth are convenient symbols for formula use, bhp, ehp, and so forth are generally used hereafter to conform to more common usage.

5.8. REFERENCES

Baker, D. W., and Patterson, C. L., Jr. (1972), "Some Recent Developments in Representing Propeller Characteristics," *Naval Engineers Journal,* **84,** 35–44.

Blount, D. L. (1966), "Smooth Water Powering Characteristics of an Experimental LCVP(T)," David Taylor Model Basin Report 2283.

Miniovich, I. Ya (1960), *Investigation of Hydrodynamic Characteristics of Screw Propellers under Conditions of Reversing and Calculation Methods for the Backing of Ships,* BuShips Translation 697, U.S. Government Printing Office.

Todd, F. H. (1967), "Resistance and Propulsion," Chapter VII of *Principles of Naval Architecture,* Society of Naval Architects and Marine Engineers.

Wennberg, P. K. (1962), "Design Considerations and Selection of Controls for Vessels Equipped with Controllable Pitch Propellers," Society of Automotive Engineers, Paper 609B.

Chapter

SIX

MATCHING ENGINE AND PROPELLER

6.1. INTRODUCTION

The matching of an engine and its propeller is a design process that seeks to establish the optimal fuel-to-thrust conversion under rated operating conditions, while ensuring that all possible operating conditions are acceptable to each component.

The process is basically one of working with the power-rev/min or torque-rev/min characteristics of engine and of propeller. Conservation of energy demands that the power produced by the engine (minus any loss in transmission) equal that absorbed by the propeller. Engine torque, multiplied by the reduction-gear ratio, likewise must equal propeller torque and must do so at the common rev/min. Since the respective characteristics are always represented graphically, the matching technique involves finding the intersections (i.e., points where torque, power, and rev/min are equal) of these curves, then adjusting engine or propeller parameters to place these intersections at the desired locations.

The process may be complicated by changes that occur in service. Among these are minor changes in hull resistance because of fouling, weather variation, or draft changes; changes in number of engines running (in multiengine installations); and possibly the gross change in resistance that occurs if a tow is taken on.

In some instances the propulsion engine also drives auxiliary loads, such as electric generator or hydraulic system pump. The effect of such loads on

general fact that the higher the reduction gear ratio—whether stock or special—the higher the cost of the gear.

With respect to the availability of engine sizes, there is little that the designer can do save design for a vessel speed that is attainable with the available power ratings. The little which he can do within the scope of engine-propeller matching is to adjust the rating a moderate amount up or down from the nominal value, accepting some penalty for doing so. If he adjusts it upward, engine life will be reduced; if downward, efficiency is lessened, and some investment is being wasted, since, in effect, a larger engine than needed is being purchased. To illustrate, suppose that in Figure 6.2 the power at point B is required, rather than at A. If the propeller is given a lesser pitch, its operation may be shifted to the curve passing through B. Since the engine controls will allow no rev/min greater than rated rev/min, B becomes the equilibrium point. A glance back at Figure 4.2 shows that the fuel rate is likely to be poorer at B. Investment is wasted, since an engine of A horsepower is to be bought to produce only B horsepower. This suggests the alternative of actually using the full capability of the engine by designing the vessel for a higher speed that may increase its earning capacity or other utility. Such wider-horizon considerations are beyond the scope of this chapter, however, so we shall not explore this path beyond the hint just given.

Now suppose that the problem of engine choices does not face the designer, but that he is constrained by cost considerations to using stock-model reduction gears. To contrive an example, let's say that the available ratios are 2:1, 3:1, 4:1, and 5:1. Suppose further that the rated rev/min of the engine is 2400, that the rev/min of the best propeller is 540 when developing the required propulsive power. The resulting situation is pictured in Figure 6.3, based on the tentative assumption that the best propeller would be used. The propeller curve, with rev/min changed by two different ratios, is shown. Note that the choice of reduction gear is either 4:1 or 5:1, and that neither allows the engine to develop full power (A) under rated conditions. If the 5:1 gear is used, the engine would reach its rated speed—and presumably maximum speed for practicable operation—before reaching rated power; if the 4:1 gear is used, it would reach the rated turbine inlet temperature before reaching the rev/min necessary for rated power. Between the two, the 4:1 gear looks to be much the better alternative, because the power reached is higher, and because it is likely to be of lower first cost. Indeed, the choice of the 4:1 gear may be satisfactory, since the difference in power between points A and C is small (about 5% in this illustration). But if the 4:1 gear is chosen, two subsequent alternatives are available to the designer in seeking the optimal match, these being:

1. Operate the engine slightly above its best rating, so that the desired pro-

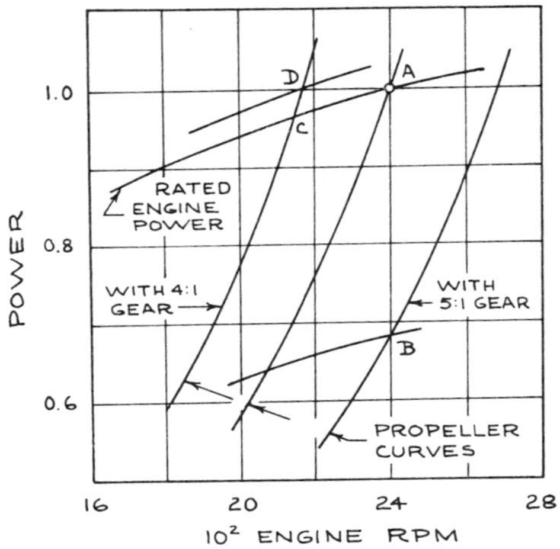

Figure 6.3. Matching engine and propeller—a choice between two reduction ratios.

pulsive power can be reached (point D rather than C). The penalty would be a slightly shortened time between overhauls.

2. Use a propeller of slightly less pitch, causing it to turn fast enough at rated power to operate at the best engine conditions, point A. The penalty would be a slightly lesser propeller efficiency.

The basic process of matching engine and propeller is thus one of shifting two curves on the power-rev/min plane (or torque-rev/min plane, if preferred) to place the equilibrium point at the optimal position. Moving the propeller curve is usually effected by pitch changes; moving the engine curve usually means requiring the engine to operate at a different rating. The propeller curve is also shifted by changes in reduction gear ratio. (When plotted in terms of engine rev/min, that is, if the abscissa of Figure 6.3 were propeller rev/min, a gear-ratio change would shift the engine curve.) As already noted, each such shift is likely to cause some change not directly shown on the plot: change in propeller efficiency, fuel rate, cost of reduction gear, or time between overhauls. The designer would weigh these factors, trading off among them in deciding which location of the equilibrium point is optimal.

B. Additional Points

Whatever propeller power-rev/min curve is chosen in design, it is unlikely to remain unchanged in service. If hull resistance increases from any cause, such as roughening of the surfaces by corrosion, a given power will require a higher rev/min under a specified set of operating conditions. Let us suppose that torque-rev/min curves are available for propeller, generator, diesel, and gas turbine and that reduction gear power loss is known for the specified condition. Reduction gear ratios for the diesel, gas turbine, and generator (all with respect to propeller rev/min) are r_d:1, r_{gt}:1, and r_{gen}:1, respectively. Proceed as follows:

1. Change the scales on the diesel torque-rev/min plot by dividing rev/min scale by r_d and multiplying torque scale by r_d.
2. Do same for the gas turbine, using r_{gt}, of course.
3. Observe that conservation of energy requires that total drive torque equal the sum of diesel torque and gas-turbine torque. Since steps 1 and 2 have adjusted their rev/min to a common basis, simply add their individual torques to get a total driver torque.
4. Divide power lost in the reduction gear by rev/min (using suitable constants for consistent units) to get the lost torque.
5. Subtract the gear-lost torque from the curve found in 3. This now represents the torque-rev/min behavior of the driver—the lumped combination of diesel, gas-turbine, and reduction gear.
6. Now work on the loads, lumping them in similar fashion. First, change the scales on the generator torque-rev/min plot by dividing rev/min by r_{gen}, multiplying torque by r_{gen}.
7. Conservation of energy requires that total load torque equal the sum of generator and propeller torques. Add their torque curves to produce the total load torque-rev/min characteristic.
8. The problem is now reduced to one of a single driver and a single load. Torque-rev/min characteristics of both are on the same scale. So simply intersect the two curves, as done in Figure 6.1.

An example illustrating this technique more specifically is given in Section 6.5.

It should be observed that a power-rev/min characteristic conveys the same information as torque-rev/min, since power is proportional to the product of torque and rev/min. The analysis discussed above could, therefore, be done as well with power-rev/min plots, and one finds in following sections that power-rev/min is used frequently in lieu of torque-rev/min. It is mostly a matter of convenience, of using whichever form of the data is furnished in the instance at hand.

6.3. MATCHING ENGINE AND PROPELLER

A. The Fundamental Problem

A certain thrust power is needed to drive the vessel at its design speed. This power is developed b, a propeller—or other propulsor, such as a jet pump—that ideally should do so with the least shaft (input) power, that is, with the highest possible efficiency. The input power is developed by an engine that ideally should operate at rated conditions suitable for the application, that is, at its rated rev/min and at a turbine-inlet temperature that will allow the optimal engine life. If a reduction gear is available with the correct ratio so that engine and propeller can each operate at its desired rev/min, then engine and propeller are perfectly matched, and there is no "problem." On the power-rev/min plane, the situation can be represented by the heavy curves of Figure 6.2, it being implied that the engine curve is the rated engine curve (e.g., perhaps the continuous-power curve of Figure 4.2), that the propeller curve represents the performance of the propeller of highest efficiency allowed by other factors (such as a limit on diameter), and that the intersection at A provides the power required for the rated service of the vessel.

But several factors may spoil easy perfection and thereby call for some effort by the designer in attaining the best possible solution under the circumstances. These are (1) the availability of only a limited number of sizes of gas turbine engine, (2) the relative cheapness of stock reduction gears as opposed to special designs, (3) the limited number of these stock reduction gear ratios, and (4) the

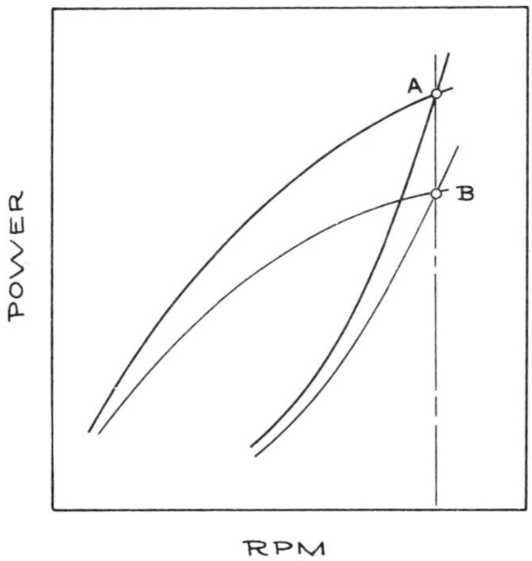

Figure 6.2. Matching engine and propeller on the power-rev/min plane.

Shaft speed can nominally go all the way to zero, but at low rev/min the power required to keep the shaft turning may fall so low that the gas generator part of the engine drops below its self-sustaining speed, and operation therefore cannot be maintained. If sustained low-speed operation is required, some additional feature in the engine-propeller drive train is needed. An example is a pneumatic clutch, partially inflated to allow slippage between engine rev/min and propeller rev/min, with the friction torque of the clutch furnishing an additional load on the engine. Another scheme, perhaps more commonly used with gas-turbine propulsion, is the controllable-pitch propeller. For low vessel speed, its pitch can be reduced to a low value while rev/min remains high. Because the propeller becomes very inefficient, its input power remains high enough to load the engine adequately for self-sustained operation.

B. Multiple Engines

The use of multiple engines connected to the same propeller shaft is common. It may be done merely to obtain the total power needed from available engines, or one of a pair may be a cruising engine, that is, one much smaller than the other. The purpose in the latter case is to improve part-load efficiency; when the vessel is at cruising speed, the small engine provides all of the propulsion power, operating at its rating.

With multiple engines, the torque and power characteristics are the sums of the characteristics of the individual engines, as illustrated by Figure 6.5 for

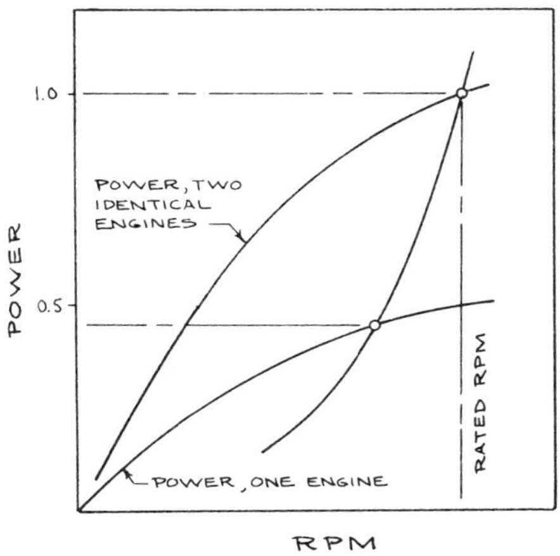

Figure 6.5. Equilibrium points for two identical engines operating together and alone.

two engines. One might say that drawing a figure to show that the whole is the sum of its parts is a trivial exercise, but the figure demonstrates a significant phenomenon that might not otherwise be obvious: note that although the two engines are identical, neither operating alone can provide one-half power when operating at its rating. As the figure demonstrates, the load curve is such that a single engine cannot reach its rated rev/min and hence cannot reach its rated power.

C. Towing Loads

If a vessel takes another in tow, its resistance at any given speed is increased, and the propeller torque-rev/min or power-rev/min curve shifts to the left. Qualitatively, the effect is the same as when any such shift occurs: the engine slows, and there is some consequent loss in power unless the engine is loaded more heavily to make up for the loss in rev/min. This very thing has been discussed in Section 6.4.A with respect to the small increases in resistance that usually occur in any service. With towing, however, a major increase in resistance can occur, and the possible difference in magnitude warrants an additional discussion.

Figure 6.6 shows on both ehp-speed and shp-rev/min planes what happens when a tow is taken that quadruples resistance (and hence ehp) at any vessel

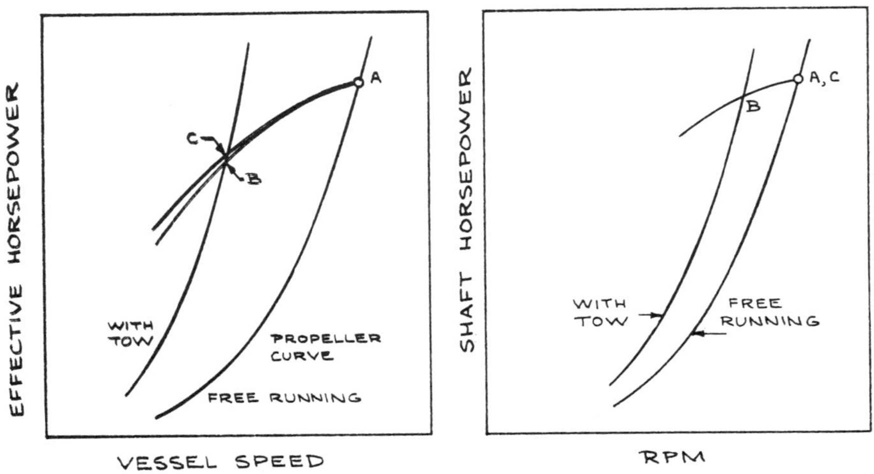

Figure 6.6. The effect of taking a tow on vessel speed, rev/min, shaft horsepower, and effective horsepower.

propeller rev/min. The propeller curve thus tends to shift to left in service. To use Figure 6.3 for illustration, if the propeller curve through A is chosen in the original design, it tends to become a curve such as the one through C after service deteriorations have occurred. The loss of power evident in the figure, therefore, must be accepted, or the engine must be operated above its rating to maintain the original power (point D).

The same phenomenon occurs with other types of engines, since it is necessary for any engine to develop full rev/min in order to develop full power without over-temperature or some other aspect of overload. For the diesel engine, the situation is as pictured in Figure 6.4. The power-rev/min characteristic is typically a straight line with slope proportional to the mean effective pressure of the engine. If the propeller curve shifts to the left in service, rev/min and power fall to point B, unless the mean effective pressure is increased. Since the thermal and mechanical loadings on vital engine parts are functions of this pressure, the increase is undesirable, just as an increase in gas-turbine temperature above the rated value is undesirable. It is, therefore, general practice in diesel propulsion to provide a margin in design. The propeller curve passing through C in Figure 6.4 would be selected. As it shifts leftward in service, the engine rated rev/min is kept the same by increases in the mean effective pressure. If the designer's margin were well chosen, average service conditions would find the engine actually operating at its rating, point A.

So much for diesel-design practice, well established over many thousands of vessels. Should the same practice carry over into the gas turbine field? The answer is a qualified no, based on the difference between the diesel charac-

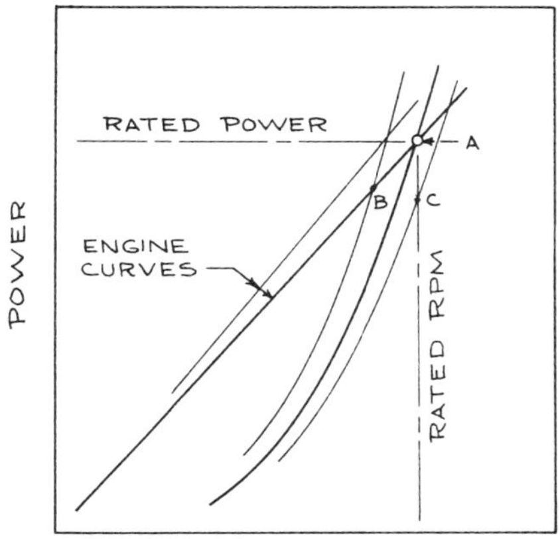

Figure 6.4. Engine-propeller equilibrium for the diesel, showing loss in power caused by shift of the propeller curve.

teristic and the gas turbine characteristic. If the turbine engine-propeller match is such that the engine operates at the peak of the power curve, that is, at the point of maximum utilization, the curve is virtually flat. A small loss in rev/min caused by a shift of the propeller curve will thus entail a trivial loss in power, one that can be recovered by a trivial increase in turbine temperature. (The shift of the propeller curve is likely to be much less than suggested by Figures 6.3 and 6.4; these are exaggerated for clarity.) Only in unusual circumstances does it seem necessary to consider a margin to allow for propeller-curve shift. If, for example, the propeller-engine match were for some reason set so that the engine operated on the steeper part of its power-rev/min curve, the situation would be much like that of the diesel engine, and the same sort of margin might be advisable.

Similar words should be said about uncertainty in matching engine and propeller. There are always small (hopefully small!) inaccuracies in testing, in design, and in manufacture, so that it is unlikely that an engine or a propeller will perform precisely as the curves predict. The point of match—the power and the rev/min—are unlikely, therefore, to fall exactly on the expected spot. Point A, for example, in Figure 6.3 (or any other such point) is not actually a point, but a disc of some finite radius. The designer may wish to allow margins for this phenomenon, but unless he has reason to be especially concerned about reaching an exactly specified power without exceeding an exactly specified temperature (and there will be uncertainties in determining these things in service), margins for uncertainty appear to be unnecessary.

6.4. OFF-DESIGN OPERATION

A. Low-Speed Operation

The torque-rev/min and power-rev/min characteristics of the gas-turbine engine have been introduced in Chapter 4, and the finding of an operating point by intersecting either of these curves with the corresponding propeller curve has been discussed in this chapter. It has been stated that the only possible operating point is at the equilibrium represented by the intersection. But operation at low speeds is an obvious necessity, and with a fixed-pitch propeller this is accomplished only by cutting back the fuel flow. The result is not a violation of equilibrium, but a new equilibrium at the point where the propeller curve intersects the engine curve that represents its behavior at reduced fuel flow. Note though, that this implies a movement of the equilibrium point along the propeller curve as fuel is cut back, so that the in-service characteristic of the engine is dictated by that of the load.

point for his attention. Although such matters are a bit beyond the scope of this text, it might be noted that increasing the loading on a propeller that is originally near its optimum inevitably decreases its efficiency, so that generally any attempt to improve efficiency at point B is likely to be a fruitless one. On the other hand, in attempting to keep engine power from declining, the designer has several options, very much like those he may otherwise face in engine-propeller matching, Section 6.3.1. He may intend that the engine be overloaded when towing, with consequent shortening of its time between overhauls, or he may design the propeller with less pitch so that when towing it turns at the rev/min intended for point A. If towing were to be the usual service condition for the vessel, the latter would be the obvious thing to do. But when running free, a power and speed loss would occur because the engine would reach its rated rpm before becoming fully loaded. Its operation would be like that of point B in Figure 6.2. Another alternative is use of a controllable pitch propeller. If pitch can be changed in service, then the same rev/min can be kept under all conditions of lading. The engine operation at constant rating would follow contour A–C in Figure 6.6 and, in fact, would do so for any type of engine. Note, though, that changing pitch does not maintain ehp, for propeller efficiency declines with loading, as previously remarked.

It is appropriate to explain here how the contours in Figures 6.6 and 6.7 are constructed. The essential ingredients are data on the torque coefficient of the propeller, a statement of the torque-rev/min characteristic of the engine, definitions of the parameters of interest, and assumptions or data on the variation of wake fraction and thrust deduction with vessel speed. For the last, it is here assumed that these two factors are constant. All of the necessary relationships used in finding the contour through A and B in Figure 6.6 (a reminder: this is the ideal gas-turbine torque-rev/min relationship mapped onto the ehp-speed plane) are set forth below. Most of these are conveniently presented as ratios, with the zero subscript representing the value of the quantity at point A.

$$\frac{Q}{Q_0} = 2 - \frac{N}{N_0} \tag{6.1}$$

$$\frac{J}{J_0} = \frac{V}{V_0}\frac{N_0}{N} \tag{6.2}$$

$$\frac{K_Q}{K_{Q_0}} = \frac{N_0^2}{N^2}\frac{Q}{Q_0} \tag{6.3}$$

$$\frac{K_T}{K_{T_0}} = \frac{T}{T_0}\frac{N_0^2}{N^2} \tag{6.4}$$

$$K_Q = K_Q(J) \tag{6.5}$$

$$K_T = K_T(J) \tag{6.6}$$

$$\frac{P_E}{P_{E_0}} = \frac{TV}{T_0 V_0} \tag{6.7}$$

where V is vessel speed, N is rev/min, T is thrust, P_E is ehp, and Q is torque. The first of these is the engine torque-rev/min relation, the next three are the definitions of the propeller advance coefficient (J), torque coefficient (K_Q), and thrust coefficient (K_T) formed into ratios, the next two denote graphical relations of K_Q and K_T to J for the particular propeller, and the last is the definition of effective horsepower.

There are seven equations with eight unknowns (J, Q, K_Q, K_T, P_E, V, N, T). It is thus possible by eliminating the unwanted variables to produce a relationship between any two of them. In plotting, it may be more convenient to calculate directly, rather than establishing the relationship in symbols. For example, in plotting the A–B contour in Figure 6.6, one proceeds as follows:

1. Choose a value of the ratio, N/N_0, say 0.9.
2. Find from this Q/Q_0 by Eq. 6.1.
3. Use N/N_0 and Q/Q_0 in Eq. 6.3 to find K_Q/K_{Q_0}.
4. Enter the appropriate propeller chart with this value, and read J and K_T. The base value of the propeller parameters must be known, of course, so that the J/J_0, K_Q/K_{Q_0}, and K_T/K_{T_0} ratios can be established or translated into chart readings.
5. With the values just found and the chosen value of N/N_0, calculate V/V_0 by Eq. 5.2 and T/T_0 by Eq. 6.4.
6. Use Eq. 6.7 to calculate P_E/P_{E_0}.
7. Plot P_E/P_{E_0} versus V/V_0.
8. Repeat with other values of N/N_0 until sufficient points are obtained to establish the contour.

If shaft horsepower is wanted, it can be found by the product of torque and rev/min, that is,

$$\frac{P_S}{P_{S_0}} = \frac{QN}{Q_0 N_0} \tag{6.8}$$

If a contour of constant shaft horsepower is to be plotted, with this condition obtained by keeping both torque and rev/min constant, the equations above are greatly simplified, for both N/N_0 and Q/Q_0 are 1.0. The propeller must be of the controllable-pitch type, and a chart showing values of K_Q and K_T for the different pitches must be used. An appropriate procedure, analogous to the

speed. The curve passing through points A and B is the ideal gas-turbine torque curve mapped onto the respective planes in the figure; at point A the turbine is assumed to be operating at its speed of maximum utilization. The increased resistance causes the rev/min to fall, so that shaft horsepower decreases. Propeller efficiency also falls, so that the decrease in effective horsepower is steeper than that of shaft power. (The figure was constructed with the aid of data for a particular propeller, taken to be operating at peak efficiency at point A.) The net result is operation at point B. The loss in vessel speed is appreciable, due mainly to the increase in resistance, but also due in part to the decline in shaft power.

The difference in vessel speed due to loss in power is demonstrated by also plotting a contour of constant shaft horsepower on the left sketch in Figure 6.6. The plot is based on the assumption that rev/min might be kept constant by decreasing propeller pitch (implying, of course, the use of a controllable pitch propeller), hence the contour is merely a point in the right sketch. In the left sketch, the equilibrium point for the towing condition is point C, so that the loss in vessel speed previously noted due to decline in engine power is the difference between points B and C as measured on the abscissa.

The difference in speed between points B and C appears to be trifling. The reason lies in the rising torque curve of the gas turbine. As noted in Chapters 3 and 4, torque ideally doubles as the engine slows from rated rev/min to 0 rev/min, and, as a result, the power-rev/min curve is nearly flat in the vicinity of rated rev/min, a behavior that is approximated quite well by practical engines. As a result, the two contours in Figure 6.6 lie very close together over the range of the plot. The result is somewhat different for the diesel engine, for which the torque curve is ideally flat, making the loss in power proportional to loss in rev/min. This loss is often regarded as significant in diesel-powered towing vessels, and for them often justifies means of keeping rev/min constant. [See discussion by Vibrans (1956).] Some idea of the difference between the ideal gas-turbine behavior and ideal diesel behavior in this situation can be obtained from Figure 6.7. The ideal gas-turbine contour from Figure 6.6 is repeated (curve A), and an ehp-speed contour for constant torque is plotted (curve C). The greater loss of power along the latter curve is evident.

It may be concluded tentatively that possible towing loads need not concern the designer in matching engine and propeller, since measures he might take to prevent or compensate for the loss in engine power produce only trivial results. Some qualifications to this conclusion should be noted, however. First, the conclusion depends somewhat on the magnitudes chosen in constructing Figure 6.6. Recall that the tow was said to increase the resistance by a factor of four. This was an arbitrary choice. If the factor were ten, or a hundred, both of which are quite possible, the difference between points B and C would be greater, for the two contours are diverging toward the left. Second, it was

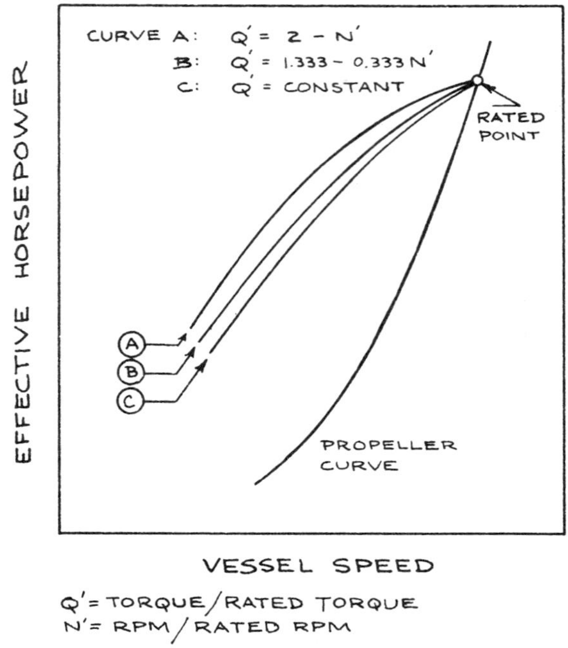

Figure 6.7. Contours on the eph-speed plane for several different torque characteristics.

assumed in constructing Figure 6.6 that the engine was to operate at its best rev/min in the free-running condition, point A, and the close approximation to constant-shp behavior derives from the near flatness of the power curve in the vicinity of that rev/min. If the engine were rated at a lower rev/min, the A–B contour would be steeper. An example is shown in Figure 6.7. The curve labeled B in that figure is calculated from a torque characteristic, $Q' = 1.333 - 0.333N'$ (Q' is torque divided by rated torque; N' is rev/min divided by rated rev/min), which states the torque for an engine whose rated power occurs at one-half its rev/min of highest utilization. The resulting contour more nearly resembles the ideal diesel contour than the ideal gas-turbine contour. The conclusion may be amended, therefore, to say that towing loads are likely to be a concern to the designer only in the face of some extreme condition of towing or of engine rating.

Suppose, nonetheless, the designer does wish to improve performance under towing conditions. A large part of the decline in ehp is due to a decrease in propeller efficiency (assuming that it was designed for high efficiency to start with), so that an attempt to keep this efficiency high seems to be the obvious

power available to turn the output shaft. Should he be confronted with a "gross" horsepower for *any* kind of engine, he must deduct the power required for the auxiliaries before proceeding to any phase of his propulsion design.

Hopefully, the designer will not be faced with the need to correct gross horsepower, but he may want to add additional auxiliary loads for ship-service purposes. A generator for the electrical needs of the vessel may be driven by the propulsion engine; this is a widespread practice in small vessels and is sometimes seen in even the largest. Circulating pumps for lube-oil cooling water, bilge pumps, and hydraulic-system pumps may also be driven in the same way. The power required for these units obviously subtracts from that available for propulsion and thus should be a factor in the engine-propeller matching process. If the power is quite small compared to propulsion power, the designer might safely neglect it, but if the auxiliary power is 20% (a rough guide only) or more of rated propulsion power, he should analyze its effect on the best match. An example, following, will illustrate.

Suppose that a gas turbine engine rated at 400 hp drives a fixed-pitch propeller whose power-rev/min characteristic is cubic. A hydraulic system is used to drive certain deck machinery, and the system pump is driven by the propulsion engine. This pump may absorb power that is a significant fraction of propulsion power; the designer is consequently concerned with the rev/min and power that will remain available for propulsion when the pump is running.

Recall from Section 6.2 that the technique advised for determining the operating point of a complex system of loads and drivers is to lump them into a single driver and a single load. Here the items involved are a piping system, a pump, a propeller, and an engine. The piping and pump for a driver-load pair that in turn is in parallel with the propeller as a load for the engine. The first step is, therefore, to find a load characteristic for the piping-pump in the same form as that of the propeller. Figure 6.9, top sketch, gives a head-flow characteristic for the piping, and head-flow characteristics for the pump with rev/min as a parameter. Each intersection of a pump curve with the piping curve is an equilibrium point for the particular rev/min. Such a pump power-rev/min curve (based on an assumed 50% pump efficiency) is plotted in the lower sketch. This curve is then added to the propeller curve to represent the total load as imposed on the engine. Equilibrium occurs where this curve crosses the engine curve; total power, power to the propeller, and power to the pump, can be read from the appropriate intersection. The power taken to drive the pump is, of course, subtracted from the propulsive power. In addition, there is a loss in power because of the slowing down of the engine. It is, however, a small loss, approximately 2% of the original 400 hp. The loss would be more serious if the engine were a diesel, for example. A dashed line labeled "ideal diesel" is included in the figure, demonstrating the constant

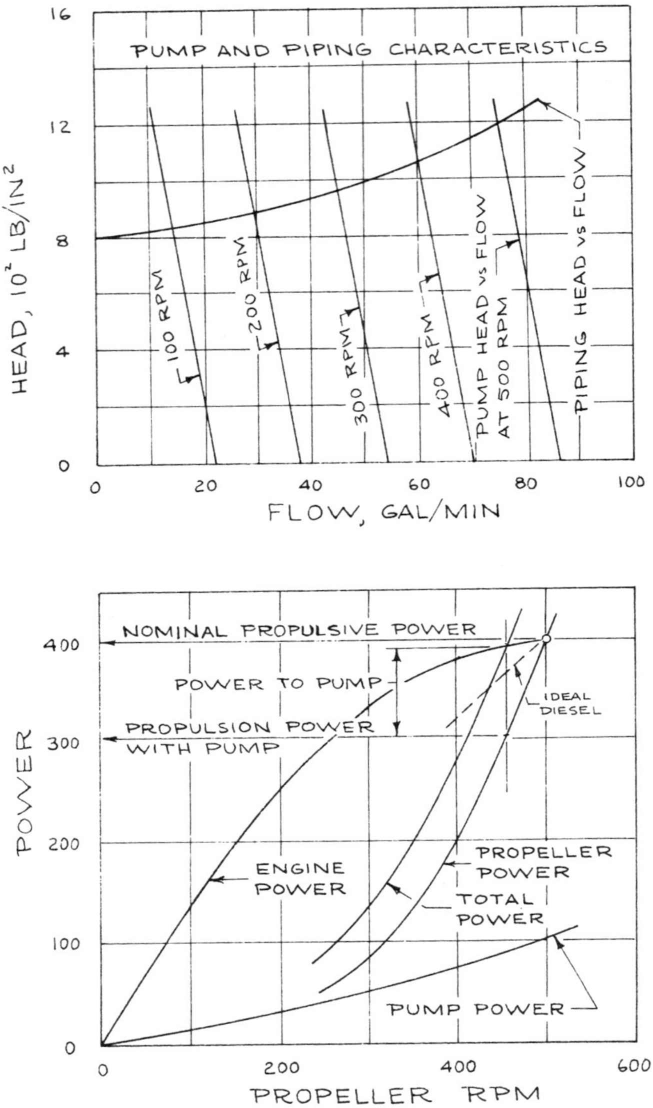

Figure 6.9. Illustrating the auxiliary load example.

one outlined above, begins with a choice of a value of J/J_0, again say 0.9. Since both N and Q are perforce unchanged, $K_Q = K_{Q_0}$ for this or any other value of J/J_0 choice. Pitch is read there, and for the same pitch and same J/J_0, K_T is read. Once K_T/K_{T_0} is obtained, the remaining quantities appear readily from the other relationships.

6.5. CODAG AND CODOG

The lightness and compactness of gas-turbine engines, especially of the aircraft-derivative type, have been exploited in propulsion plants that must operate efficiently at low power for long periods of time (cruising mode), and at high power levels for short periods (boost mode). The gas turbine provides the latter with little penalty in weight and volume, while a diesel engine—having the advantage of high efficiency—provides the cruising power. The scheme is commonly designated CODOG, from *co*mbined *d*iesel *o*r *g*as turbine.

The plotting of driver and load curves for CODOG illustrates a significant point, Figure 6.8, left. The figure shows that the diesel and gas-turbine engines cannot both be on the line at the same time, at least if either is operating at its full-power rating. This is a handicap, since it means that the propulsion plant cannot take advantage of the output of both for maximum

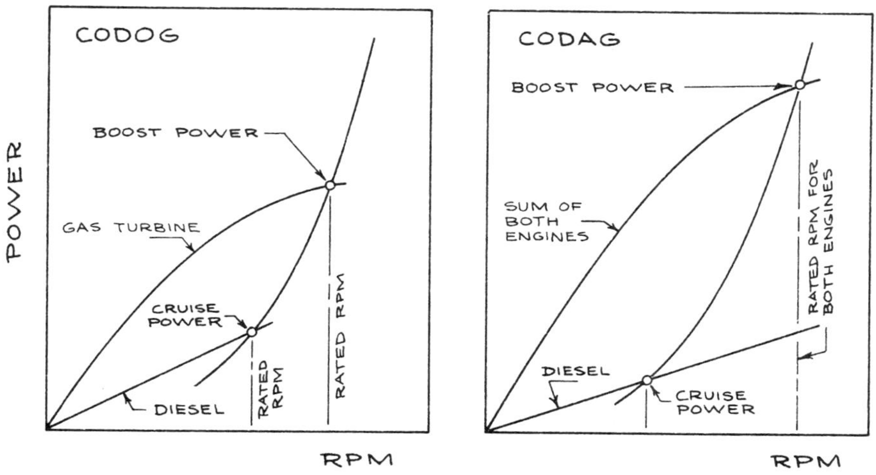

Figure 6.8. Rated conditions for CODOG and CODAG propulsion plants.

speed. The reason is this: if the engine-propeller matching is such that each engine operates along at its rated rev/min, then the sum of their powers would move the equilibrium point to a higher rev/min, overspeeding both engines.

The apparent remedy is to match the propeller to the sum of the engines, that is, design the propeller with a higher pitch so that rated rev/min is not reached until the rated power of both engines is applied. This is the CODAG scheme, the word coming from *co*mbined *d*iesel *a*nd *g*as turbine. The matching essentials are shown in the right half of Figure 6.8. The solution also has a drawback, illustrated by the figure. Note that the diesel must be matched to the same propeller rev/min, but when it is running alone in the cruising mode, the intersection of its characteristic with the propeller characteristic is at a much lower rev/min, and this engine cannot then develop its rated power. The phenonmenon is the same as that discussed in Section 6.4.B.

It is apparent from the preceeding discussion that if two engines are to be used to drive a common propeller shaft, a propeller pitch that is optimal when both are driving is a poor match when only one is driving, and vice versa. The solution to the dilemma is a propeller whose pitch can be changed in service, and such a propeller—the controllable-pitch propeller—is frequently used. The contribution of this device, and the special aspects of it relative to matching engine and propeller, are treated in Section 6.7.

6.6. AUXILIARY LOADS

Engines of all types require certain auxiliary service for their operation. Fuel pumps, lube-oil pumps, cooling-water pumps, valve gear, distributors, feed pumps, condensate pumps, electrical generators, superchargers, and so forth, are among the auxiliary devices found on engines of one type or another. Most often in the case of internal combustion engines, the auxiliary equipment is driven directly by the engine itself, so that the engine is essentially an independent unit, independent of external power sources, except perhaps for starting. In the case of the gas-turbine engine, its fuel pump and lube-oil pump are almost certainly driven directly by the engine.

Since the auxiliary devices, when driven by the engine, are in effect no more than internal engine parts so far as the marine designer is concerned, they are not involved in matching engine and propeller. The engine torque and power curves are taken at the engine-output coupling, and any power tapped off to run auxiliary devices is automatically deducted from the power produced within the engine. Nonetheless, the designer should be aware of the practice of quoting "gross" horsepower of the typical automotive engine. This is power *before* the requirements for auxiliaries are deducted and is definitely not the

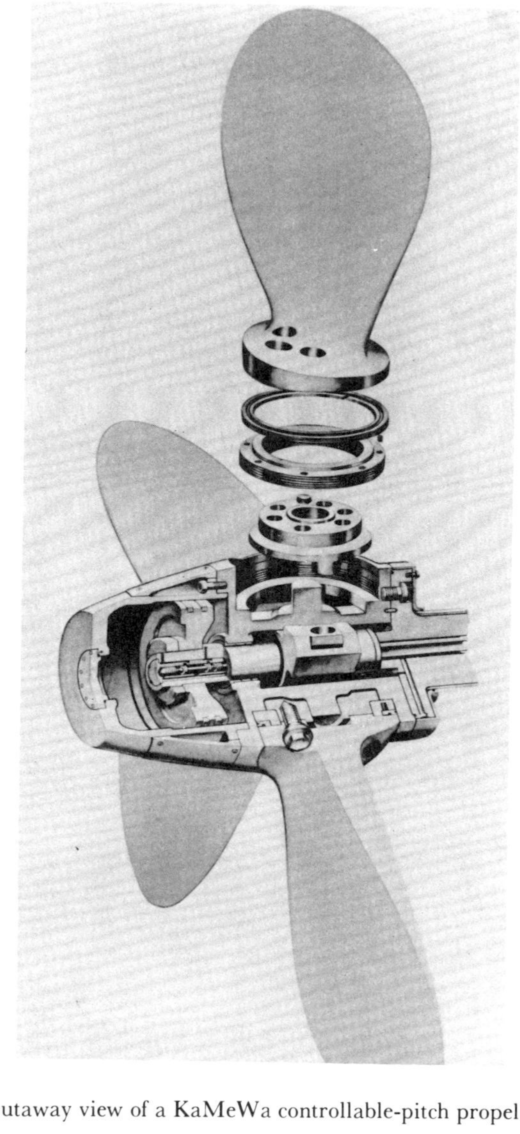

Figure 6.10. Cutaway view of a KaMeWa controllable-pitch propeller.

TABLE 6.1. SOME COMPARATIVE STOPPING TIMES, FPP AND CPP

Ship	TDW	Power, BHP	Starting Speed, knots	Stopping time, seconds	Type	Condition[a]
M/S Andorra	12000	12000	19.8	183	CPP	B
M/S Azuma	13150	15000	22.0	160	CPP	B
M/T Esso Fawley	16700	10080	17.0	199	CPP	B
Tanker	18000	8000	15.0	534	FPP	F
M/S Columbialand	24850	11400	15.8	240	CPP	B
M/S Holtefjell	35500	12600	15.5	366	CPP	B
Tanker	33000	11000	15.0	558	FPP	F
Tanker	35000	12500	16.3	582	FPP	F
M/T Sinclair Venezuela	51300	2 × 8400	16.5	286	2 CPP	B
Tanker	47000	15000	16.6	560	FPP	F
Tanker	48500	16000	15.8	630	FPP	F
M/S Nuolja	72500	17600	17.0	420	CPP	B
M/S Nikkala	72500	17600	16.3	426	CPP	F
Tanker	65000	17500	17.0	690	FPP	F
Tanker	79000	22000	15.9	750	FPP	F
SS Fort Henry	12000	6000	20 miles/hr	293	FPP	B
M/V Fort Chambly	12000	6000	20 miles/hr	163	CPP	B
SS Murray Bay	25000	10000	17.75 miles/hr	296	FPP	B
M/V Saguenay	25000	9500	17.75 miles/hr	285	CPP	B

[a] B = ballast; F = full load.

mean effective pressure characteristic of a diesel. Its intersection with total power lies about 13% below 400 hp, a magnitude much more likely to concern a marine designer than 2%. Whether or not he wishes to do something about it in either case depends upon the possible significance to the application of losses in propulsive power when the pump is operating. The "something" would be installation of a lower pitched propeller to allow the engine to turn at full rev/min when driving the pump, or installation of a controllable-pitch propeller so that the same rev/min could be maintained with and without the pump.

The loss in rev/min may be more significant to the auxiliary load than to the propulsive load. Note that the flow through the piping system is nearly proportional to pump speed. Thus the decline in engine—and hence pump—rev/min from 500 to 460 represents about an 8% loss in flow. If the auxiliary load is important, it may dictate an adjustment in propeller matching, or in the speed ratio between engine and pump, to compensate for this loss. The designer may find it expedient to draw a diagram such as Figure 6.9 to check the auxiliary operating conditions, even though he knows that the effect on propulsion load due to rpm change is trivial.

It is also appropriate here to note how the auxiliary power can be taken from the engine. In the case of a large ship propulsion engine (typically 20,000 hp or more), this load is likely to be driven by a mechanical connection to the reduction gear. Otherwise, especially in small vessels, the engine itself is provided with auxiliary power take-off shafts by the manufacturer. Typically there are two, one geared to the power-turbine shaft, and the other to the gas-generator shaft. Auxiliary items essential for operation of the engine (e.g., fuel pump) and provided as part of the engine are likely to be driven by the latter, since it must be running at least at self-sustaining speed whenever the engine is operating. The marine designer usually prefers to drive ship-service auxiliary loads from the power-turbine shaft, as was implied in the numerical example just above. The reason is that the slowing of the gas-generator shaft caused by the additional load there may cause the engine fuel control to impose a further cutback in power, since it may be designed to provide a fuel flow proportional to gas-generator speed. In a typical instance, the reduction in power at the engine output shaft is 1.5 times the power actually absorbed by the auxiliary load driven by the gas generator. But since this phenomenon is dependent on the programming of the fuel controller, the designer should check with the engine builder if he contemplates tapping off power from the gas generator.

6.7. CONTROLLABLE-PITCH PROPELLER

A. Introduction

The pitch of a controllable-pitch propeller is adjusted in service, either automatically, or at the command of the vessel operator, to change rev/min, power, thrust, or direction of thrust. Figure 6.10 is a cutaway view of such a propeller, showing the mechanism by which pitch is changed. The ability to change pitch gives a second degree of freedom (the other being fuel control) to the designer or operator in fixing engine-propeller operating points, particularly at part load. It permits optimal matching of the two components under more than one condition, in ways that have been mentioned in several places earlier in this chapter. But it also places an additional burden on the designer, for advantage can best be taken of this second degree of freedom only if it is restricted by linking fuel control and pitch control in a way that he must specify. The reason is this: if freedom to control pitch allows the operator to optimize operating conditions, it also allows him to make them worse than they would be with fixed pitch. Although his intentions may be the best, lack of operational information, or ignorance of its significances, may defeat his intentions. The second degree of freedom, therefore, is used in the design stage, with the result being an operational program of pitch and fuel settings. The operator is supplied with only a single control handle. the circuitry that it actuates has built into it the program selected by the designer, and the operator must in consequence follow the intentions of the designer.

But before delving into the designer's responsibilities, it is helpful to summarize the features of the controllable-pitch propeller that may lead to its being chosen in the first place.

B. Merits of the Controllable-Pitch Propeller

The most clear-cut advantage of the controllable-pitch propeller over fixed-pitch is its ability to reverse its thrust through reversal of pitch rather than reversal of rotation. It thus obviates the need for a reverse train in the reduction gear, or for an additional engine for reverse rotation. Greater astern thrust can usually be developed by pitch change during rapid maneuvering, since pitch can be adjusted to keep the engine at its full-power rev/min throughout the maneuver. This feature is important from the safety standpoint, since it usually results in shorter stopping time and distance. Table 6.1, taken from a paper by Ridley and Midttun (1970), demonstrates the generally better stopping performance of vessels equipped with the con-

him, however, because pitch can be changed to do this in service as the need arises. The question may rather be whether he wishes to allow these pitch changes and how they are to be accomplished. The likely answer is discussed in a later paragraph.

A bigger question involves the choice of part-load operating points. It is usual, of course, for any vessel to operate over a range of speeds, even though one (typically the highest) is the speed for which it is primarily designed. It has been noted earlier above that the lower speeds may be obtained with a range of engine fuel and propeller pitch combinations, and that some combinations may be better than others. Since the operator cannot or may not distinguish the good from the bad, they must be determined in the design stage and specified by the "program" previously alluded to. Figure 6.10 shows the plots of several programs on the power-rev/min plane.

Two considerations usually influence the choice of the program: (1) possible requirements of auxiliary loads, and (2) efficiency of propulsion at part load.

The most likely demand by an auxiliary load is constant rev/min for an a.c. generator. In such a case, all part load operation would be accomplished by pitch change alone, and, indeed, the operator's control handle would set only pitch, while the engine ran under control of its governor. The program as determined by the designer would consist only of a statement of the pitch to be set by each position of the control handle.

If there is no restriction to a particular rev/min, a combination of pitch and rev/min that gives best conversion of fuel to thrust can be sought. Figure 4.2 has demonstrated the variation of fuel rate over possible operating range of a gas-turbine engine, and the essential part of the figure is repeated here as Figure 6.12, with the addition of a fixed-pitch cubic propeller curve. Let's suppose that the engine is rated as 36,000 hp, 4000 rev/min, and then examine, say, operation at 20,000 hp. Fuel rate there is distinctly higher than at rating, but the magnitude of the increase depends strongly on the rev/min at which the power is developed. With the same pitch as at the rating, and the consequent 3150 rev/min, fuel rate is 0.515 (interpolating between the 0.50 and 0.525 fuel-rate contours). If pitch could be decreased to raise the rev/min to about 3600, the rate might decline slightly to about 0.510; if pitch could be increased sufficiently to move the operation to the normal power curve (about 1400 rev/min), the fuel rate would be 0.70. This last possibility points out why the choice of pitch and rev/min is not to be left in the hands of the operator.

Propeller efficiency is also an element in the fuel-to-thrust conversion, so Figure 6.12 doesn't give the complete picture. However, if high propeller efficiency is obtained at the rated point, the path of best propeller efficiency at part load lies close to the fixed-pitch propeller curve in the figure, so that best fuel rate and best propeller efficiency coincide approximately. Hence if the

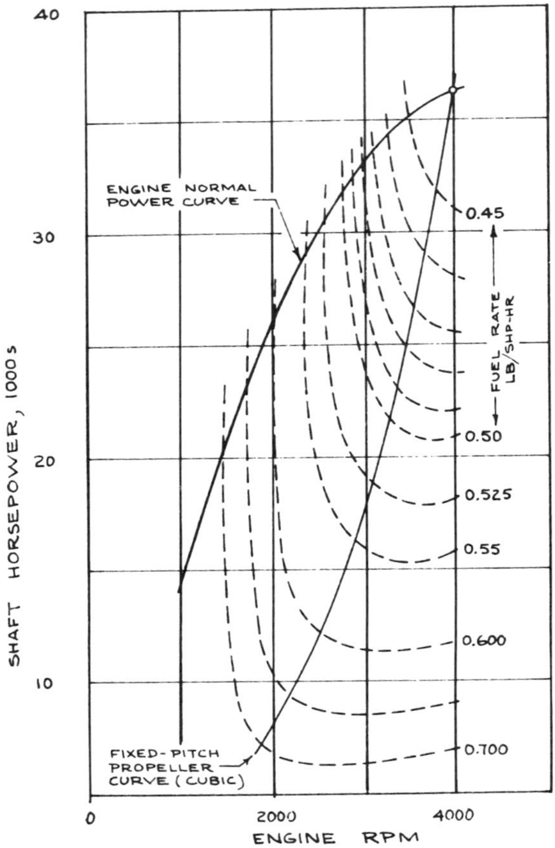

Figure 6.12. Engine fuel rate contours, showing propeller-load curve.

designer chooses a program that gives best part-load fuel rates, he will be quite close to highest overall efficiency.

Now for the basic technique of constructing the program. Figure 6.13 is a repeat of Figure 6.12 with information added as necessary to provide the program information. It is assumed that the rated pitch ratio (pitch/diameter) is 1.0, and propeller power-rev/min curves are added to the figure for several other pitch ratios. It is also assumed that rated vessel speed is 20 knots, and contours of vessel speed are added, based on a cubic relation between shp and speed. These contours show minima along the line of pitch ratio 1.0, indicating that the best propeller efficiency lies along this line. These curves, obtained from hull and propeller data that must be available, provide the input

trollable-pitch propeller. (None of the ships listed is powered by the gas turbine, but that is immaterial here.)

Low-speed maneuvering is facilitated by the controllable-pitch propeller, since a reduced pitch allows a low thrust to be developed at high rpm. Any vessel speed down to zero can thereby be maintained steadily with the engine running above its minimum self-sustaining speed and power. A significant shaft power is required because of the very poor efficiency of the propeller at low pitch and high rev/min.

There may also be need to maintain a constant shaft rev/min over a wide range of vessel speeds, even outside the low-speed maneuvering range. The most likely such situation would involve an auxiliary electrical generator driven by the propulsion engine. If it were an a.c. generator, constant rev/min might* be essential to maintain the design frequency. Vessel speed would be varied by pitch change with rev/min remaining constant, perhaps over the full range of powers.

Since a given thrust power can be obtained over a range of rev/min when pitch is suitably adjusted, the controllable-pitch propeller confers the ability to operate the engine at any desired combination of power and rev/min within its capability, rather than along a single line dictated by the characteristic of a fixed-pitch propeller. Figure 6.11 illustrates several paths of operating points that are possible. One of these is the fixed-pitch path, another is the constant-rev/min path mentioned in the paragraph just above, a third lies along the engine characteristic, and the fourth is an arbitrary one. If both the engine curve and the propeller curve can be shifted (by fuel change and by pitch change), then the equilibrium point can be placed anywhere within the range of capabilities of the two components. One such equilibrium point is shown in the figure, in addition to the rated point.

When service deteriorations, towing loads, or starting of auxiliary load or boost engine (CODOG or CODAG) forces the engine to deviate from its rated rev/min, a pitch change can restore that rev/min and so allow the engine to continue running at its rated power. This feature has been mentioned earlier in this chapter, though it is generally been said that small changes in rev/min are inconsequential to the gas turbine because of its almost flat power characteristic near the rating. The controllable-pitch propeller is consequently less justified for this engine on grounds of rev/min adjustment under varying load than it is for the diesel, say. CODOG- and CODAG-propulsion plants, therefore, benefit more from this particular feature of controllable pitch than do pure gas-turbine plants.

* "Might" because means are available for producing a steady frequency from a variable-speed generator, as is discussed subsequently.

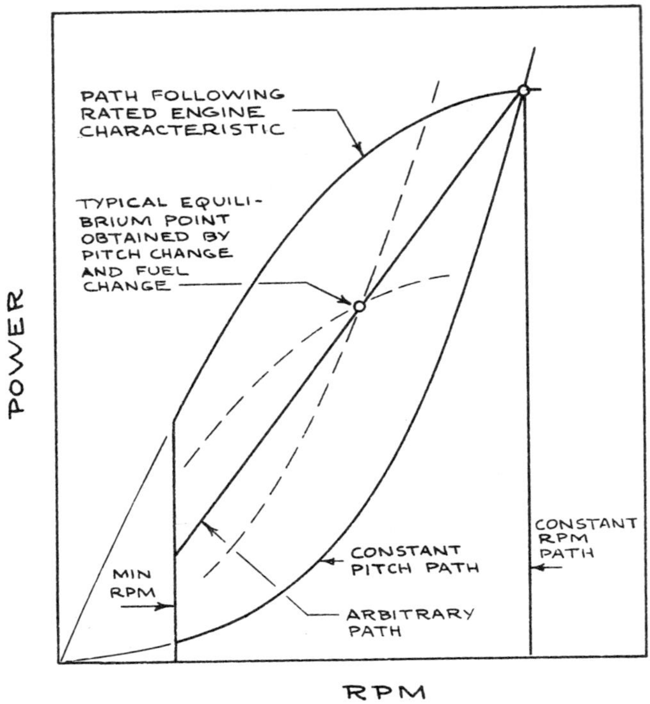

Figure 6.11. A sampling of possible paths of operating points on the power-rev/min plane, as permitted by controllable pitch.

C. Matching Engine and Propeller

Here we shall extend the meaning of the term "matching" to include the selection of propeller pitch and engine fuel flow at any possible operating point of the vessel. This was not necessary in the earlier discussion of fixed-pitch matching, because once the match was established at the rated power, all other operating conditions were automatically established (i.e., along the propeller characteristic).

At the rated point, the matching problem is essentially the same as that facing the designer in the fixed-pitch situation. He attempts by best choice of reduction gear ratio to obtain the rated propulsive power of the vessel with the propeller of highest feasible efficiency, and with the engine operating at the rating appropriate to the service. The possible need to make compromises in rev/min to give margin against service deteriorations and uncertainties, to obtain full power when auxiliary loads are started, and so forth, does not face

needed. Now suppose that the designer's intention is to lay out a program based on equal increments of vessel speed (i.e., speed of the vessel is to be proportional to position of a control handle), and that provides the lowest attainable fuel rate at any speed. His first step is to draw a path through the regions of lowest fuel rates. Such is the heavy line in Figure 6.13 labeled "program line." His remaining step is to read the necessary data at equally spaced speeds along this line. Table 6.2 shows the results if readings are made at 2-knot intervals between 20 knots and 12 knots. This table contains the essential information needed to fix the designer's intentions on the fuel-pitch control system. This system will subsequently be designed so that movement of the control handle sends signals to the pitch controller specifying the pitches

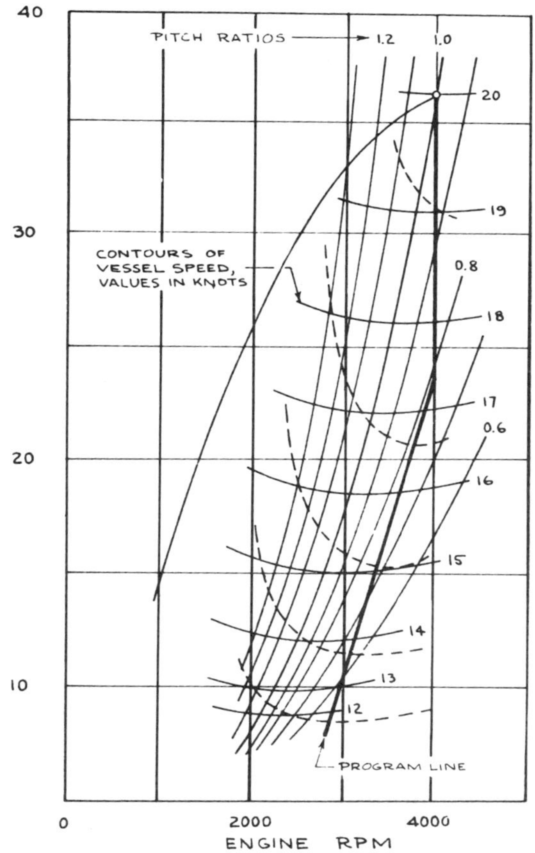

Figure 6.13. An example controllable-pitch propeller program.

TABLE 6.2. DATA FOR ILLUSTRATIVE ENGINE-PROPELLER CONTROL PROGRAM

Speed	Engine rev/min	Shaft Power	Pitch Ratio
20	4000	36,000	1.00
18	4000	26,000	0.85
16	3600	18,500	0.76
14	3130	11,500	0.66
12	2870	9,000	0.50

required by this program and simultaneously sends signals to the fuel controller specifying the rev/min required. The example above outlines the procedure by which a fuel-pitch program may be derived. There is no intention, however, of recommending the program demonstrated. In fact, such a program would probably not be used in practice because one that merely followed the design pitch ratio would be almost as good in terms of fuel rate and would have the advantage of a simpler control system, since pitch change would not be required over the range of speeds shown. Actual programs have preponderately followed the fixed-pitch path in the higher speed ranges, such as the 12- to 20-knot range here.

But what of the remainder of the speed range, down to 0 knots? The usual practice at low speeds is to maintain the engine rev/min at a constant value over this range and allow the control handle to set a pitch directly, ranging from zero to full pitch. Such a program may not give the best efficiency in its speed range, but this consideration matters little because absolute fuel consumption (i.e., wt/hr) is low, and most vessels operate in this range only during brief maneuverings.

Astern operation must also be considered, and this may be essentially a mirror image of the ahead situation: as the control handle is moved through a sector representing astern propulsion, astern pitch is first increased directly while rev/min is held constant, until a point is reached at which rev/min must be increased along a program similar to the ahead program. Or maneuvering power only may be required astern, in which case the latter part is omitted.

Figure 6.14 shows some aspects of a program incorporating several of the features discussed above. Pitch, rev/min, and vessel speed are plotted against handle position. The program features a low-speed range in which rev/min remains constant at a minimum value while pitch is varied from zero in either ahead or astern direction, plus a higher-speed range over which pitch remains constant while rev/min is increased up to rated value. Ahead and astern operations are identical, though the result—vessel speed—is obviously different in two different directions. The program is designed to produce pitch changes proportional to handle position in the maneuvering range and rev/min changes proportional to handle position in the higher range.

As a more concrete example, let us take a brief look at the engine-propeller control program used by the *Euroliner* class vessels of Seatrain Lines [Carpenter, Hoburn, and O'Neill (1973)]. These are large, fast, container ships designed for North Atlantic service, with propulsion by twin FT4A-12 (Turbo Power & Marine Systems) gas turbines rated for 30,9000 hp at 3600 rev/min (135 propeller rev/min), and by LIPS (a Dutch builder of propulsion equipment) controllable-pitch propellers. A significant feature of the machinery plant is production of ship-service electrical power by a 1000 kW generator geared to each engine. Although they are a.c., 60-Hz power is produced over a range of speeds. This is accomplished by rectification into an intermediate d.c. loop, followed by inversion to a.c. by a bank of thyristors. In this way, a constant 60-Hz frequency is maintained with engine speed as low as 2000 rev/min (75 propeller rev/min). The necessity for a constant-rev/min program above the maneuvering range is obviated. The program actually used is this: up to 12,000 hp, engine rev/min is kept constant at 2000 while increasing pitch is set directly by command of the control handle; when full pitch is reached, further movement of the handle increases rev/min over the 2000–3600 range, while pitch is unchanged. Figure 6.15, taken from [Carpenter, Holburn, and O'Neill (1973)], is a pictorial sketch of the control

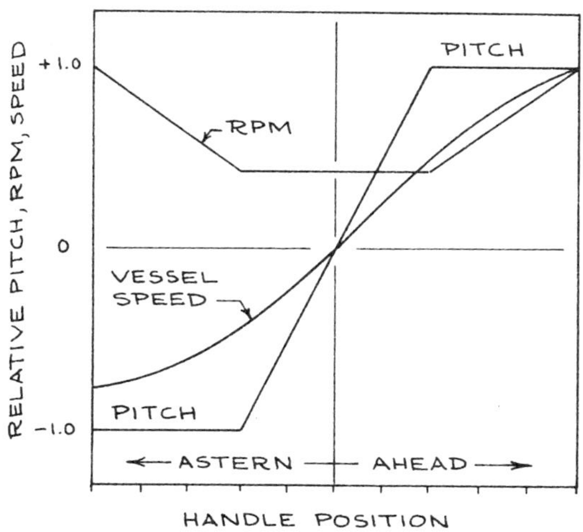

Figure 6.14. Representation of a sample fuel-pitch program.

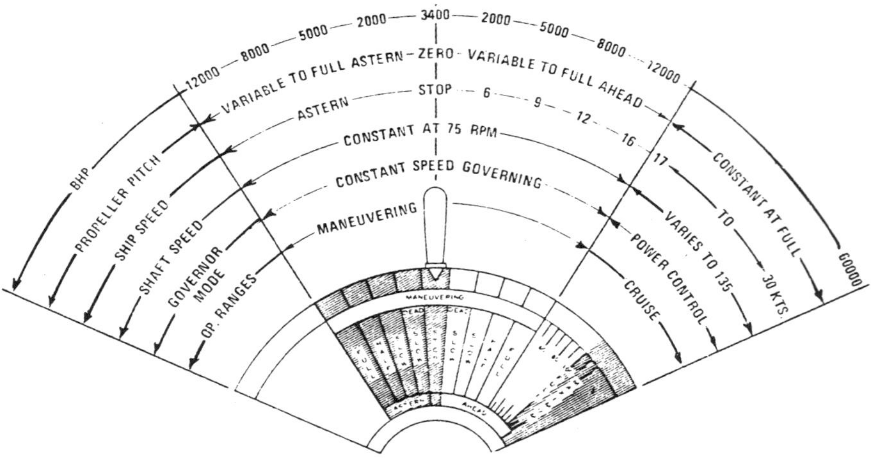

Figure 6.15. Propulsion control handle for Euroliner vessels. [Carpenter, Holburn, and O'Neil (1973)]. From *Marine Technology,* January 1973, published by the Society of Naval Architects and Marine Engineers.

handle, showing the values corresponding the handle position of several parameters of interest. Note that only a maneuvering range is provided in the astern direction.

Now let's go back to the question of pitch changes intended to compensate for service deteriorations, starting of auxiliary loads, towing, and so forth. Such changes might be accomplished automatically if the pitch commanded by the program were maintained by a governor. (A governor is a control device that acts on some parameter—pitch in this case—to maintain a speed constant.) The tendency of a load increase to slow the shaft would then be automatically countered by a reduction in pitch until rev/min was restored. Engine power would thus not suffer the loss caused by decline of rev/min discussed at several earlier points in this chapter. This feature is often included in diesel-propulsion plants that drive controllable-pitch propellers, for the loss in power may be significant with that type of engine. With its flatter power characteristic, the gas-turbine engine is another matter; the constant-rev/min feature has less appeal because power remains nearly constant for small-rev/min changes. The preponderance of practice is against providing automatic pitch changes. On the *Euroliner* ships, for example, the parameter maintained constant by governing at each handle position in the 2000- to 3600-rev/min range is engine power; pitch is constant at the full-pitch value, and rev/min establishes itself at the driver-load equilibrium point.

D. CODAG Considerations

A CODAG propulsion plant with controllable pitch is inevitably more complicated than if either diesel or gas turbine is used alone, but presents little additional conceptual problem. Presumably the single control handle sends simultaneous signals to both engines, so that the designer must decide and specify how the load is to be split between them. Figure 6.16 demonstrates one

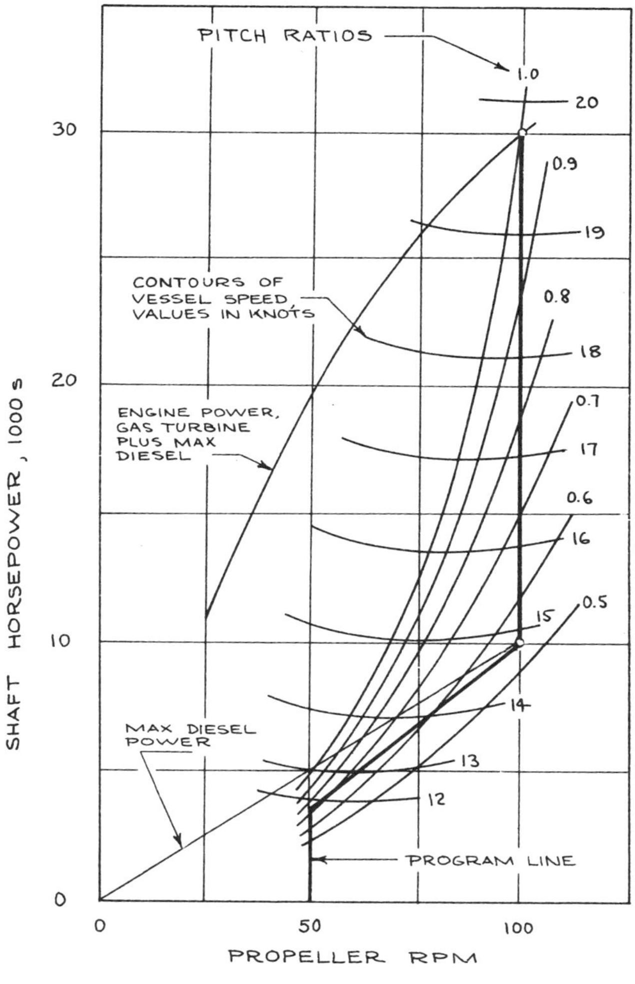

Figure 6.16. An example CODAG program.

possibility. The figure is similar to Figure 6.13, but now it is assumed that power up to 10,000 hp is provided by a diesel engine, with a gas turbine put on the line only to provide power above 10,000, up to 30,000 hp. A program line is shown, fixing rev/min, power, and pitch for the diesel. Up to approximately 12 knots, the diesel rev/min is held constant while pitch is changed; above this speed rev/min and pitch are varied—as can be read from

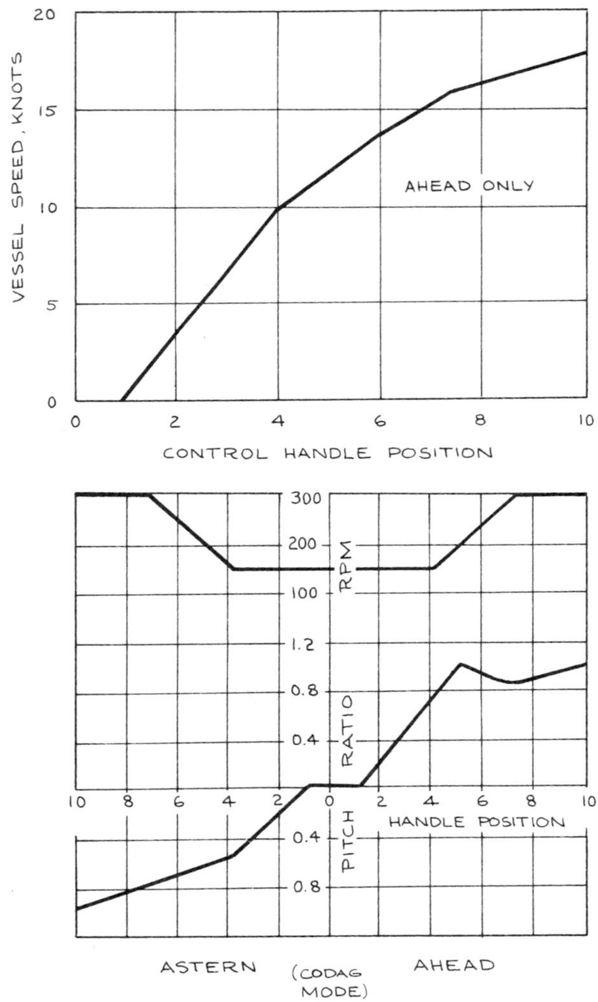

Figure 6.17. An example CODAG program [Adapted from Hamilton (1967)].

TABLE 6.3. CONTROL-PROGRAM POINTS FOR U.S. COAST GUARD VESSELS [Hamilton (1967)]

		Turbine			Diesel			CODAG		
Handle Position	Vessel Speed	hp	rev/min	Pitch Ratio	hp	rev/min	Pitch Ratio	hp	rev/min	Pitch Ratio
2	3	250	150	0.25	250	150	0.25	250	150	0.25
4	10	500	150	0.7	500	150	0.7	500	150	0.7
5	12	1300	225	0.8	1300	195	1.1	1300	195	1.1
6	14	2000	300	0.58	1900	240	1.1	1900	240	0.85
$7\frac{1}{4}$	16				3000	300	0.88	3000	300	0.85
10	18							5000	300	1.03

the program line—up to the full power and rev/min of the engine. Further advancement of the control handle presumably starts and clutches in the gas turbine, then commands increments of power and pitch along the program line. Meanwhile the diesel engine operates at a fixed rev/min and power.

A feature likely to be needed with such a propulsion plant is the ability to operate the gas turbine alone, since this would provide a valuable backup capability should the diesel be out of service. In effect, a separate program would be required for this mode of operation. Doubtless, the same control handle would be used, but the control circuitry would require a switching device to actuate the hardware that generates the commands necessary for this program. Such a scheme is used on a class of U. S. Coast Guard vessels [Hamilton (1967)]. Diesel alone, gas turbine alone, or CODAG modes of control are provided, with the programs designed so that vessel speed as a function of handle position is the same in all three modes, at least over the power range that any of the three can cover. Figure 6.17, taken from [Hamilton (1967)], shows pitch ratio, rev/min, and vessel speed as functions of handle position in the combined (CODAG) mode. Table 6.3, taken from the same source, shows the same information for all three modes.

6.8. REFERENCES

Carpenter, D. B., Holburn, J. G., and O'Neil, D. A. (1973), "System Integration of the GTS *Euroliner* from Conception to Operation," *Marine Technology* 10, 38–50.

Hamilton, R. (1967), "Control Design and Experience on CODAG-Powered 210-ft *Reliance* Class Cutters," ASME paper 67-GT-28.

Ridley, D. E., and Midttun, O. H. (1970), "Economic Considerations of Controllable-Pitch Propellers with Diesel Main Propulsion," *Naval Engineers Journal* 82, 71–83.

Vibrans, F. (1956), "Variations in Marine Engine Performance with Departures from Design," *ASME Oil and Gas Power Proceedings*.

Chapter

SEVEN

CONTROL OF THE GAS TURBINE PROPULSION PLANT

7.1 INTRODUCTION

The fundamental output of a propulsion plant is the thrust developed by its propulsor; this thrust must be regulated to give the speed of the vessel ordered by its operators. The fundamental input to the plant is its fuel; its flow to the engine, therefore, is to be regulated to control the thrust. Such, in its most simple terms, is the control of a combustion-engine propulsion plant.

Complications to the simple picture abound, however, and must be carefully regarded in the design of a practical propulsion control system. Most of them originate in the control requirements of individual plant components, most specifically the engine. Some of these requirements are based on inherent qualities of the component quite independent of the type of service. A ready example is the necessity for limiting temperatures and rotational speeds of the gas turbine, irrespective of whether it is in marine service. The engine is likely, therefore, to be provided by its builder with a control subsystem that must subsequently be integrated into the overall propulsion plant control system. Therefore, the discussion in this chapter applies in some sections to the control of the gas-turbine engine as such and elsewhere to the control of propulsion plants built around this engine. But it should be observed that the division is somewhat artificial, for control of plant and engine cannot be totally categorized. Shaft rev/min, for example, is a vital parameter that is common to both.

In the beginning the discussion by looking first at the plant requirements, let's note that it is seldom that the operator's prime interest is in thrust. In most types of vessels the result of a certain thrust, the speed it produces, is most meaningful. Although speed is, to be sure, a function of thrust, speed is also a simple function (nearly linear, typically) of shaft rev/min. Shaft rev/min is easier to measure than thrust and also is a more significant parameter to the engine. Therefore, shaft rev/min is in practice the manifestation of output that is most often controlled in the traditional propulsion plant. The gas turbine, however, presents a unique rev/min situation in that it typically possesses two significant rev/min: its output rev/min and that of its gas generator. There are reasons noted later for primarily controlling the latter, giving rise to what is known as "power control," since establishing a gas-generator rev/min fixes its output to the power turbine. Power control is often preferred for the propulsion gas-turbine engine, but when used, the operator's control signal may be modified by the control circuitry so that the result is a ship speed proportional to the position of his control lever.

If the propulsor is a controllable-pitch propeller, further amendment must be made to the original statement about thrust being controlled, for a certain thrust or ship speed can be produced by a range of pitch-rev/min combinations. This is discussed in Chapter 6, wherein the concept of a *program* of these combinations is introduced. The principal function of the propulsion control system hence may be said to be regulation of propeller pitch and fuel flow such that the programmed combination of pitch and rev/min (or power) produces the desired speed through the water. The fixed-pitch propeller may be regarded as a special, simplified case.

Means of thrust reversal is an essential feature of the propulsion plant, and the control system must provide the desired direction at the command of the operator. Pitch reversal, or engagement and disengagement of clutches, or switching of electrical transmission, is required, integrated in all cases with control of the engine as its load is suddenly changed.

Another vital stricture on propulsion plant control is the requirement that it protect plant components from operational hazards. If clutches must be engaged and disengaged to effect shaft reversal, for example, these actions must be integrated with rev/min and torque to prevent damage to the clutches. If the engine can impose torques on the transmission above rated value (as the gas turbine can because of its torque characteristic), then the propulsion control is required to monitor torque and order a fuel cutback when it reaches the limiting value. And if a windmilling controllable-pitch propeller can overspeed the engine during a rapid reversal of thrust, a limit must be imposed on the rate at which pitch can be changed.

A seemingly simple, but highly important, function of the propulsion plant control is to provide the operator with appropriate means of effecting his commands from a distance, and from several different operating stations. Control from the bridge of the vessel, and from a centralized engineering station near the machinery, is typically required. Some of the complexity in control systems stems from the duplication of control stations, from the need for communication between them, and from the extensive monitoring equipment that remoteness of the human operator makes necessary.

The engine-control subsystem receives orders from the propulsion-control system to produce a certain rev/min or a certain power. It may do this simply by ordering the appropriate rev/min of either the gasifier or the power turbine. Or, if the power turbine nozzles are adjustable, it follows a program of nozzle angles and rev/min to produce the ordered output. In this it is acting analogously to the propulsion control that may have options of rev/min and pitch with which to produce a required power or ship speed.

Protection of plant components from hazardous operating conditions has been mentioned as a function of plant control, but part of this may devolve on the engine-control subsystem. For example, cutback of fuel in the event of overspeed or tripping of a fuel cutoff valve in the event of severe overspeed almost certainly are functions of the engine controls. Other hazards are more uniquely engine problems, such as loss of lube-oil pressure, and protection is also a responsibility of engine controls.

Starting and stopping of the engine is likely to be done independently of propulsion-plant control; typically the engine is not turned over to propulsion-plant supervision until it is brought to a steady idling condition under the command of a starting sequencer. This device orders the correct sequence of events that must be followed for starting, all the while protecting the engine against hazards specifically associated with startup. Other transient conditions are likely also to require special control action by the engine subsystem. For example, the fuel controller may translate an order for instantaneous increase in power into a gradual (ramped) increase in fuel flow to prevent overshooting the ordered rev/min, temperatures, and torque.

To summarize the discussion to this point, the combined requirements of propulsion-plant control and engine control may be listed as

1. Provide means for remote management by the human operator.
2. Set and maintain the plant output parameters at the commanded level.
3. Manage transitions from one operating condition to another, including
   a. Startup and shutdown of the engine.
   b. Accelerations and decelerations.
   c. Thrust reversal.
   d. Switching engines.
4. Protect the propulsion plant from hazards, including
   a. Protection from externally imposed hazards, such as loss of load or improper command from the operator.

b. Protection of any component from unacceptable stresses caused by action of any other component.

c. Protection of components against internal malfunctions, or against failures of their auxiliary equipment (e.g., loss of engine lube-oil pressure).

Put in such general terms, these requirements could easily apply to any type of propulsion plant. Each type may have unique needs for emphasis in certain areas, and unique ways of supplying the control functions. Discussion of the particulars for the gas turbine, of course, follows in this chapter, but one additional generality regarding this type of machinery should be noted before getting to the particulars: the gas-turbine propulsion plant is typically more automated than other types. The term *automated* implies human control from remote locations, but more importantly, means that control functions are carried out in *closed-loop* fashion by nonhuman equipment.

The control from a remote location is a necessity because of high temperature and high noise level in the immediate vicinity of the engine. A gas-turbine propulsion plant is usually located either in an unmanned machinery space, or the engine itself will be totally enclosed in isolation from direct human supervision. Another incentive for remote control stations is simply operator convenience; centralization of control and monitoring functions at one location reduces the need for human effort, and so likewise does placing an operating station on the bridge. (This second incentive is equally applicable to other types of propulsion machinery, so that centralized control stations and direct bridge commands to the engine are also common with steam and diesel plants.)

Automatic control is also widely used in steam and diesel propulsion, but it is more extensively used with the gas turbine, and this from necessity. The gas turbine responds more rapidly to changes in load and fuel supply, and its rated conditions—especially with respect to temperatures—are closer to the hazard level. Human responses just aren't adequate to cope with the possibly rapid excursions from safe to unsafe conditions. The steam plant may be contrasted, its boiler serving as a good example. Modern boilers are typically automated; fuel, air, and water are automatically regulated to meet the varying steam load demands without human intervention. But the control functions can be carried out manually if the automatic equipment fails. The gas turbine, on the other hand, is not provided with means of manually regulating its fuel. Although some propulsion plants will be seen to have so-called manual controls, these work through the engine-control subsystem. The human operator can manually order a power or speed, but the order must pass through this subsystem so that his manipulations cannot inadvertently send rev/min and temperature into dangerous ranges.

The term *closed loop* introduced three paragraphs back refers to the fundamental characteristic of the usual automatic control device, namely, its ability to sense errors in the controlled variable (error is the difference between the value ordered and the actual value), and to impose corrections to reduce the error. Now, the analysis and design of closed-loop control components and systems is a subject far too great in scope for this chapter. But let's give brief attention to the problems to be overcome by the control designer. Beginning with generalities, he must contend with the same requirements that apply to technical components and systems of any nature, namely, such things as reliability, compatability with the environment, cost and weight restrictions, and so forth. More specifically, he seeks two characteristics of controller and of the device controlled: adequate speed of response and stability.

Speed of response is essentially a self-defining term; if a controlled temperature, say, is rising toward the danger level, it is obvious that the controller must respond quickly enough to arrest the trend in time. In reacting rapidly, however, the controller may cause the controlled variable to overshoot its set point, in which case action in the opposite direction is required. If response to the overshoot then produces an overshoot of equal or greater magnitude in the opposite direction, the controlled variable will never settle to its set value, and the condition of instability exists. If these oscillations do converge toward the set value, the action is stable, but in practice may be unacceptable, for the overshoot may be of dangerous magnitude, or the convergence may take an excessive time.

The successful resolution of these problems calls for high technical skill and constitutes a field of engineering in itself. The treatment in this chapter is perforce only descriptive; we look qualitatively at what must be done for the propulsion plant and its engine, and describe some actual methods and equipment. The reader should realize, though, that behind these briefly depicted items lies a major engineering effort, comparable in scope to the design of the propulsion plant itself.

7.2. SOME BASIC CONSIDERATIONS

The control of a propulsion plant and its engine is fundamentally a matter of governing, logic, sequencing, and timing. As the preceding section has promised, the quantitative theory and technique used in design of the control components and systems that carry out these functions is beyond our present scope. But qualitative discussion, appealing to the reader's basic understanding of things mechanical, are reasonable and enhance the subsequent descriptions.

A. Governing

A *governor* is a device that acts automatically to maintain a speed at (or near, as will be seen) an ordered set point. Since the set point can usually be changed, the governor also acts to change speed upon command. It is the traditional device for controlling the output of an engine, having been applied even to the early steam engines and continuing into contemporary gas-turbine technology. In the marine field it is also often used to control the rev/min of a propulsion shaft via pitch change. Controlling an engine, the governor manipulates some parameter—valve position, fuel-pump stroke, or fuel-pump speed—that modulates fuel flow rate. Controlling pitch, it usually manipulates a pilot valve that admits hydraulic fluid to the pitch-actuating cylinder.

The governor of recent technology is usually a mechanical-hydraulic device; it senses speed mechanically by balancing a spring force against a centrifugal force and applies the force necessary to move the control element (e.g., valve) via a hydraulic cylinder. The most recent gas-turbine fuel controllers are electric (or electronic, if you prefer that term) and are indeed sometimes referred to by that more general name—fuel controller—because they incorporate features other than simple speed control, such as protection against over temperature, and supervision of fuel increases and decreases during operating transients. The basic function is that of governing, nonetheless, and the principles of governing are essentially the same no matter what the energy medium employed. The explanation to follow is based on the mechanical-hydraulic governor, since the mechanical and hydraulic actions can be outlined without the need for a specialized technical background on the part of the reader.

The governor consists of three elements, (1) a speed-sensing device, (2) an amplifier to increase to usable magnitude the small forces generated by the sensor, and (3) a servomotor to position the device that moderates fuel flow. All of these functions can be carried out in several different ways. The universal way of mechanical speed sensing is the balancing of a spring force against the centrifugal force generated by rotating weights or "flyballs." (See Figure 7.1.) The device shown would be rotated by an auxiliary shaft connected to the shaft whose speed is to be sensed. Inspection of the figure shows that centrifugal force in the flyballs produces couples about the pivot points that are balanced by couples produced by the spring force. Position of the base of the spring is a function of initial spring force, and of speed. For a given initial spring force, which is the operator's way of setting a desired speed, there is a unique position for every speed. The output force is reduced by friction. Nonetheless, this force is sometimes used in simple governors to set the fuel directly. In such cases, sensor, amplifier, and servomotor functions are combined into a single device. Alternatively, the output force moves a hydraulic pilot valve (the amplifier) that in turn directs oil under pressure to an

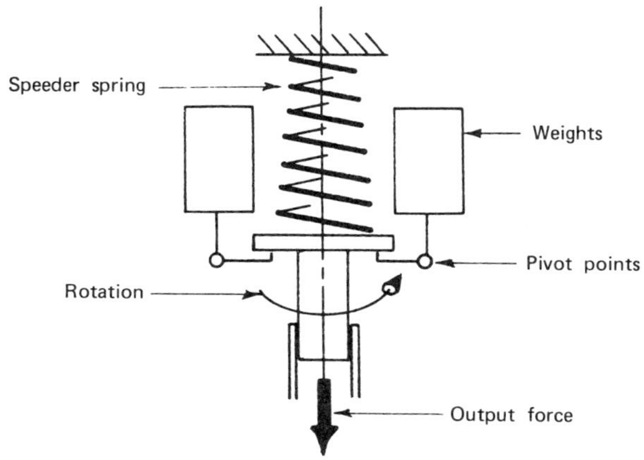

Figure 7.1. Elementary governor [Drake (1951)].

actuating cylinder (the servomotor), thus providing what is usually called a hydraulic governor. The sketches in Figure 7.2 show these elements clearly. The flyball output force is quite small compared to that needed by the mechanical governor, since the amplifier makes a large sensing force unnecessary.

Three qualities should be expected of a governor, (1) force or torque output adequate to position the fuel setting, (2) adequate speed of response, and (3) stability (i.e., the engine should not "hunt" about the set rev/min). These, especially the last two, involve engine characteristics as well as governor characteristics, and especially the dynamic characteristics of both. A complete understanding of the problems must be based on knowledge of at least the basics of automatic control theory. However, considerable intuitive insight may be offered by the sketches of mechanical-hydraulic governors in Figure 7.2. The first (*a*) is the simplest device. The sketch shows the flyball sensor positioning a hydraulic pilot valve that controls the flow of oil to an actuating cylinder. By adequate sizing of pilot valve and cylinder, and by provision of adequate oil pressure, force and speed of response can be provided, but the device is unfortunately unstable. A moment's reflection on how it works should show that when an error in speed causes the pilot valve to be opened, it stays open until after the error is removed. A speed adjustment always results in large overshoots, and they are never reduced to zero, hence the instability.

Sketch (*b*) shows a more complicated arrangement in which the instability is eliminated by a lever connecting the fuel valve and speeder spring. This lever provides a proportional action. As the fuel valve opens, say, to increase speed, the lever moves to relieve the speeder spring and to allow the pilot valve to move toward closure. The speed thus creeps in on the set value, stably. The

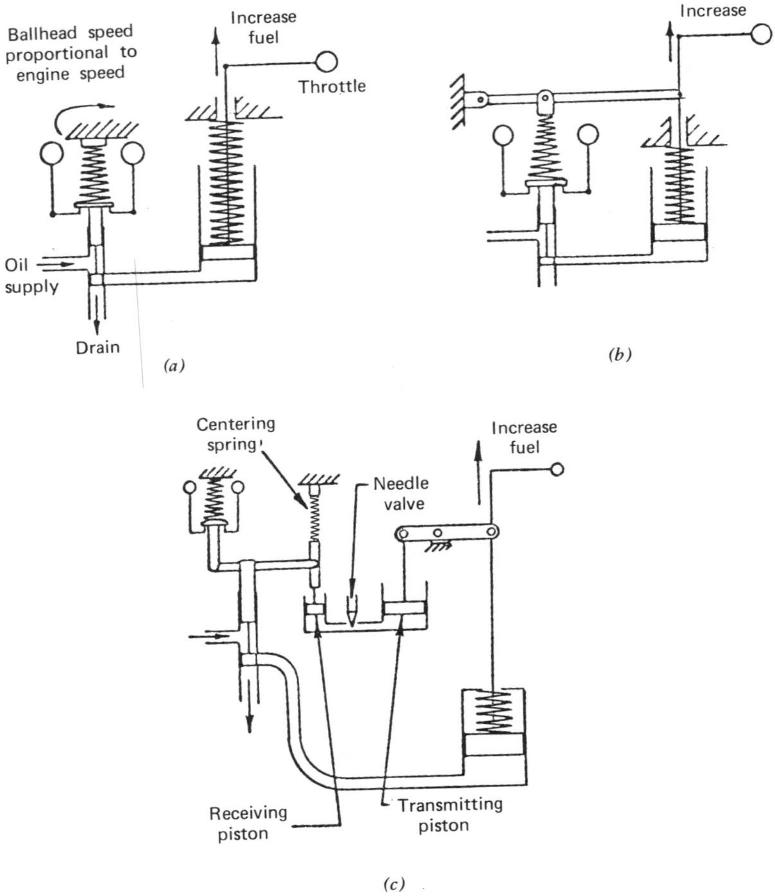

Figure 7.2. Principles of mechanical-hydraulic governors [Drake (1951)].

consequence, however, is *droop,* since the set point (top of the speeder spring) is moved in the process. The resultant speed is less than the set value by some small amount that increases as load on the engine increases. There is, therefore, a load-speed characteristic that droops, hence the terminology. The third sketch shows a yet more complicated mechanism that is both stable and *isochronous,* the latter term meaning that it maintains the set speed without droop. Stability is obtained via transient droop. It can be seen that if the needle valve is closed, the assembly of transmitting piston, receiving piston, and so forth, must act as a solid lever, and the action is the same as in the second sketch. However, the needle valve does allow oil to leak out so that the centering spring can slowly restore the receiving piston to its original position, thus uncoupling the set point from the throttle position.

The meaning of droop might be amplified by a look at Figure 7.3. The formal definition of the term is given in the top sketch of the figure. As shown, the rev/min-load characteristic curve that expresses droop is usually linear. The second sketch in this figure shows how droop is exploited to establish load sharing among engines driving the same load, as in a multiengine propeller shaft. The sketch shows mismatched droops enforcing a very unequal distribution on two engines; as can be seen, identical curves are required for equal distribution.

B. Logic, Sequencing, and Timing

Governing is carried out continuously and under steady-state conditions. Obviously, though, orders to change from one steady state to another must be accepted and carried out free of hazard. Orders to start or stop an engine, to

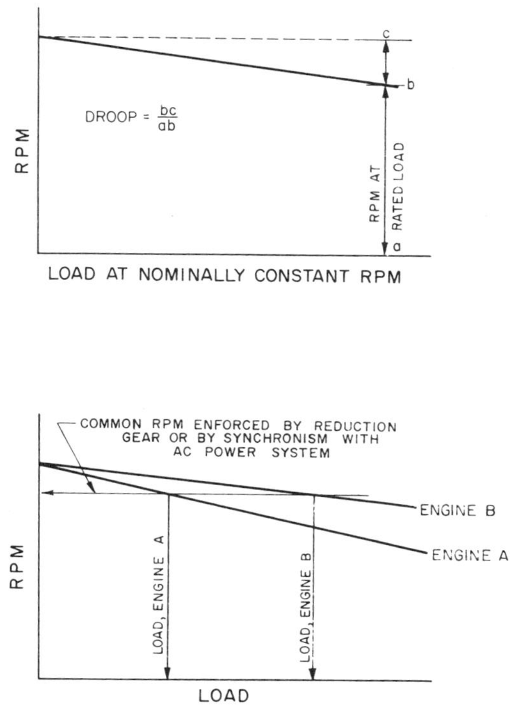

Figure 7.3. Definition of governor droop and example of its application.

change shaft rev/min or propeller pitch, and to engage or disengage clutches or shaft brakes are ready examples. Some of these are merely changes in operating level, but others are discontinuous changes from one operational state to another. In the latter case, a safe transition may be dependent on certain prior conditions being met—transition from state A to state B to be permitted only if conditions B1, B2, B3 exist, say. (Example: an engine to be clutched into the propulsion shafting only if the shaft brake is disengaged, if the engine is running under control of its governor, and if no other engine is engaged to the same shaft. The array of such conditions constitutes the *logic* of the transition, and this logic must then be built into the control circuitry that carries out the transition.

Often a transition consists of a series of steps that must occur in a certain sequence. Starting an engine provides a good example, since ignition cannot be effected until fuel flow is established, and fuel flow must not begin until adequate air flow is established. The device that manages these and other necessary steps is usually called the starting sequencer, since it orders the initiation of the steps in the proper sequence. The sequence may be simply a consequence of the logic. For example, if orders were given simultaneously to activate the starter (a motor that rotates the compressor and thus provides a starting air flow), open the starting fuel valve, and energize the ignitor, proper sequencing would follow if logical circuits permitted fuel valve opening only if a certain compressor discharge pressure were present, and permitted ignition only if the fuel valve were open.

Timing may also be important. The starting sequence again may be used as an example. After ignition has been ordered, a decision must be made whether to continue the sequence toward normal idling conditions if ignition has actually occurred, or to terminate the sequence and begin again if ignition has failed. But time must be allowed for ignition to occur, and perhaps for the temperature rise to indicate that it has, before the decision is made. A suitable time delay, therefore, is built into the sequencing circuitry.

Logic and sequencing can be illustrated further in the overall propulsion-plant control. Let's use as an example the clutches and shaft brake that may be found in a system with fixed-pitch propeller and a reverse-reduction gear. Clutches and brake are typically pneumatic, that is, they are engaged when inflated by air to say 15 lb/in^2 or above, and are disengaged when the air is exhausted. In the example system, each clutch and the brake are connected to supply or exhaust by a three-way solenoid valve, the engaged condition being obtained when the solenoid is energized. Generally speaking, a clutch may engage safely only when the other clutch, the brake, its input shaft, and its output shaft, meet certain conditions. The conditions that might be considered are listed following, each symbolized by x_1, x_2, and so forth:

x_1 other clutch disengaged

x_2 brake disengaged
x_3 engine speed zero
x_4 shaft rev/min less than 50
x_5 input and output shafts rev/min less than 50 apart
x_6 engine fuel in idle position

The combination of conditions for which clutch engagement is safely permitted is shown by a logic diagram in Figure 7.4 (top sketch). The corresponding electrical pathway from a selector switch to the solenoid is shown for one of the clutches in the lower sketch. For the clutch to engage, at least one path must be established by the closing of contacts. In the top diagram this means that A and B and C must provide input to the largest AND gate for there to be an output, x_2 or x_3 must provide input to the B OR gate, and so forth.

Each contact is presumably actuated by a magnetic coil that is energized when the prescribed condition is met. For example, pressure switches on the other clutch and on the brake will energize the coils of x_1 and x_2 whenever air pressure falls below the engaging pressure.

Sequencing and timing are also accomplished by the arrangement shown. If, for instance, the selector switch is moved from BRAKE to AHEAD, the necessary sequence that the brake disengage before the clutch engages is maintained by x_1, since it will not close until the pressure in the brake has fallen to its disengage value. The time required to engage and disengage depends on the rate at which air can flow in and out, hence on the size of the valves, and on the size and length of the piping. If the action must be fast or slow for some purpose of plant operation, these features may be designed to suit. Or, if the actuation were electronic, analogous timing features would have to be built into the control circuits.

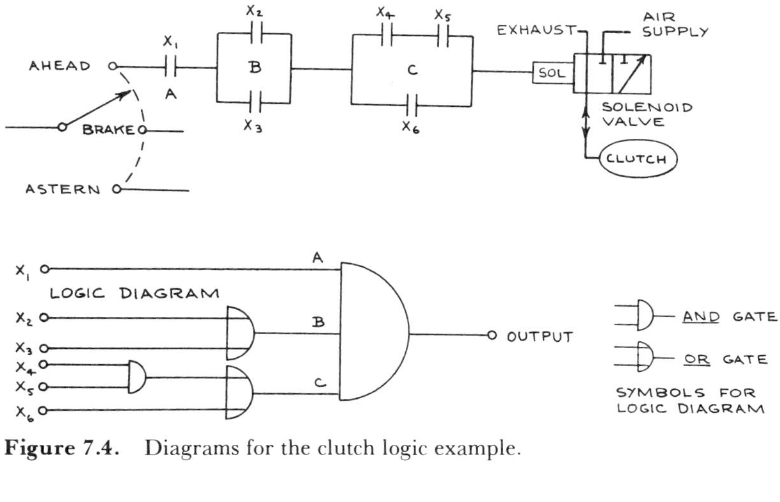

Figure 7.4. Diagrams for the clutch logic example.

7.3. CONTROL OF THE GAS-TURBINE ENGINE

A. Control of Output during Steady-State Conditions

During normal steady-state operation of a two-shaft gas-turbine engine, the parameters to be controlled are the power-turbine speed, the compressor speed, and the compressor-turbine nozzle temperature (i.e., the highest temperature in the engine). The first of these is the only one of significance to the load, and if traditional power machine practice is followed, a governor senses power-turbine speed and adjusts fuel flow to maintain this speed at its ordered value. A second governor—known as the *topping governor*—senses the gas-generator rev/min and overrides the other governor to order a fuel cutback if its speed passes a safe value. The temperature is likewise sensed and may override normal control to protect the engine.

It is also quite feasible to apply the prime control to gas-generator speed, with the topping governor sensing and protecting the power turbine speed. This scheme is known as "power control," since maintaining a set gas-generator speed assures an essentially constant engine power output. The power turbine is left free to find the equilibrium speed with its load, subject to the limits imposed by the topping governor.

Power control has an advantage for marine propulsion in that it protects the gas generator, fuel-system components, and control system from excessive wear that may occur with the first-mentioned alternative, called "speed control." With speed control, small fluctuations in shaft rev/min caused by a following sea, say, are sensed by the prime governor, with the consequent attempts to correct them by manipulating the fuel. Control and fuel system components, therefore, may be in continuous motion. But if this rev/min is not controlled, its small fluctuations may cause no significant change in power (recall the "constant power" characteristic of turbines), and hence no response by the prime governor in control of gas generator speed. Speed control nonetheless has its adherents and is used in some instances. It may be necessary with electrical transmission, or if the propulsion engine drives an a.c. generator for ship service power. Most frequently, engine control in marine propulsion is a combination of power and speed control. Typically, speed control is used in a maneuvering (low power) range over which ship speed is regulated by adjusting propeller pitch (or perhaps by slipping a clutch with a fixed-pitch propeller), and power control is used for powers above the maneuvering range.

The temperature used for topping control is also subject to alternatives. The hottest temperature, that of the gas entering the compressor-turbine nozzles, is the one that should be limited, but exhaust temperature is often the one actually sensed. The reason is that thermocouples in the exhaust stream pose no hazard to downstream turbine blades should they break off. (The speed-of-response requirement may preclude the use of a protective well for the thermocouples.) When exhaust gas temperature is used in this way, the limit must of course be set to allow for the difference between this temperature and the one for which protection is primarily required. And this difference must be adjusted in service to correct for operating conditions that can change it. Typically, ambient temperature is sensed and used as the correcting factor, since this temperature affects temperature levels throughout the engine.

If the engine is of the "twin-spool" design (see Chapter 1), a topping governor may be required to protect the low-pressure compressor shaft from overspeed. This feature is omitted on some twin-spool engines, however, for the close thermodynamic and hydrodynamic coupling between the two compressor-turbine pairs makes it almost certain that an excess speed in one will cause an excess speed in the other.

If the power-turbine nozzles can be adjusted to meet load changes, automatic control must be provided to integrate control of this parameter with that of fuel (somewhat analogous to the integration of fuel control and pitch control in a propulsion plant with controllable-pitch propeller). Recall from Chapter 4 that the purpose of the variable nozzles is to improve engine efficiency at part loads by maintaining combustor outlet temperature at its rated value. Ideally, an automatic nozzle control would adjust nozzle position to keep this temperature (or a closely related temperature, such as exhaust temperature) constant as the speed or power governor adjusts fuel flow to meet changing load. But recall also that the compressor surge limit precludes adjustments wide enough to keep temperature constant over the full range of possible loads. This control must in consequence be programmed otherwise for low powers. A typical arrangement maintains compressor speed constant at 80% of its rated value while the load is in the 0–50% load range. Temperature varies with load, and reaches its rated value at 50% load. Thereafter nozzle manipulation maintains it constant, while compressor speed is allowed to vary up to its rated value at 100% load. This is the scheme followed by the General Electric industrial-type engines. Further details on their control appear later in this chapter.

B. Control During Transients

The rate at which fuel flow can change must be limited in both the increasing and decreasing direction. An order to accelerate carries with it the danger that a sudden increase in rate of fuel flow will lead to overshoots in rev/min and temperature before the normal protective features can act to prevent them. Therefore, acceleration limiting is commonly employed. This may take the form of a definite limit on the rate at which fuel flow can be increased, or this rate may be tied to the rate at which air flow increases. Compressor discharge pressure is often sensed, and used as a measure of air flow; fuel flow is limited to a value that will not cause excess temperatures with the resulting air/fuel ratio. Increase in fuel is thereby limited to the rate at which the gas generator can accelerate.

Deceleration presents the opposite hazard: a sudden cutback in fuel can momentarily produce an air/fuel ratio so high that combustion is extinguished ("flameout"). Rate of fuel reduction is therefore usually directly limited to a rate known to be safe from this standpoint.

C. Safety Trips

A *trip* is a device that acts to stop an engine by cutting off its fuel supply when a hazardous operating condition occurs. It may be contrasted to a topping governor or other limiting device that protects the engine by reducing fuel flow when protected parameters reach limiting values. It may serve as a backup for these normal controls, or may protect against a condition that the normal operating controls cannot affect. An example of the former is an overspeed trip that would be set to act at a slightly higher speed than the power turbine topping governor. (Ships classed by the American Bureau of Shipping, for instance, must have overspeed trips set at not more than 115% rated rev/min.) An example of the latter is a low lubricating-oil pressure trip; it should be obvious that the normal fuel control can do nothing to ameliorate a defect in the lube oil system.

Over temperatures are also often used as a tripping condition.

Some engines are equipped with vibration sensors that initiate a trip if a safe level is exceeded. Excessive vibration is an indication of a loss of balance by the rotating part of the engine, a condition that is caused by either a serious mechanical defect, such as loss of a compressor or turbine blade, or by heavy fouling within the engine.

Loss of flame is a dangerous condition that usually actuates a trip. The consequences of not taking quick action following flame loss is flooding of fuel into the hot parts of the engine. Failure to ignite during starting is similarly the cause for terminating a starting sequence. On the other hand, the low lube-oil pressure trip is usually bypassed during a start, since this pressure is abnormally low until the gas generator comes up to operating speed.

D. Starting and Stopping

Starting involves a sequence of steps necessary to bring an engine from rest up to idling speed in a self-sustaining condition, whence it may be turned over to the normal operational controls. The sequence is commonly initiated by a simple pushbutton command, "start," from the human operator. The start is then managed by a device (electronic, in most cases)—the *start sequencer*—that orders the starting steps in sequence, and by its built-in logic protects the engine against hazards that may be inherent in these steps. By sequencing, logic, and timing, the start sequencer typically manages the following steps:

1. Insures that the engine is in a condition to start. (Examples: shaft brake disengaged, engine compartment ventilation in operation.)
2. Actuates the starting motor (spins compressor up to self-sustaining speed).
3. Bypasses any steady-state control that would interfere with the starting sequence.
4. Admits fuel to the combustion chambers when suitable air flow is reached.
5. Actuates an ignitor.
6. Confirms that ignition has occured after allowing necessary time.
7. Deactivates ignitor and starter.
8. Surrenders control to the normal operating system when idling speed is reached, or alternatively, terminates the sequence if ignition fails.

Stopping is similarly managed following a pushbutton "stop" order, though the sequence is much shorter. The essentials are cutback of speed to the idling level, followed by closure of a fuel valve. In some cases this may be followed by a timed run on the starting motor to promote even cooling and to purge the engine of fuel vapors.

A heavy-duty industrial-type engine may burn residual fuel, but usually only at steady loads: starting, stopping, and perhaps maneuvering, are done on a lighter fuel. The starting sequencer for such an engine may, therefore, include logic to prevent starting on the wrong fuel, and the stop sequence may include the steps of switching to the light fuel and the time of a run on this fuel sufficient to purge the residual from the fuel system.

E. Monitoring

Monitoring is the surveillance of operating parameters, with information displayed or recorded for the benefit of the human operator. Many uses may be made of the information, but the immediate operational use is to allow the operator to take corrective actions before some abnormal condition escalates to the point of causing a trip. The parameters whose abnormalities should cause a trip are usually monitored with a visual or audible alarm to alert the operator; the alarm values are set somewhat below the trip value.

Engine monitoring is usually combined with monitoring of the propulsion plant. Propeller pitch and rev/min, and status of clutches and shaft brakes are likely, therefore, to be displayed at the monitoring location. Also, parameters not directly associated with control of either engine or plant may be monitored. An example is the pressure of reduction gear lubricating oil. Moreover, such a parameter may cause a trip via a stop order generated by the monitoring system.

F. Examples of Engine Control

The engines of each manufacturer have their unique means of putting into practice the generalities just discussed. Here as examples are described the controls furnished with two of the engines that are finding prominent use in marine propulsion.

The first example is the General Electric heavy-duty industrial-type engine described in Chapter 1 (Figure 1.5). Its most significant feature in the present context is its use of adjustable power-turbine nozzles; a key function of the engine controls, therefore, is integration of nozzle angle with fuel setting to produce the optimal combination of engine speed and temperature.

The engine is controlled by the proprietary "Speedtronic" (General Electric trademark) system, a system that has been described widely in the company's promotional literature, and in technical papers. See, for example, Drake (1951) and Rowen (1970). Figure 7.5, taken from Rowen (1970), is a simplified block diagram that serves to introduce our discussion. Fuel is furnished by a fuel pump whose delivery to the engine is fixed by the position of a bypass valve. Valve position is set by the output of a comparator/amplifier that compares actual fuel flow to that commanded. The command signal is a voltage (designated VCE) imposed by the control system. An increase in output is obtained by directing the control system to increase VCE; limiting conditions, therefore, must be able to prevent an increase in VCE, or reduce it as necessary. This feature is accomplished by a minimum-value gate that sets VCE equal to the lowest of several voltages offered. These voltages are the command voltage and the voltages generated by each of the limiting parameters. The latter must be inverse voltages, that is, they decrease as exhaust temperature increases. If this temperature exceeds its limits, the voltage becomes lower than those generated by the operator's command, and by other limited parameters, and hence is selected by the VCE gate. In this way exhaust temperature becomes controlling and limits fuel supply to stay within the limiting temperature.

The variable power-turbine nozzles are used to hold compressor speed

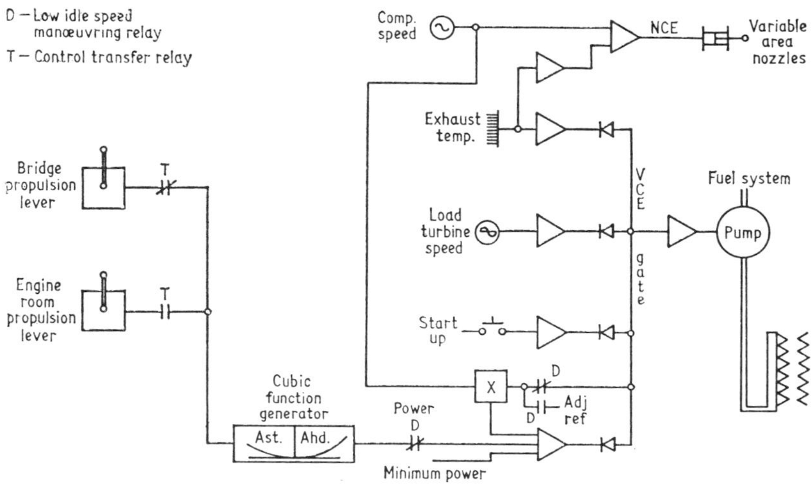

Figure 7.5. General Electric SPEEDTRONIC control system.

constant over the 0–50% load range. This speed is sensed by the automatic nozzle angle control, and held at 80% of rated value by angle changes. In the low-load range, none of the limiting voltages should be reached, so that the operator's commands for increased power are met by increases in fuel flow, while the nozzle angle is automatically adjusted in response to keep compressor speed steady. Meanwhile, exhaust (and other) temperature increases, and reaches limiting value at 50% power. When an increasing load passes 50%, exhaust temperature transiently exceeds its limit, causing the nozzles to adjust to allow a compressor-speed increase, and this action continues upon further load increases until 100% compressor speed is reached at 100% power. Temperatures are essentially constant over this range, but a command for power increase above 100% causes a further rise, since compressor speed has reached its allowed limit. The VCE gate will thus find exhaust temperature to be limiting and admit it to control of the fuel pump, thus preventing the power increase.

Figure 7.6 shows the variation of nozzle angle (α), exhaust temperature (T_X), combustor temperature (T_F), and gasifier speed (N_{HP}). (The circles enclosing Y, B, G, or R represent indicator lights showing the control mode status to the operator.)

The exhaust temperature limit is adjusted to suit ambient conditions via sensing of the compressor discharge pressure, with the adjustment being programmed so that the combustor temperature (gasifier-nozzle temperature, T_F, in Figure 7.6) is kept within a constant limit in the face of changing condi-

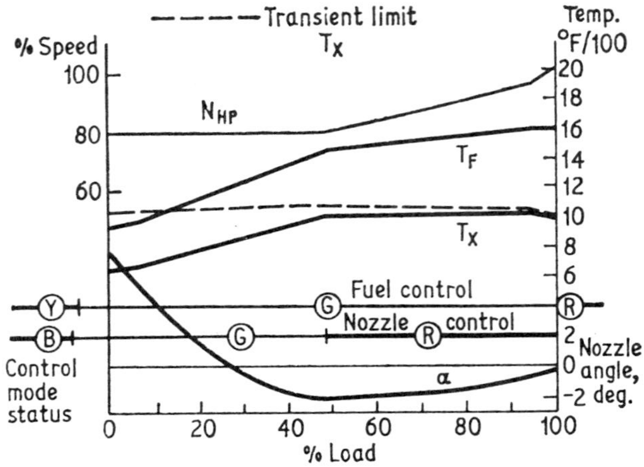

Figure 7.6. Temperatures, power-turbine speed, and nozzle angle as programmed by the General Electronic SPEEDTRONIC system [Rowen (1972)].

tions. Accleration and decleration are limited by the maximum rates at which fuel supply rate can be changed. Torque limiting is also applied; this is discussed in the subsequent examples of propulsion plant control.

The second example is the General Electric LM2500 engine (recall Figures 1.1 and 1.2), an aircraft-derivative gas turbine. The major difference between this engine and the one just treated, from the standpoint of control needs, is the absence of the variable-power turbine-nozzle feature and the attendant speed-temperature program. A somewhat analogous feature, however, is adjustment of the compressor inlet guide vanes and the first six rows of compressor stator vanes to suit inlet temperature and compressor rev/min, this being done to give compressor acceleration and deceleration stall protection and to maximize part-load efficiency of the compressor.

Control throughout the operating range is basically power governing, with fuel control responding to a power command by setting the appropriate gas generator speed. A power-turbine-topping governor prevents speed of the output shaft from exceeding 104% of its rated rev/min by cutting back on throttle-command signal to the main fuel control.

Acceleration is governed by compressor discharge pressure, compressor-inlet air temperature, and compressor speed. These are sensed by the fuel control, and fuel flow is limited to that rate which these parameters indicate can be burned without exceeding acceleration limits.

Compressor-inlet temperature is sensed by the fuel control and used to adjust the acceptable ratio between compressor discharge pressure and fuel flow.

The compressor-inlet guide vanes and stator vanes are also positioned by the fuel control. This device senses compressor speed and inlet temperature for the other purposes mentioned, and is programmed from them to position the vanes via a mechanical linkage.

Overspeed shutdown is provided by two solenoid shutoff valves in series in the fuel system. Each may be deenergized and thereby closed by a speed switch that is tripped when a signal from a magnetic speed pickup senses power turbine speed of 110% of rated rev/min. The overspeed shutdown is a backup in the event of a load drop that is more rapid than can be controlled by the speed topping governor.

7.4. PROPULSION PLANT CONTROL

A. General

Control of the propulsion plant involves coordinated management of the engine and of the other active items in the plant, principally those in the power-transmission system. If, for example, the fixed-pitch propeller is used, ahead and astern clutches are doubtless present as adjuncts to a reverse-reduction gear. If a controllable-pitch propeller is used, the equivalent item is the propeller-pitch actuator. With either type of propeller, there may be a shaft brake, and if more than one engine drives a propulsion shaft, each engine is provided with a disconnect clutch. An example of a component not part of the propulsion transmission system, but one whose control may be integrated with that of propulsion, is an electrical generator driven by the propulsion engine.

It is almost essential that the propulsion plant receive only a single command from the operator so that their actions are properly integrated by the programmed responses of the propulsion-plant control system. As an example of the importance of this requirement, consider the possible consequences of disengaging a clutch in the transmission without regard to status of the engine. If the engine were, say, at full power, the clutch would doubtless be subjected to excessive wear and heat generation during the transient interval when it is disengaging; the engine would react as if to a sudden loss of load, with the likely consequence being an overspeed trip. To prevent such disasters, and even minor operational upsets, all practical propulsion control systems are indeed integrated. They receive a single command from the human operator, which contains both the direction and intensity (i.e., power and speed level) desired, and by their programs, logic sequencing, timing, and automatic responses, carry out the functions which have been outlined for propulsion plant control in Section 7.1. To make further use of the clutch example just cited, the system reduces engine speed to idling rev/min before instituting a disengage order to the clutch, checks other logical requirements for clutch action, and also supervises the engagement of the other clutch if, say, a reversal has been ordered.

B. Example—The *Callaghan* System

The roll on/roll off ship *Admiral William M. Callaghan* presents an interesting and unique case of propulsion-plant control since it is the only U.S.-built gas-turbine ship to use reversing reduction gears with fixed-pitch propellers (twin screw) and is our only example of an essentially pneumatic control system. A simplified diagram of this system is given in Figure 7.7. The figure is taken from Richardson (1969), and this reference is the basis of my description following.

The clutches are of the pneumatic design typical of those used with diesel reverse-reduction gears. The driving and driven elements are connected by an inflatable part resembling a pneumatic tire. When it is deflated, it lies flaccid between these elements, and consequently no torque can be transmitted. When

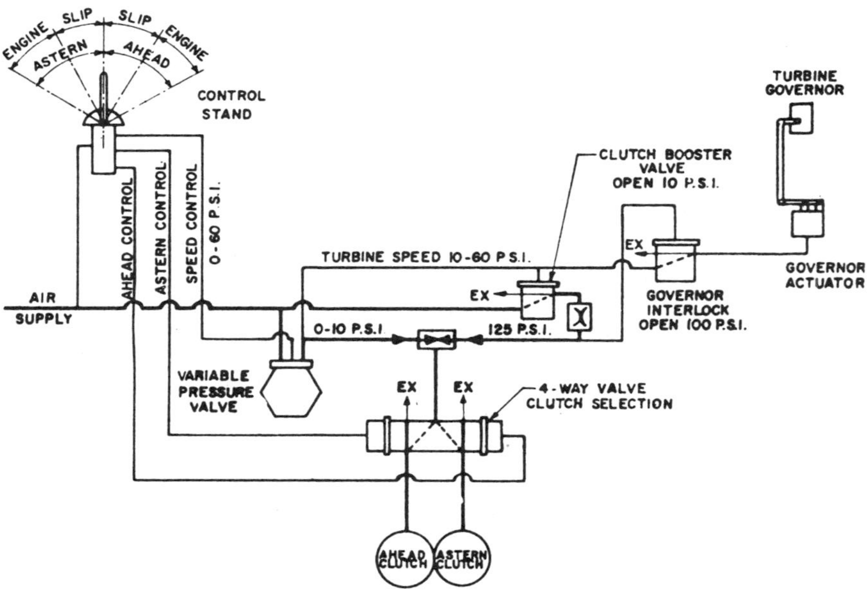

Figure 7.7. Simplified diagram of pneumatic propulsion control system, *Admiral William M. Callaghan* [Richardson (1969)].

it is fully inflated, it presses against both elements and thereby makes a rigid connection by which torque is transmitted. When partially inflated, some torque can be transmitted, but above a limiting value of torque, the clutch slips. This action is undesirable, since it means that transient slippage with resulting wear and heat generation occurs upon every engagement and disengagement. But since engine control is integrated with clutch control, torque to be transmitted during these transient intervals is small enough and the time is brief enough that wear should be trivial, and the heat generated well within the capability of the clutch to absorb and dissipate it. And in some instances—as on the *Callaghan*—the slip phenomenon provides a way of reducing propeller rpm for slow-speed maneuvering. The "lock up" pressure for the *Callaghan* clutches is 10 lb/in^2 at engine idle speed. If a pressure (P) below this is maintained, the clutch slips continuously at a slippage roughly proportional to $P/10$ (slippage = difference in input and output rev/min divided by input rev/min). Because the torque being transmitted is low, the continuous-wear rate and heat-generation rate are not matters of concern.

The control handle is shown prominently at the left of Figure 7.7. This handle mechanically actuates, via cams within the control console, a pressure regulating valve that establishes actuating pressure for the clutch directional selector valve, and a modulating pressure (0–60 lb/in^2) that controls degree of clutch inflation and output of the engine. When this signal is between 0 and 10, the modified pressure to each component is as follows:

1. For the engine, the pressure signal is constant at 10 lb/in^2. This is interpreted by the engine governor actuator as a command for idling rev/min.
2. For the clutch, the supply pressure is set equal to the primary signal pressure. Since "lock-up" (engagement without slip) pressure is 10 lb/in^2 at idling rev/min, the clutch does not fully engage, and the resulting slippage reduces propeller rev/min. Low-speed maneuvering of the ship is thereby enhanced, since rev/min is continuously variable down to 0 rev/min.

When the control handle is pushed beyond the slip range, the clutch booster valve opens to admit 125 lb/in^2 air directly to the clutch, and it is thereafter rigidly engaged. The engine governor actuator receives air pressure equal to the signal pressure, and this calls for engine power proportional to that pressure. As originally built, engine-speed control was used, but was soon changed to power control. It had been found with speed control that gas-generator power at times fluctuated as much as 3000 hp in responding to shaft-speed changes caused by pitch and roll of the vessel. The change to power control eliminated the power fluctuations. A benefit of steadier power was an apparent improvement in fuel consumption [Zeien, Smith, and Hirst

(1969)]. Logic is provided by the governor interlock valve; this prevents the increase-power signal from reaching the engine governor until the clutch has inflated to a pressure sufficient to ensure lock-up and returns the governor signal to idle if clutch pressure should be lost.

Several refinements to the control action are added by components not shown in Figure 7.7. One, for example, involves a bias signal supplied to the boost valve at low speeds so that it acts faster, speeding up maneuvering at those speeds. Also, the reader should realize that the *Callaghan* has served somewhat as an experimental vessel and that changes have been made in its propulsion plant since it first went to sea. One instance was the replacement of engines by those of other manufacture. This change required no basic change in the control system. The system as described was retained, the only modification being the addition of a delay network in the line to the engine governor in order to reduce the rate at which the engine could be ordered to accelerate.

C. *Euroliner* System

The *Euroliner* is the first of a class of high-speed North Atlantic container vessels built in Europe for Seatrain Lines of New York. The power plant is twin-screw, each shaft being driven by a Turbo Power & Marine Systems (United Aircraft) FT4A-12 aircraft-derivative gas-turbine engine (Chapter 1, particularly Figure 1.4). Propellers are controllable-pitch, by LIPS of Holland. A feature of note that has a bearing on engine and propeller control is an auxiliary generator coupled to each engine. The propulsion plant and its controls have been described in several publications [Müller and Stasiowski (1970); Carpenter, Holburn, and O'Neill (1973)], which are the major sources here.

The engine-propeller control program has been discussed briefly in Chapter 6. Recall that the speed requirements of the generators limit the minimum engine rev/min to 2000, a speed that corresponds to 75 propeller rev/min and 5000 hp per engine. Maximum engine rev/min is 3600, and although the generators are a.c., speed variation over the 2000–3600 range is acceptable to them because of the frequency-conversion equipment used on the electrical side of the system. While the engine is at its minimum rev/min, ship speed is controlled only by pitch adjustment, and this constitutes the maneuvering range of ship operation. Above the 2000 rev/min lower limit, pitch is held constant at its maximum value, and engine speed increases as a function of power demand up to 3600 rev/min and the corresponding 26,000 hp. The power-turbine rev/min is the controlled parameter for the engine fuel control in the maneuvering range; above 2000 rev/min, fuel control responds to input command from the engine telegraph by setting the gas generator speed that corresponds to the ordered power. Both "speed control" and "power control," therefore, are used in this plant. Note that were it not for the variable-speed capability of the electrical system, speed control would be required over the entire range.

Figure 7.8 is a block diagram of the propulsion-plant control system, and Figure 7.9 is a pictorial view giving an idea of how the physical equipment looks overall. A central feature of Figure 7.8 is the DATAZENT 100 (a trade name of AEG-Telefuncken). This is an extensive monitoring system that displays and prints several plant parameters, and gives alarms when required. It also is able to intervene in propulsion plant control in several instances where it may detect unsafe operating conditions outside the scope of the control system. An example of the last is low lube-oil pressure at the reduction gear; this condition initiates engine shutdown upon detection by the DATAZENT system.

Figure 7.10 is a block diagram of the starting sequence part of fuel control. The simple stop sequence is also included in the diagram.

Two significant features of the propulsion plant control fall within the category of actions that protect propulsion plant components against damaging conditions imposed by another component. One of these imposes torque-limiting; the other is the anti-windmilling feature.

The anti-windmilling feature protects the engine from overspeeds that might result from low pitch at high vessel speed during rapid maneuvering at high power. Since the overspeed would be the consequence of an input from the output shaft, normal overspeed protection via fuel cutback or fuel trip is ineffective. The situation that could produce windmilling overspeed is therefore avoided by forbidding propeller pitch reduction until engine rev/min has fallen below 3000 rev/min. Logic circuits interrupt that part of the reversing sequence that reduces pitch until 3000 rev/min is reached by means of the normal decay of ship and shaft speeds after engine fuel is reduced to idle (2000 rev/min) value. This procedure was found by simulation in the design stage to provide adequate protection against the windmilling hazard.

There is a brake on each propulsion shaft, but for emergency use only. Control is independent of the integrated propulsion plant control system, except for protective logic that prevents the brakes from being engaged during normal operation.

D. Propulsion Systems Using the General Electric Heavy-Duty Engine

The General Electric heavy-duty gas-turbine engine is being installed during the early 1970s in several classes of merchant ship built in Europe, the United States, and in Australia. The description here covers the basic system of pro-

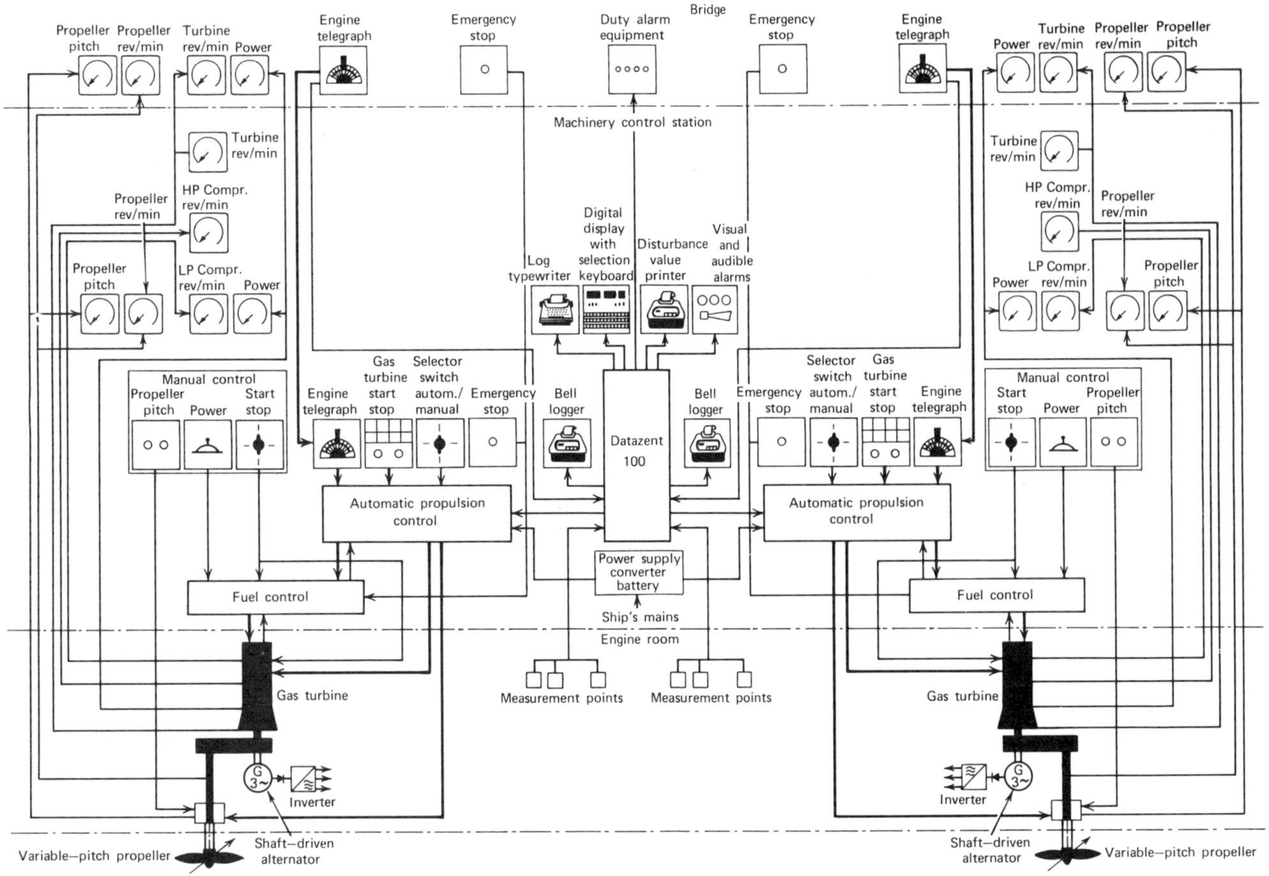

Figure 7.8. *Euroliner* propulsion-control block diagram.

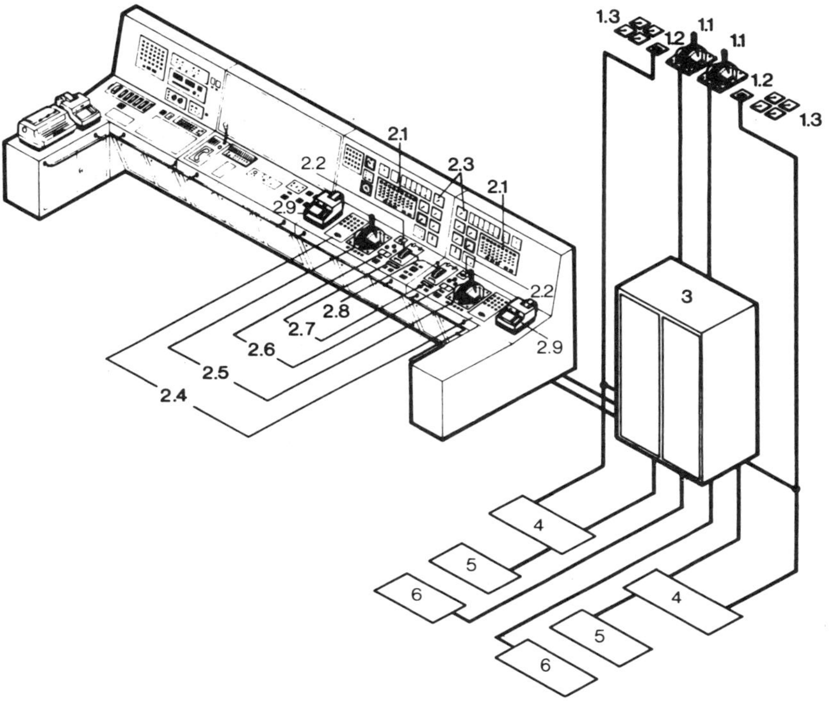

Bridge

1.1 Engine telegraphs
1.2 Emergency stop
1.3 Indicators for
rpm's
power,
propeller pitch

Maneuvering position in the control console

2.1 Automatic start/stop sequence of gas turbines with indication of operational conditions
2.2 Selector switch automatic/manual
2.3 Indicators for rpm's, power, propeller pitch
2.4 Start/stop sequence, manual
2.5 Engine telegraphs
2.6 Emergency stop
2.7 Manual setting of propeller pitch
2.8 Manual setting of power
2.9 Bell logger

3 Pt. and stbd. automatic propulsion controls
4 Fuel control
5 Fuel valve of gas turbine
6 Hydraulic actuator of controllable pitch propeller

Figure 7.9. *Euroliner* propulsion-control equipment.

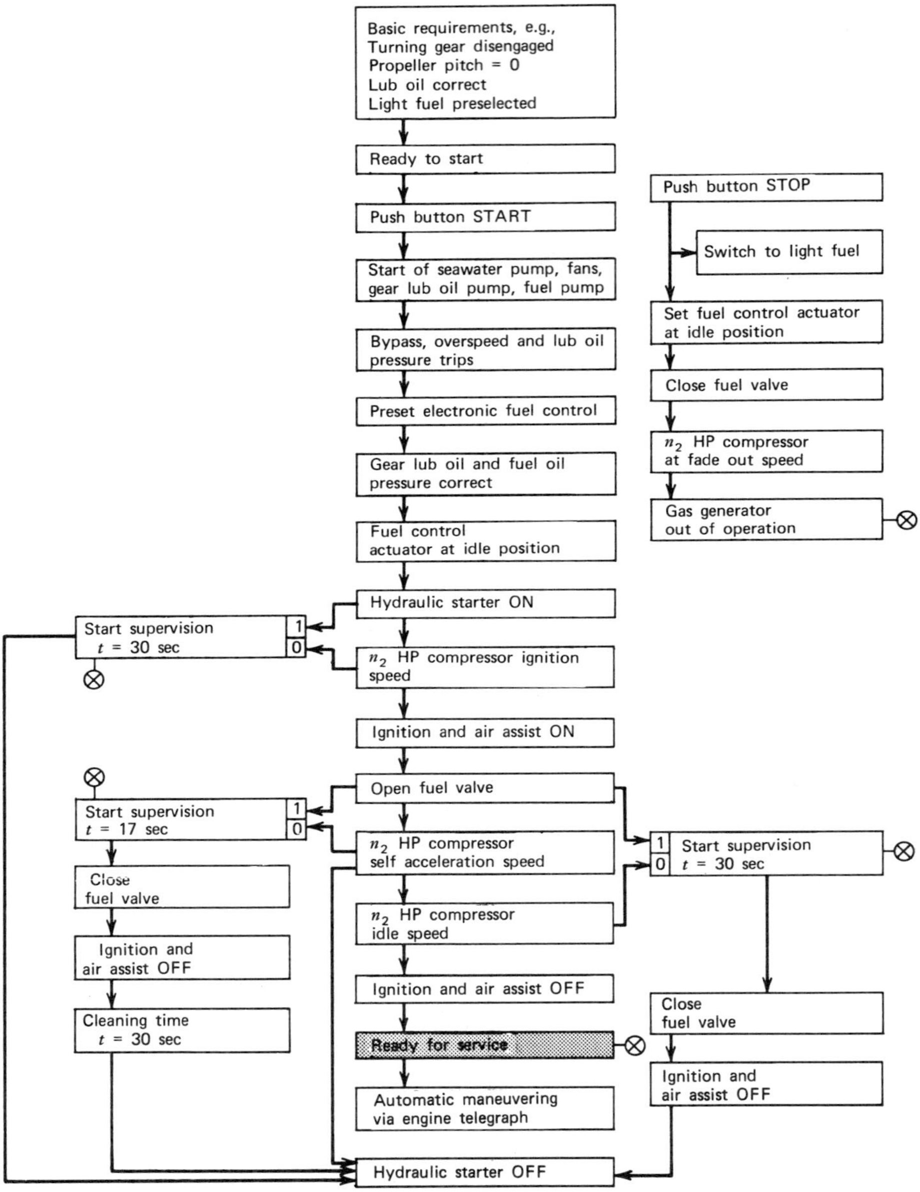

Figure 7.10. *Euroliner* starting-sequence block diagram.

pulsion control from which the individual systems are adapted [Rowen (1970, 1972)]. (Short descriptions of two particular applications appear in following sections.)

Figure 7.5 illustrates the principal features of the engine control. The only item shown there specifically related to marine propulsion is the cubic-function generator. This device produces an output signal proportional to the cube of the input signal. Since the power-speed characteristic of any displacement ship is likely to be at least approximately cubic, the result is an effectively linear relationship between propulsion lever position and ship speed.

Figure 7.11 shows in similar fashion the added circuits required to integrate engine control into that of a particular type of propulsion plant that shall serve as the main example here. The significant features of this plant are a controllable-pitch propeller and a constant-speed auxiliary generator driven by the propulsion engine. (The line labeled "to gas turbine power control" is the input to the cubic function generator of Figure 7.6.) Three modes of control are indicated: *docking, maneuvering,* and *at-sea.* The desired mode is chosen by closure of the proper contacts shown in the figure (e.g., closure of contacts M, D, and D × 1 puts the system into the docking mode).

The at-sea mode is used to maintain constant power turbine rpm while the engine is driving the attached a.c. generator. Note that a pitch governor senses this speed; it maintains speed by adjusting pitch. The engine meanwhile is controlled by the SPEEDTRONIC components as previously noted and consequently is subject to power control even though the output shaft rev/min is required to stay constant for the benefit of the generator. Direction of rotation is sensed by the pitch governor and compared with the ordered direction. This feature permits emergency reversal to be accomplished while the plant is in the

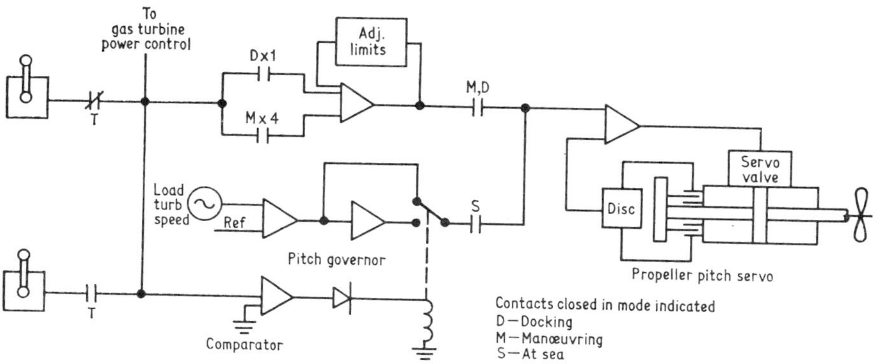

Figure 7.11. Propulsion control as integrated with General Electric SPEEDTRONIC System [Rowen (1972)].

at-sea mode (though doubtless at the expense of some transient upset in the generator frequency).

The maneuvering mode covers the full range of powers without regard for generator speed (ship service power would presumably be furnished by a standby unit during maneuvering). Over about the first 40% of the propulsion-control lever-handle movement, pitch and ship speed increase linearly while power-turbine rev/min is constant at its minimum value. Pitch reaches its maximum value at 40% speed and remains there at higher levels, so that powers above 40% are reached by rev/min increases at constant pitch.

The low-speed maneuvering, or docking, mode covers only the power outputs attainable with the engine operating at its minimum rev/min. The signal reaching the pitch control from the propulsion control lever is one-quarter its magnitude in the M maneuvering mode. Full pitch is therefore reached only at the end of the propulsion lever travel with the engine still operating at its minimum speed. The effect is to permit the vessel operator to make much finer adjustments in vessel speed. He has only about 6% power available at the limit of handle travel, but he may switch modes if greater power should unexpectedly become necessary.

Torque limiting can be included by adding a low-value gate between the cubic-function generator and the power amplifier (triangle symbol). This gate is furnished a second signal proportional to shaft rev/min. If rev/min signal is low while power signal is high—the combination that produces excessive torque—this gate selects the former signal, effectively overriding further increases in power until the rev/min increases sufficiently to lower the torque to a safe level.

Electrical transmission of propulsion power demands some additional control features, since a generator and a motor take their place in the transmission system. Now, there are several possible types of electrical transmission; for instance, both d.c. and a.c. systems have been used, and rectified a.c. (a.c. generator, d.c. motor) seems practicable with contemporary equipment. Here we limit discussion to just one type that has been offered by General Electric for use with its heavy-duty gas turbine. It is an a.c. system based on a synchronous propulsion motor and a fixed-pitch propeller. Description here is based on Rowen (1970, 1972) and Ekstrom (1973).

A synchronous motor is characterized by unity power factor and hence by quite high efficiency during normal operation. To achieve this, however, it must remain locked in step electrically with the propulsion generator. Speed of the propeller, therefore, is changed only by changing the rev/min of the engine. (In d.c. systems, propeller speed can be varied independently of generator rev/min.) This characteristic presents problems during startup and during low-power operation when the engine cannot reduce its rev/min to the level required, for unless the synchronous motor can remain in step, it produces no torque whatever. The solution is to remove the excitation normally supplied to the rotor windings and reconnect them so that the motor acts an an induction motor. This type is characterized by slip between its rev/min and that of the generator, with high torques being produced at all slips up to 100% (0 rev/min). Power factor is high and efficiency is low at high slips, but this is of little consequence at low powers and for brief intervals.

Figure 7.12 is a simplified block diagram of the propulsion plant and of the major control components. In normal at-sea operation, the engine power is controlled by the SPEEDTRONIC system previously discussed, and the propulsion motor rotates in synchronism with the generator. Power flows from generator to motor via the ahead circuit breaker (AH in the figure) in the propulsion switch gear (SW GR). Generator voltage is proportional to generator rev/min, controlled in this way by automatic regulation of its excitation. Motor excitation is regulated to maintain unity power factor.

In the maneuvering range, that is, at propeller speeds that would be below the lowest allowed rev/min of the engine, the propulsion motor is operated as an induction motor. Torque in this mode is proportional to the square of applied voltage, so that voltage regulation provides the means of controlling motor torque. The engine is meanwhile governed at constant power-turbine rev/min so that the frequency remains constant. As less voltage is supplied to reduce motor torque, and it slows in consequence, rev/min of motor and of generator diverge and the slip of the motor increases. High slip means high power factor and the associated high current for a given torque. This in turn implies a high heat-generation rate, but at the low powers where this occurs, the normal cooling capacity of the motor is sufficient to dissipate the heat.

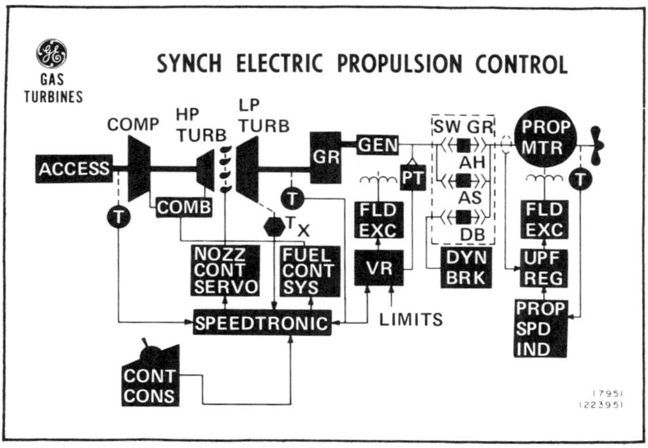

Figure 7.12. General Electric Synchronous electric-propulsion control [Carpenter, Holbrun, and O'Neil (1973)].

During rapid maneuvering, the propeller shaft is slowed by dynamic braking. The windmilling propeller drives the motor as a generator loaded by a bank of resistors (DYN BRK in Figure 7.12). Upon rapid reversal from ahead, say, the motor is switched from the generator to the resistors so that mechanical energy in the moving hull and the rotating shaft can be rapidly degraded to heat. After the shaft slows sufficiently, the dynamic braking circuit breaker is opened and the astern breaker closed so that the motor—now acting as an induction motor—can accelerate the shaft in the reverse direction.

E. Chevron Tankers—an Electric Drive Example

The Chevron (Chevron Oil Company) ships are a class of 35,000-ton (deadweight) single-screw tankers built in the early 1970s by FMC Corporation-Marine & Rail Equipment division. (See Figure 10.2). The propulsion engine is a General Electric heavy-duty industrial gas turbine with a.c. electric drive. Discussion here is abstracted from General Electric Technical Proposal MDS-11349, which was kindly furnished by that company.

The propulsion system differs somewhat from the electric drive described just above in that the propeller is controllable-pitch. This makes possible the use of the propulsion generator to supply ship service power, since its rev/min can be maintained constant, even during emergency maneuvering from ahead to astern (or vice versa). It also makes unnecessary the operation of the propulsion motor in the induction mode during low-speed maneuvering.

It is of interest to note the starting sequence for the propulsion motor. A prior necessity is that the engine and propeller generator be running and ready to accept load. A second condition is that propulsion shaft auxiliaries be operating correctly. These are the stern-bearing lube-oil pump, and the thrust-bearing and motor lube-oil pump; correct operation is manifested by adequate lube-oil pressures being developed. This step may be taken by manual start of the pumps, or be by an automatic sequence. When the requisite conditions are met, motor start may follow, provided that the following preconditions are also met:

1. Propeller pitch hydraulic servo pressure is correct.
2. The propulsion control lever is set in the zero-power position.
3. Propeller pitch is in the zero-thrust position.
4. Shaft turning gear is disengaged.

Motor start is initiated by a manual pushbutton, but is managed by an automatic sequencer that performs the following steps:

1. De-energizes the motor space heaters.
2. Starts the motor vent and exhaust fans.

3. Starts the bearing lube-oil pumps.
4. Connects the motor windings in wye.
5. Energizes the motor windings.
6. Reconnects the motor windings to the delta configuration when motor speed reaches 97% of the rated value.
7. Energizes the motor field to pull the motor into synchronous speed (100 rev/min at 60 hz).
8. Stops the bearing lube-oil pumps.

The motor operates during this sequence as an induction motor until step 7, the period in which it is accelerating from 0 to 100 rev/min. (The switch from wye to delta winding configuration is a common practice with large induction motors, intended to reduce the starting current surge.)

In normal propulsion operation the propulsion generator and motor run at constant rev/min, with ship speed controlled by pitch variation. As the propulsion control lever is moved from *stop* in the ahead or astern directions, pitch is increased directly in an approximately linear pitch-lever angle relationship over the full power range. Up to about 40% full pitch, the engine is controlled in the speed mode; its power turbine is governed at the constant rev/min. Above this point, power governing is used, but propeller-shaft speed is also controlled by a pitch governor that adjusts the pitch ordered by the control lever. In reducing power the transition from power to speed governing takes place at about 30% pitch, the difference being intended to prevent cycling of the program when steady operation in the neighborhood of the transition point is desired.

The auxiliary ship-service generator, which is intended for use in port when the propulsion generator is secured, can be used as an alternative source of power when the main propulsion generator is unavailable. Revolutions per minute is kept constant by the auxiliary turbine-fuel control, and the propulsion control lever sets pitch as with the normal propulsion mode, except that the range is limited by the comparatively low power rating of the auxiliary machine. The ship-service electrical load is meanwhile transferred to the emergency generator, since the auxiliary generator operates at reduced speed while furnishing propulsion power.

F. Reksten LNG Carrier

This ship, built in Norway in the early 1970s, is powered by the General Electric heavy-duty engine, using mechanical drive with a controllable-pitch propeller. Discussion here is abstracted from General Electric Specification MDS-11381, which was kindly furnished by that company. The engine and propeller are controlled by a single propulsion-control lever, located either on the bridge or at an engine room console. As the lever is moved from the neutral position in either ahead or astern direction, pitch is increased while shaft rev/min remains constant until full pitch is reached at about 40% ship-rated speed. Thereafter, pitch remains constant while rev/min is increased up to its full value. Governing of the gas turbine is in the power mode throughout.

A low idle mode of operation can be selected as an alternative for maneuvering. In this mode the engine remains at a low idle rev/min over the full range of handle movement, a speed at which it produces no more than about 1% of its rated power. Propeller pitch varies as a function of handle position over the full range of movement.

A shaft brake is also a feature of the propulsion system, although it is used only to stop and lock the shaft, and not as an aid to normal maneuvering. Its setting and release is nonetheless strongly dependent on the operational state of the propulsion plant at the time. The logic essential to setting the brake is expressed by the necessity for all of the following conditions being met:

1. Engine mode selector switch in the *low idle* position.
2. Engine operating in the low idle mode.
3. Propeller at neutral pitch.
4. Propulsion lever in the neutral position.
5. Power turbine shaft below 50% rated rev/min.
6. Propeller shaft turning gear disengaged.
7. Engine compressor rpm below 55% rated rev/min.

Further logic forbids switching the engine from the low idle mode while the brake is engaged, and the brake cannot be released unless the propeller pitch is at neutral.

The nominal fuel for the heavy-duty gas turbine is residual oil, but distillate fuel is used during startup, shutdown, and during sustained low-power operation. The ship discussed here, being an LNG carrier, also has available cargo "boil-off" (gas) as fuel. Since the rate at which this gas is supplied is strictly fortuitious, it seldom matches the demand of the engine for fuel. As a result, capability of burning gas and an oil fuel simultaneously is required. As might well be expected, this requirement complicates gas-turbine fuel control, as well as the fuel and combustion systems themselves. The simultaneous burning of gas and oil fuel in the General Electric engine, and the control of the process, is described in the chapter on fuels and so is not attempted here. Let us only note the several choices of fuel required aboard the Reksten ship, with a few words on the implications of each. They are (1) residual oil, (2) normal, (3) distillate/gas, (4) gas, and (5) distillate. Each of these is selected by a five-position selector switch.

The residual position restricts fuel suppy to residual fuel. Normal position permits either residual and gas or the distillate and gas and, as the name implies, is the normal mode of operation. However, switching the engine to the low idle mode causes fuel to be limited to distillate and gas only. Distillate/gas position restricts use to distillate and gas only. Gas position permits only gas to be used and, in effect, limits power output to that corresponding to the available gas flow. Distillate position likewise permits only the fuel named to be used.

Normal startup and shutdown sequences take place only in the distillate mode. Transfer to this mode takes place automatically when the operator orders either startup or shutdown.

G. A Naval System—The DDH-280 Canadian Destroyer Class

DDH-280 designates a class of twin-screw Canadian destroyers. Each shaft is driven by a Turbo Power & Marine Systems (United Aircraft) FT4 engine in its full-power mode, or by a smaller FT12 engine in the cruise mode. The propellers are controllable pitch [Benn (1969); Sachs (1969)].

The engine-pitch program is similar to those described for other vessels: there is a maneuvering range in which pitch is varied from zero to maximum in either direction while the engine runs at its maximum rev/min; the power range follows with rev/min being increased up to its maximum value at full power. This program is depicted in Figure 7.13, plotted against control-lever angle. Note there that vessel speed is proportional to control-lever position.

Control is normally affected either from the bridge or from a machinery control console located in a machinery control room on the deck above the machinery space. A third control station, the Local Operating Position (LOP), is in the compartment with the engine, but is normally not manned, being used only for testing or emergency control.

Control of the propulsion plant is illustrated in block form by Figure 7.14. A ship speed command (i.e., handle angle, as shown by Figure 7.13) is sent pneumatically from either the bridge or the machinery control console to the shaft speed programmer, a device that translates this command into an engine rev/min signal. The same command is given to the pitch controller, which sets pitch in accord with the command. The signal from the shaft speed programmer nominally passes directly to the engine fuel control (F.C.U. in the figure) by means of the LOP actuator. However, the signal passes through several intermediate devices that also contribute to the fuel setting, as seen by following the heavy line in the figure. These contributions are as follows:

1. On-line controller: This is in effect a governor that acts during the constant-rev/min portion of the program to maintain that rev/min by fuel adjustment.

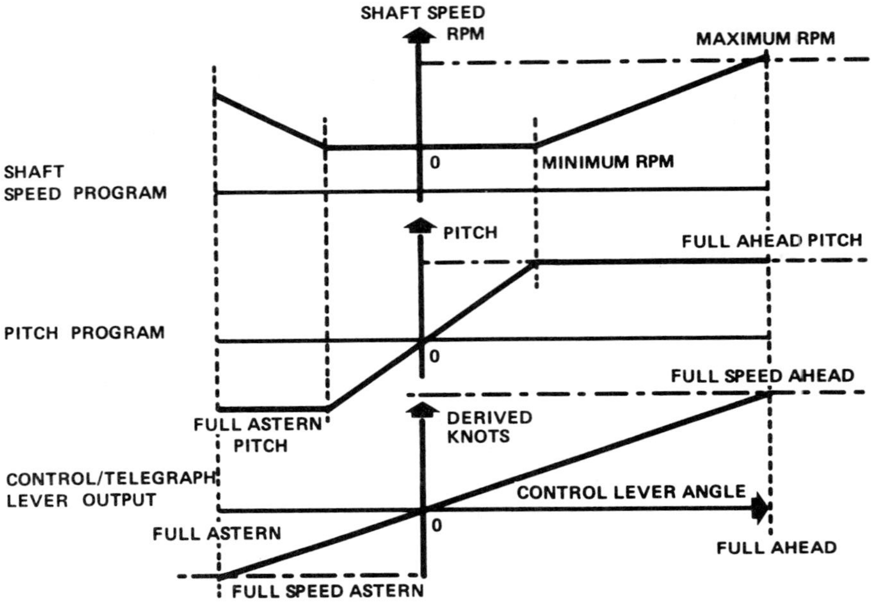

Figure 7.13. DDH-280 pitch-rev/min program, [Sunley and Patterson (1969); Sachs (1969)].

2. Off-line controller: This is in effect a governor that acts during the constant pitch part of the program to maintain the rev/min ordered by the shaft speed programmer.
3. Turbine inlet temperature compensator: This device senses ambient temperature, and adjusts the command so that power produced at a given lever angle is always the same as that produced on a 59 F day.
4. Sea-state adjustment: This device is a gain adjustment for the off-line controller. When gain is reduced, it lessens the fluctuations in engine power that could be caused by propeller-shaft torque variations induced by heavy seas.
5. Torque limiter: This limiter is furnished with a torque signal by a torquemeter, and cuts back the signal from the shaft speed programmer if a safe level of torque is exceeded.
6. Ramp: Noted subsequently.

Sequencing equipment is also provided for engine starting, stopping, and for transfer of power between the main engine and the cruise engine. The starting and stopping sequences are much like those described for the *Euroliner,* and indeed are typical of gas-turbine engines in general. If a transfer of power is ordered, the starting sequence is accomplished for the engine assuming power, it is then "ramped" to the speed commanded (i.e., power level is increased at a limiting constant rate, and not instantaneously), and the proper clutches are engaged and disengaged. The offgoing engine is ramped to idle, then maintained at idle speed for five minutes before being stopped.

The several safety trips, including those possible during starting and transfer sequences, are listed below [here quoted from Sunley and Patterson (1969)].

1. Power turbine overspeed of any engine shuts down that engine.
2. If a cruise engine clutch does not disengage or on power transfer to a main engine and the cruise engine power turbine is thereby accelerated

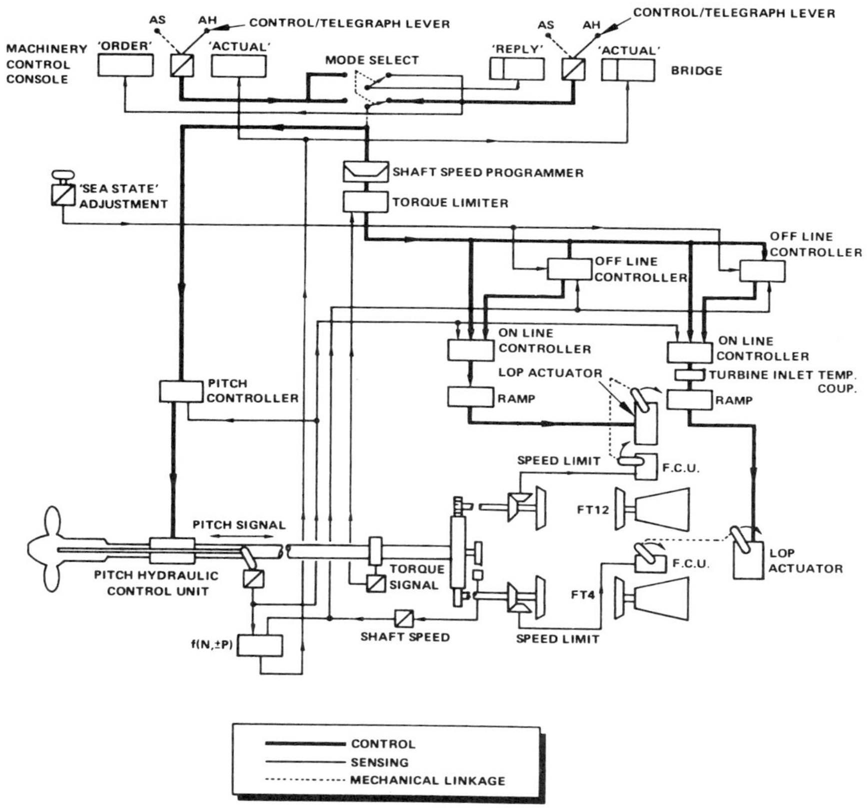

Figure 7.14. DDH-280 propulsion-control block diagram [Sunley and Patterson (1969); Sachs (1969)].

to its overspeed trip level, the main engine also trips through the sequencing equipment.

3. During automatic operation, if the engine vibration exceeds preset limits, the engines will be shut down. A vibration trip override is provided on the machinery control console for battle conditions or periods of critical maneuvering. When the override is set on the vibration trip, the operator relies on an independent alarm at trip levels.
4. During automatic starting, until the idle speed condition is reached, a preset power-turbine inlet temperature limit trip is in effect. Above idle speed, alarms only function.
5. During automatic starting, the following sequences must be completed within preset times or the engine is automatically shut down.

a. On starting the main engines, the cruise engine clutches must unlock within approximately 3 sec or the main engines will be shut down.

b. On starting, the gas generator must reach ignition speed within a preset time period.

c. After ignition, the gas generator must reach a preset speed within a preset time. This acceleration check ensures that satisfactory ignition has been achieved.

d. After ignition, the gas generator must reach idle speed within a preset time.

The control system is supported by an extensive monitoring system. Value of approximately 150 plant parameters are automatically recorded at 1-hr intervals. Many of these are indicated visually at the control stations, and some trip alarms.

H. The United States Coast Guard Icebreaker *Polar Star*

The *Polar Star* is a triple-screw diesel-electric icebreaker, with gas-turbine boost power for heavy ice breaking. Each shaft is normally powered in cruising and light icebreaking by a d.c. motor, energized through rectifiers by twin a.c. diesel generators totaling 6000 engine horsepower. Propellers are controllable-pitch. Under heavy icebreaking conditions, each shaft is powered by a single Turbo Power & Marine Systems FT4A-12 gas turbine driving mechanically through a reduction gear. Gas-turbine engine rating is 20,000 shp, with a maximum output of 25,000 shp. Operation is strictly CODOG, that is, diesel-electric and gas-turbine engines cannot drive a shaft simultaneously. During gas-turbine operation, the propulsion motor remains in the shaft line, but only as an inert rotating member.

Figure 7.15 depicts schematically the propulsion plant and the control

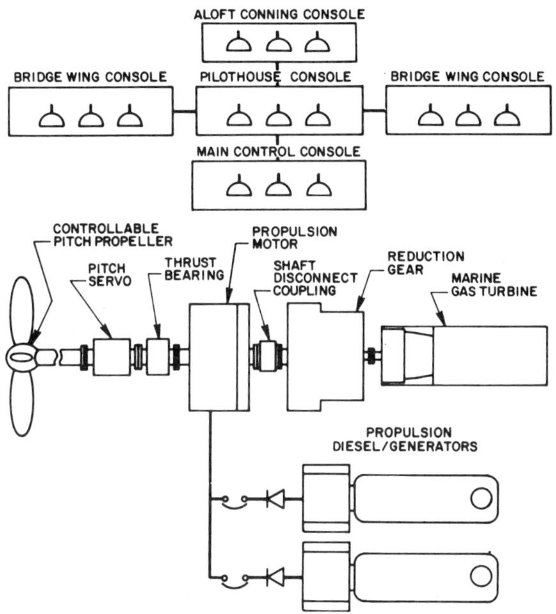

Typical throttle control schematic

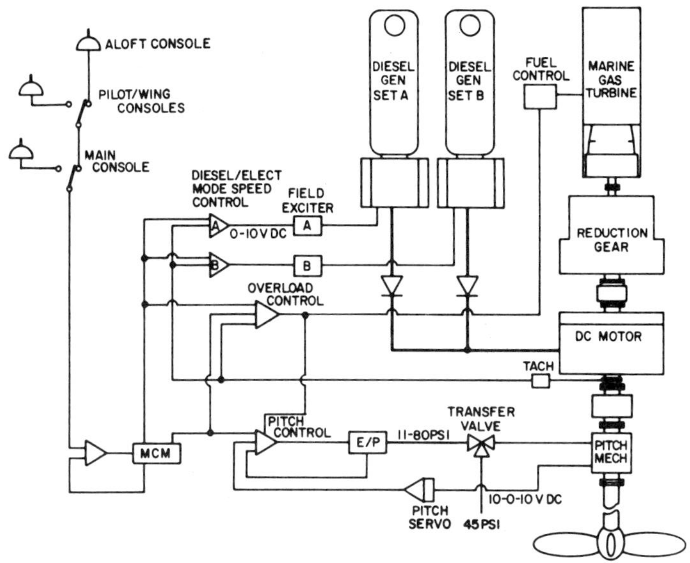

Basic control system schematic

Figure 7.15. *Polar Star* propulsion-plant and propulsion-control block diagram [Finizio (1973)].

system basics. Here, the description is limited to the gas-turbine control aspects and is abstracted from Finizio (1973) and Johnson and Stasiowski (1973).

The heavy icebreaking operational mode with gas turbines driving is conducted at a constant 175 propeller shaft rev/min; propulsion control lever position is therefore translated directly into a proportional pitch position while the engine fuel controller maintains the constant rev/min (speed control mode being used). However, heavy ice breaking usually entails a wide range of load fluctuation. The propulsion control system responds automatically to such fluctuations in these ways:

1. If increased load reduces shaft speed, a switch to power mode governing is made and the power is increased as necessary, up to 20,000 shp.
2. If shaft speed is still low when 20,000 shp is reached, pitch is reduced, down to the minimum allowed for the particular control handle position.
3. If the minimum permitted pitch does not restore rev/min to within 10 of 175, then engine power is increased to 25,000 shp.

The pitch reduction is accomplished as quickly as possible, but a slower integration rate is used in the control amplifier to restore pitch when the load (e.g., chunks of ice impeding the propeller blades) reduces this to ameliorate the rapid cycling that might occur.

When a gas turbine is to be placed on the line for this service, it is first started by a sequencer typical of those mentioned earlier in this chapter. The shaft-disconnect coupling, which can be engaged and disengaged only with the shaft stopped, normally will have been engaged earlier, with a clutch in the reduction gear meanwhile allowing independent rotation of the engine and propeller shaft. When a power-mode selector switch is placed in the transfer position, this sequence occurs:

1. Propeller shaft rev/min is changed to 105 (by diesel plant control).
2. Propeller pitch is set to neutral.
3. The clutch is engaged.
4. Propulsion generator fields are reduced to zero.
5. Diesel rev/min is set at its idle value.
6. The propulsion motor field is switched off, and brushes are lifted.

The transfer switch is then placed manually in the gas-turbine position, which automatically brings shaft rev/min to 175, provided the logical conditions implied just above have been met. Transfer back to diesel-electric drive is essentially a reverse procedure.

7.5. PROPULSION DYNAMICS

Acceleration and decleration of a ship place demands upon the engine for rapid changed in ouput and in turn may impose hazards on the engine and drive train. The latter include danger of overspeeding by a windmilling propeller during rapid reversals, of overtorque from the need for high accelerating torques at low rev/min, and generally of any engine parameter deviating from a safe range during transient conditions. These possibilities of hazard must all be denied by adequate control functions, and the control system must also direct the engine in responding to the need for rapid change in ouput.

The features of control systems thereby made necessary have been mentioned in the general discussions and examples preceding. Examples are acceleration-limiting, torque-limiting, and antiwindmilling features. The design of these is based on knowledge of magnitudes and times in propulsion transients. Therefore, it is necessary in the design stage to compute the values of all pertinent parameters as functions of time over all likely transient conditions. This is largely a new requirement in marine engineering, something that has traditionally not been done in the design of diesel and steam-propulsion plants. These engines are of higher inertia and, therefore, less rapid to respond than the gas turbine, and they do not operate as close to conditions of hazard (thinking principally of the maximum temperatures in each). In most cases, propulsion controls based on steady-state requirements, aided by experience with success on previous plants, has sufficed. But for the reasons implied above, the gas turbine propulsion plant demands analysis by its designer of its behavior during transient conditions.

The analysis is founded on the propulsion dynamic equations, which are

$$m\frac{dV}{dt} = T_{\mathrm{p}} - R_{\mathrm{T}} \tag{7.1}$$

$$2\pi I\frac{dn}{dt} = Q_{\mathrm{d}} - Q_{\mathrm{f}} - Q \tag{7.2}$$

where m is mass of the hull, including entrained water; V is speed of the vessel; t is time; T_{P} is propeller thrust; R_{T} is hull resistance; I is moment of inertia of the rotating components, including water entrained by the propeller; Q_{d} is engine torque; Q_{f} is friction torque; Q is propeller torque, and n is shaft speed. The first of these expresses the relationship between ship acceleration and net force causing it; the second expresses the relationship between shaft acceleration and the net torque causing it. As shown, they are but the skeleton of the equations actually to be solved, for the terms on the right-hand side are

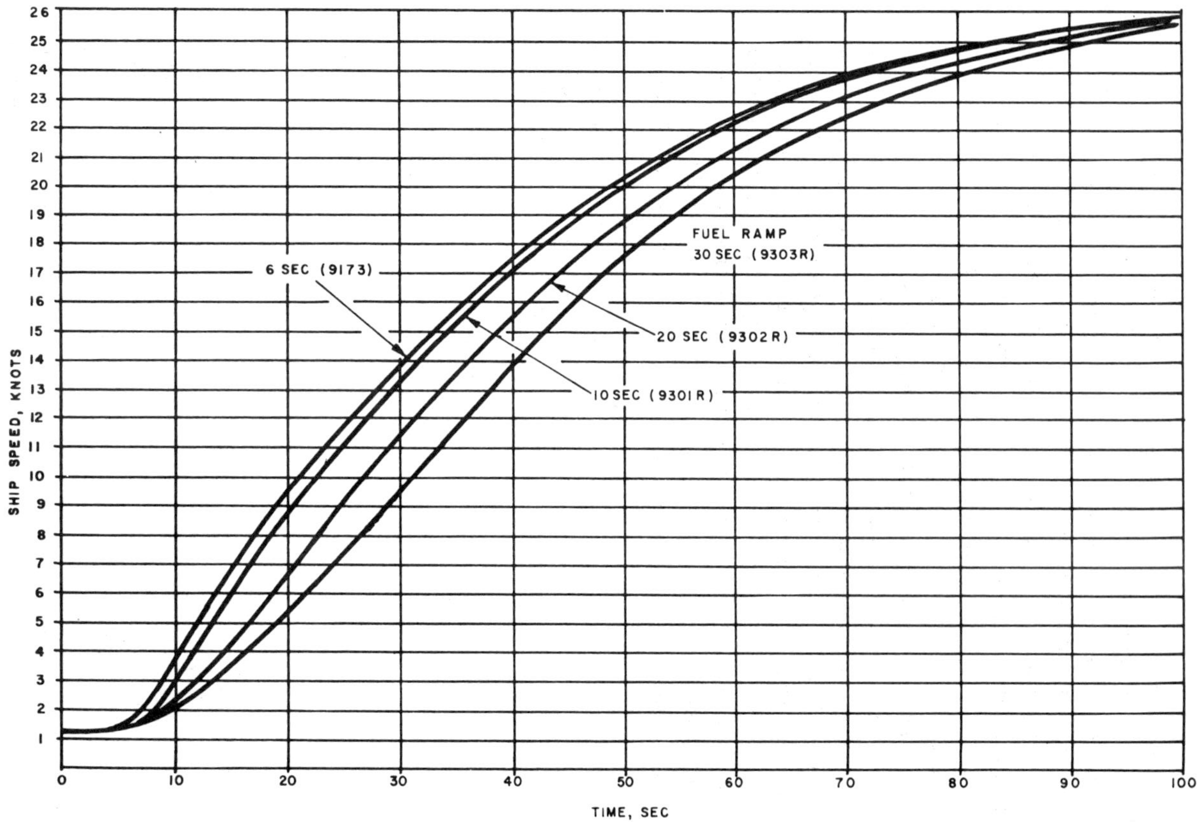

Figure 7.16. Simulated acceleration of a destroyer escort [Rubis (1972)]. From *Transactions,* 1972, published by the Society of Naval Architects and Marine Engineers.

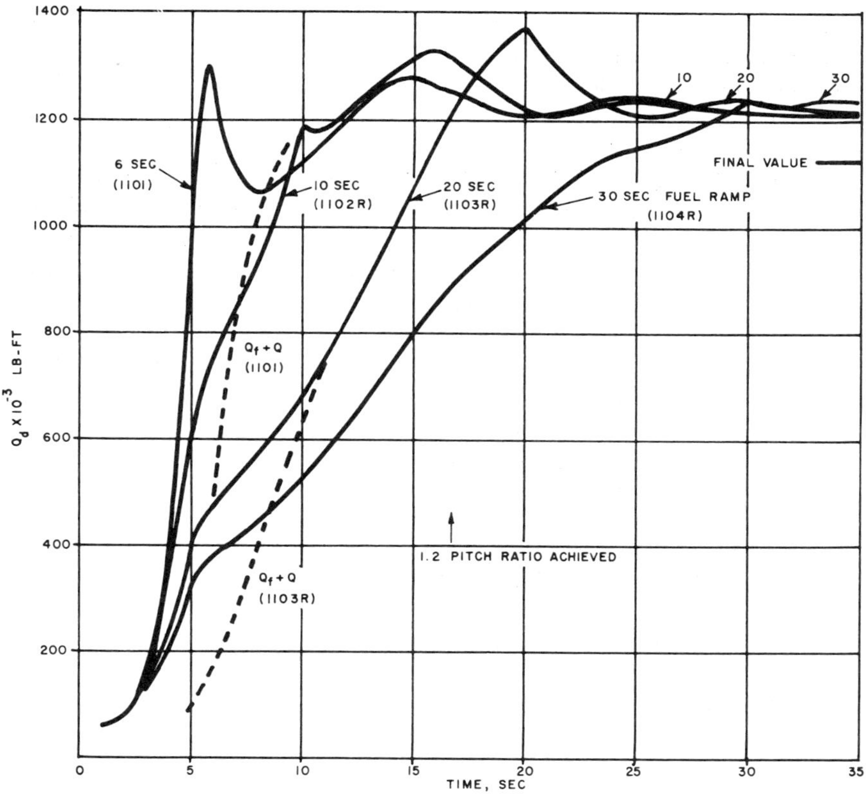

Figure 7.17. Simulated torque during acceleration of destroyer escort Rubis (1972). From *Transactions,* 1972, published by the Society of Naval Architects and Marine Engineers.

generally functions of vessel speed and shaft rev/min, and complicated functions at that. The outstanding examples are the propeller terms, T_P and Q, since these are expressible with any accuracy only in graphical form via the K_T and K_Q characteristics (Chapter 5, Figures 5.1 and 5.6). Solution—if done by hand—therefore requires a laborious iteration technique. In most contemporary cases, the calculation is done instead by an electronic computer, digital, analog, or hybrid of the two, with the solutions being plotted by the computer as continuous curves. Since the equations presumably behave in the same way as the physical systems they represent (at least when supplied with accurate data on those systems), the plots should show how the hull, propeller, and machinery behave in reality, hence the physical systems are said to have been *simulated.*

The designer also includes in the simulation equations representing the behavior of any parameter that is to be treated by the controls, and also equations representing the behavior of the control functions. The simulation then demonstrates the adequacy of the control-system design, and this design may be altered easily in the simulation stage.

Simulation has been a necessity in contemporary gas-turbine propulsion design, since each vessel, or class of vessel, has been unique. The graphical plots produced by the simulating computer are widely found throughout the references that have been cited previously in this chapter. For extensive exposition of results, and of the methods of analysis used, the reader is referred especially to the works of Rubis [Rubis (1970, 1972); Bodnaruk and Rubis (1971)]. His work treats the simulation of acceleration, deceleration, and reversal of moderate-sized naval vessels powered by gas turbines, with both fixed and controllable-pitch propellers.

Figures 7.16 and 7.17, taken from Rubis (1972), are examples of what simulation can show. (The vessel is a twin-screw destroyer escort of about 4000 tons displacement.) The first of these figures is a plot of ship speed during a large acceleration transient, showing the effect of the fuel "ramp," that is, the number of seconds over which the fuel controller advances the fuel flow rate. The second is a plot of engine torque developed during the same acceleration, also showing the effect of the fuel ramp used. This figure is a guide to the control-system designer, showing him, for example, whether there will be need for overtorque protection, and if so, at what speed it must respond.

7.6. REFERENCES

Benn, D. H. (1969), "The Propulsion Plant for the DDH-280," *Canadian Shipping and Marine News,* **40,** pages 56–60.

Bodnaruk, A. and Rubis, C. J. (1971), "Acceleration Performance Analysis of a Gas Turbine Destroyer Escort," American Society of Mechanical Engineers Transaction, *Journal of Engineering for Power,* **93,** pages 49–60.

Carpenter, D. B., Holburn, J. G., and O'Neil, D. A. (1973), "Systems Integration of the GTS *Euroliner* from Conception to Operation," *Marine Technology,* **10,** 1.

Drake, F. (1951), "The Governing of Internal Combustion Engines," *Proceedings, American Society of Mechanical Engineers, Oil and Gas Power Division.*

Ekstrom, T. E., (1973), "T2 Repowering: An Economic Process," General Electric 18th Gas Turbine State of the Art Seminar.

Finizio, A. F. (1973), "USCG *Polar Star*—Combination of a Diesel Electric Propulsion Plant with a Geared Gas Turbine Boost Plant," American Society of Mechanical Engineers, Paper 73-GT-73.

Johnson, F. W., and Stasiowski, C. (1973), "The Control System for the U.S. Coast Guard Icebreaker Propulsion Plant," American Society of Mechanical Engineers, Paper 73-GT-62.

LM2500 Marine Gas Turbine Installation Design Manual (1973), General Electric Report MID-IDM-2500-2, Marine and Industrial Projects Department.

Müller, C., and Stasiowski, C. (1970), "Automation of Jet Engines and Gas Turbines for the Propulsion of High-Speed Vessels," American Society of Mechanical Engineers, Paper 70-GT-122.

Richardson, W. S. (1969), "The Friction Clutch Reverse-Reduction Gears for the GTS *Admiral William M. Callaghan,*" American Society of Mechnical Engineers Transactions, *Journal of Engineering for Power,* **91**, pages 245–252.

Rowen, W. I. (1970), "Controlling the Gas Turbine in a Shipboard Environment—A Study in Propulsion Dynamics," American Society of Mechanical Engineers, Paper 70-GT-121.

Rowen, W. I. (1971), "The Marine Gas Turbine Control System," *Marine Engineers Review,* July, pages 25–30.

Rubis, C. J. (1970), "Braking and Reversing Ship Dynamics," *Naval Engineers Journal,* **82**, pages 65–76.

Rubis, C. J. (1972), "Acceleration and Steady-State Propulsion Dynamics of a Gas Turbine Ship with Controllable-Pitch Propeller," *Society of Naval Architects and Marine Engineers Transactions,* **80**, pages 329–360.

Sachs, R. M. (1969), "Description of Propulsion System for DDH-280 Class Gas Turbine Destroyer," American Society of Mechanical Engineers, Paper 69-GT-26.

Sunley, A. N., and Patterson, G. A. (1969), "DDH-280 Control System," *Proceedings, Second Ship Control Systems Symposium,* Annapolis, Md. (not publically distributed).

Zeien, C., Smith, H. F., and Hirst, F. W. (1969), "The Gas Turbine Ship *Callaghan*'s First Two Years of Operation," *Transactions, Society of Naval Architects and Marine Engineers,* **77**, pages 344–371.

Chapter

EIGHT

FUELS AND FUEL SYSTEMS

8.1. INTRODUCTION

As in all types of marine engines, the principal fuels for the gas turbine are petroleum-based liquids. This chapter discusses the properties of the several fuels that fall within this category, the significance of these properties to the gas turbine, and the shipboard systems that supply the fuel to the engine. The only likely alternative fuel type is gaseous, specifically boil-off from liquified gas cargoes, and a brief discussion of this fuel is also included.

Fuel oils have many properties of significance, but an introductory discussion can be based on just two. They are the viscosity and the impurity content. Viscosity is the property of greatest significance in fixing the pumping and atomizing behavior of the fuel, and is the most commonly used in simple classification of fuels: a "heavy" fuel is one that has a high viscosity at ordinary temperatures, while a "light" fuel has a low viscosity. Impurities of several kinds are troublesome because of clogging of filters and burner nozzles, deposition of their combustion products within the engine, and because of the accelerated corrosion of hot parts that these products may promote.

Fuel properties can be improved to some extent to suit the requirement of the engine. Viscosity is readily reduced by heating, and some impurities can be removed, or can be neutralized by additives. More generally, though, fuel quality must be chosen to suit the requirements of the engine, and the qualities required of gas-turbine fuels are higher on the whole than those of other engines. The major reason is that the highly stressed and precisely machined turbine nozzles and blades are exposed continuously to the hot products of combustion, a situation not found in the domains of steam and diesel. Another reason is the closeness of the combustion temperature to a level that can damage the exposed parts. A clogged fuel nozzle can cause local overheating beyond safe levels at other nozzles, since the other nozzles must pass excess flow to compensate for that lost at the stoppage. Although clogged nozzles are also detrimental to the performance of steam and diesel plants, there they do not so directly nor so quickly cause degradation of the machine itself.

The petroleum fuels used widely in marine service may be classified as either *medium distillates* or as *residuals*. These terms refer to the key characteristics of the refining process that produces them. Refining is fundamentally a process of distillation by which the many constituents of crude petroleum are separated according to their unique boiling points (although any commercial product, such as gasoline or a fuel oil, is always a mixture of many compounds covering a range of boiling points). Any product produced wholly in this way may be called a distillate; the residue of the refining process—the constitutents that have not distilled because their boiling points are impractically high—are the residuals.

A "medium" distillate is one whose temperature range of distillation is intermediate among the products normally used as fuels. Two such distillates have long been available from marine-fuel suppliers for commercial diesel use; they are commonly known as *marine gas oil* and *marine diesel oil.* The gas oil is the lighter of the two (i.e., lower distillation temperature, lower viscosity) and is used by high-speed diesel engines. The other is suitable for medium-speed engines and is used by them without heating or treatment.

Residual fuel oil (often called "bunker C") is the highly viscous remnant of the refining process, and is the traditional boiler fuel for commerical ships. Heating is always required to reduce its viscosity for handling and atomization, but steam boilers usually burn it without other treatment, even though it retains most of the impurities that were in the original crude oil.

A range of marine fuels known as *intermediates* is prepared by blending distillate with residual, the purpose being to produce a low-price fuel that is of acceptable quality for low-speed diesels. These fuels are designated by a number representing their viscosity at 100 F. For example, intermediate 15 has a viscosity of 1500 on the Redwood 1 scale at this temperature. The lower-viscosity intermediates are also used by medium-speed engines. (See Figure 8.1.)

The U.S. Navy has traditionally used two fuels, "navy special" fuel oil for steam boilers and "navy diesel" oil for diesel engines. The first of these corresponds approximately to intermediate 4, and the second is close in properties to marine diesel oil (i.e., to the commercial product mentioned previously).

All of these fuels are usable by heavy-duty gas turbines, though treatment is required for residual or fuels containing some fraction of residual. Only the distillate fuels are usable in the aircraft-derivative engines, and there must be

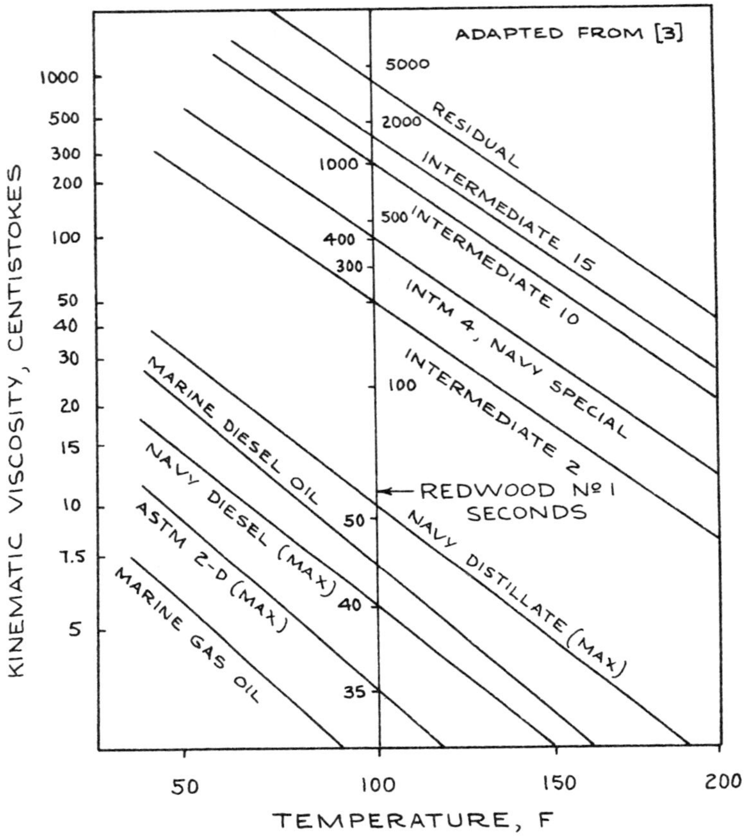

Figure 8.1. Viscosities of several fuels.

restrictive specifications that the normally available supplies cannot always meet. The traditional fuels for these engines—in aircraft service, that is—are *light distillates,* either JP-4 or JP-5 (designations given them by military specifications; there are commercial equivalents). JP-5 is less viscous than marine gas oil, and corresponds approximately to ordinary kerosine. JP-4 is yet lighter, falling between kerosine and common automotive gasoline. Since many marine gas turbines are aircraft derived, these fuels have been used in marine service, particularly JP-5.

Figures 8.1 and 8.2 should give a general idea of the major relationships among the types of fuels mentioned. The first of these shows viscosity as a function of temperature for several fuels; the second shows distillation curves; indicating the range of temperatures over which the fuels are distilled from the crude petroleum. These curves fundamentally determine what the characteristics of the fuels are, but by no means tell the whole story. The many properties, some of which depend upon the source of the crude oil, or upon the handling it receives after refining, are individually important. Those of significance to the gas-turbine application are discussed in the following section.

Price of fuel is an inverse function of viscosity; the more viscous the fuel the lower, generally, is its price. There is, therefore, an economic incentive to use the heavier fuels, an incentive that is strongly resisted by the tendency of the heavier fuels to contain higher percentages of harmful impurities. Many of the impurities that are present in the original crude oil remain in the residual, so it, or any fuel containing a residual component, is likely to be high in impurities. But as noted, residual fuel can be used in heavy-duty engines after certain treatments, and indeed the pioneer gas-turbine ship *John Sergeant* burned this fuel with apparent success [McMullen (1957)].

The techniques used to offset poor fuel quality are treatment of the fuel before firing, redesign of combustors to accomodate higher viscosity and poorer buring qualities (e.g., longer flame), protective coating of nozzles and turbine

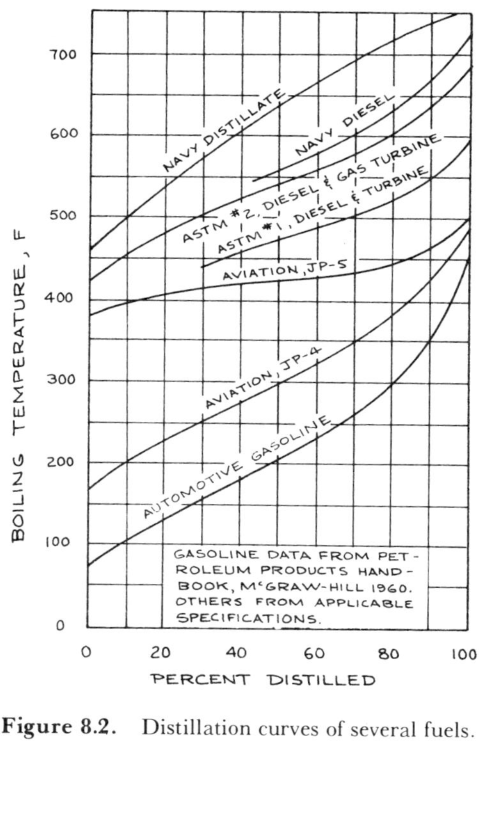

Figure 8.2. Distillation curves of several fuels.

blades, and use for these parts of alloys resistant to the unique corrosive attacks of the fuel impurities. All of these have their associated costs, and none—or all in sum—appear able to make all fuels usable by all engines. It seems unlikely, for example, that residual oil will become an acceptable fuel for aircraft-derivative gas turbines.

A partial alternative to residual oil as a cheap fuel is a *heavy distillate,* that is, a fuel of higher viscosity than marine diesel oil, but one containing none of the residual component found in the traditional intermediates. The basic idea is that impurities should not distill, but remain in the residual. This is an ideal not completely realized, and in fact some impurities may originate in the refining process and in subsequent handling, but nonetheless, impurities are sharply reduced when residual oil is kept out of the fuel. The price, however, cannot be as low as residual, hence the label "partial alternative" used a few sentences hence.

The heavy distillate is a relatively new fuel in marine experience, its specifications having been developed in the late 1960s as a consequence of the Navy's need to replace navy special fuel oil with a fuel of better quality, its desire to have a single fuel that could be used in all types of power plants, and the previously cited need for a cheaper fuel for aircraft-derivative gas-turbine engines [Sigal (1971)].

8.2. FUEL PROPERTIES

A. Physical Properties

Viscosity is that property that determines the resistance to flow of a fluid. Because of its importance to fuel technology, and because a number of systems of units are used for quantifying it, an appendix appears at the end of this chapter in which a more precise definition is given, and in which several common systems of measurement are explained.

High viscosity makes pumping difficult, and more importantly, makes atomization for combustion difficult; a viscosity of about 10–20 cSt is usually the limit for acceptable atomization (the particular figure depends on the type of atomizer, hence the range) [Brubaker and Calvino (1972)]. Fortunately, viscosity of liquids is a strong inverse function of temperature, so that moderate heating usually suffices to reduce viscosity of any fuel to the necessary value.

A low viscosity may be troublesome if it is used in a piping system designed for a higher viscosity liquid. Seals that are adequate for the more viscous liquid may leak, and pumps that should be lubricated by the fluid pumped may not be so well served by the lower viscosity liquid. But trouble of this sort is not expected when the system is designed for the lowest viscosity to be encountered.

Distillation temperatures are usually specified by the temperatures at which 10%, 90%, and 100% (end point) will have evaporated from the crude oil sources. (Specification practices vary, though; other percentages may also be listed, or perhaps only the end point.) These temperatures are of no direct concern to the engine using the fuel, but the temperature range essentially fixes the list of compounds that will appear in the distilled product, and so indirectly fixes all properties. Burning properties, in particular, are closely related to the distillation temperatures. A low 10% temperature indicates the presence of highly volatile compounds, making the fuel easy to ignite; a high 90 or 100% value indicates the presence of long-burning compounds that may cause flame impingement of the turbine nozzles and blades unless the combustors are given a size to suit the fuel.

Flash point is mainly a safety criterion, being the temperature at which fuel will ignite momentarily, but will not continue to burn when the source of ignition is removed. For example, the flash point of navy diesel oil is required to have a minimum value of 140 F by the applicable military specification (MIL-F-16884).

Pour point is the temperature at which the fuel just begins to flow, and so determines handling characteristics under cold conditions. It is obviously closely related to viscosity. Cloud point is also a low-temperature characteristic, being the temperature at which a waxy constituent begins to congeal in the fuel. This constituent can clog filters when filtration of a cold fuel is attempted. Heating is the usual remedy for unsatisfactory pour point and cloud point.

Carbon residue percentage is an indication of the amount of unburned carbon that may appear in the combustion process. In a gas-turbine combustor, a fuel of high residue value may produce carbon deposits on the nozzles that can clog them or distort the spray pattern. Carbon residue percentage is typically quite low in all fuels except the residual oil.

Thermal stability is necessary to avoid breakdown of the fuel as it passes through the high-temperature zone immediately preceeding the combustor nozzles. A breakdown would produce solid carbon constituents with results much like to those from excessive carbon residue.

B. Contaminant Properties

Contaminants may be normal constituents of the fuel, that is, minor elements or compounds present in the crude oil, or they may be things such as water and rust scale picked up in handling and storage. In the former category are substances containing perhaps several dozen elements in detectable quantity. Those of significance to the gas turbine are more limited in number; those usually regarded as significant are sulfur and vanadium. Other metals, such as sodium, potassium, calcium, lead, and copper are important also if present in sufficient concentration. Although the items on this latter list usually do not appear to significant degree in petroleum, contamination from handling and storage often provides them. Sodium, for example, is abundantly provided by sodium chloride if the fuel becomes contaminated with sea water.

Sulfur is usually the element of highest concentration in crude petroleum after the carbon and hydrogen. Most of it ends up in residual fuel, where it may be 2–4% by weight, but smaller percentages also appear in distillates. It has been identified as a deleterious component of fuel in both diesel and steam-turbine propulsion plants, especially when heat-recovery equipment (waste heat boiler with diesel, boiler air heater or economizer with steam) is used, since the resulting low exhaust-gas temperature can bring the gas below its dew point. Water condensing from an exhaust gas produced by a sulfurous fuel is in effect a dilute solution of sulfuric acid, and consequently metal surfaces are rapidly corroded. The same hazard faces heat recovery equipment that may be used with a gas-turbine engine. The simplest remedy in any case is to keep operating temperatures above the dew point (in the neighborhood of 250 F).

Sulfur is also the principal culprit in a hot-corrosion phenomenon known as sulfidation, though metallic elements such as sodium appear to be necessary also for the corrosive reactions of sulfur to occur. Since in a practical sense it is impossible to remove sulfur in either refining or in shipboard treatments (navy diesel fuel is allowed to contain up to 1% sulfur, for example), marine gas turbines of necessity are adapted to the sulfidation problem by use of coatings on hot parts and by use of alloys that resist the corrosion satisfactorily. Gas-turbine fuel specifications, therefore, are not in general especially restrictive of sulfur, but low sulfur may be specified by the manufacturers of a particular gas turbine. For example, Ford specified a maximum sulfur of 0.4 percent for its marine engine.

Other solid contaminants of noncombustible nature are often referred to as ash and are principally metals or metallic compounds. Those that are usually regarded as being harmful in a gas-turbine fuel are vanadium, calcium, sodium, potassium, copper, and lead. Vanadium is the only one of these that originates in the crude oil; the others most often can be blamed on contamination that has occurred in handling or refining. Lead, for example, may be present if the fuel has been stored in a tank that previously held leaded gasoline. Lead and copper both may be contaminants in the refining process [Brubaker and Calvino (1972)]. Sodium, potassium, and calcium most commonly originate from sea-water contamination and are likely to occur if the fuel is stored in tanks used alternatively as ballast tanks. (Even if the ship has segregated ballast tanks, this contamination can occur, since the fuel may have been handled in tankers and in a fueling barge before coming aboard.)

Vanadium is universally regarded as the most serious of these contaminants, since it is the key constituent in a deleterious vanadium-sodium(or potassium)-oxygen compound that is formed in the combustion process. This compound is liquid at temperatures normally present in the gasifier turbine and its nozzles. In this state it adheres to these turbine parts and is highly corrosive. Fuel specifications for gas-turbine fuels, therefore, generally place severe restrictions on vanadium content, although there is disagreement on permissable levels. For instance, specifications for navy distillates allow 0.5 ppm (parts per million) vanadium, while Turbo Power & Marine specifications for an equivalent fuel allow only 0.2 ppm [Morrow (1972)]. The 0.5 ppm limit, at least, is usually met by all of the usual gas-turbine fuels except residual. The use of residual fuel in a gas turbine, therefore, always requires tha addition of a vanadium inhibitor to the fuel. Magnesium added in about three parts to one part vanadium by weight is an effective inhibitor and fortunately is readily available in the form of inexpensive compounds (e.g., magnesium sulfate, or epsom salts).

The other metals listed also form corrosive compounds, or form deposits on nozzles and blades. Calcium, for example, appears in a tarlike substance that coats the turbine parts [McMullen (1957)]. Since these contaminants are not naturally occurring constitutents of petroleum, they are effectively countered if care in refining, handling, and storage is enforced to prevent their entering the fuel. Those that most likely have come from seawater—sodium, potassium, and calcium—are principally in the form of water-soluble compounds and, therefore, can be removed by a washing process (Section 8.4.B).

Water is usually present in fuel in small fractions and, if small, is not of itself harmful. But there are several secondary effects. One has already been mentioned, namely the contribution of metallic contaminants when the water comes from the sea. Another is its service as a culture medium for bacteriological growth; certain microorganisms feed on hydrocarbon compounds that may have dissolved in the water from the fuel, or are present at oil-water interfaces. The product of this activity is slime that clogs fuel filters, not only restricting fuel flow, but reducing the effectiveness of the filter to remove other solid matter. The problem usually occurs only with fuel that has been stored a long time, since the growth process proceeds slowly under normal circumstances. When conditions are favorable for this phenomenon, biocides can be added to inhibit the growth of the bacteria.

Sediment in fuel or particulate matter—dirt, to give it a plain name—is almost certain to be present, having originated in a number of different ways.

Scale from the surfaces of tanks and pipes is an obvious source for gritty stuff, as is dirt remaining from the building of the tanks. And the sediment produced by biological activity has just been mentioned. It is for these stray solid materials that filters are universally required in fuel systems.

8.3. FUEL SPECIFICATIONS

Confronted with a sample of fuel, one may ask "which of the several fuels is this?" The answer comes from which *specification* it meets, a specification being a listing of limits (upper, lower, or both) on every property significant to the intended use of the fuel. Specifications are issued by several different bodies; in the United States the American Society for Testing and Materials is the prime issuer of specifications for a wide range of products and is also the prime specifier of tests by which compliance with specifications is determined. Also of significance in the marine field is the set of specifications issued by the U.S. Navy, and some builders of gas-turbine engines issue specifications for the fuels to be used in their products.

The American Society for Testing and Materials (known as ASTM) publishes its specifications annually in 33 hard-bound volumes; volume 17 contains the fuel specifications [*1973 Annual Book of ASTM Standards*]. Specification D975 is the one for diesel fuels. It lists three fuels designated 1-D, 2-D, and 4-D. Gas-turbine fuels are specified by D2880, which lists four fuels designated 1-GT, 2-GT, 3-GT, and 4-GT. The diesel and gas-turbine specifications are reproduced here as Tables 8.1 and 8.2. An inspection of the tables shows that in categories 1 and 2, the D fuels and the GT fuels are identical, except tha a vanadium limit is specified for the latter. Within the GT classification, categories 3 and 4 are identical, except that the vanadium limit is higher for the latter.

In line with earlier discussion, the 1 and 2 categories are medium distillates, while 3 and 4 are intermediates.

An interesting point appears in the discussion of ASTM specifications, namely that a particular fuel may meet more than one specification because of identity or overlap in properties. Thus, for example, a fuel might be sold as both 2-D and 2-GT, since the items of specification tend to be identical, rather than exclusive. And the same fuel might be sold as navy diesel oil, since the specification for the latter is not mutually exclusive with ASTM. (But it does not follow that all ASTM 2 fuel samples meet the navy specification, or vice versa, for some of the limits are indeed different.)

Naval fuel is described by military specifications (known as "mil specs").

TABLE 8.1. DIESEL FUEL SPECIFICATIONS, ASTM D975-68 [*1973 Annual Book of ASTM Standards*]

Grade of Diesel Fuel Oil	Flash Point, deg Fahr (°C)	Pour Point deg Fahr (°C)	Water and Sediment, percent by volume	Carbon Residue on 10% Residuum, percent	Ash, percent by weight	Distillation Temperatures, °F (°C), 90% Point		Viscosity at 100 F (37.8 C), Kinematic, cSt (or Saybolt Universal, sec)		Sulfur, percent by weight	Copper Strip Corrosion	Cetane Number[c]
	Min	Max	Max	Max	Max	Min	Max	Min	Max	Max	Max	Min
No. 1-D: A volatile distillate fuel oil for engines in service requiring frequent speed and load changes	100 or legal (37.8)	[b]	Trace	0.15	0.01	—	550 (287.8)	1.4	2.5 (34.4)	0.50	No. 3	40[f]
No. 2-D: A distillate fuel oil of lower volatility for engines in industrial and heavy mobile service.	125 or legal (51.7)	[b]	0.05	0.35	0.01	540[c] (282.2)	640 (338)	2.0[c] (32.6)	4.3 (40.1)	0.50[d]	No. 3	40[f]
No. 4-D: A fuel oil for low and medium speed engines.	130 or legal (54.4)	[b]	0.50	—	0.10	—	—	5.8 (45)	26.4 (125)	2.0	—	30[f]

[a] To meet special operating conditions, modifications of individual limiting requirements may be agreed upon between purchaser, seller, and supplier.
[b] For cold weather operation, the pour point should be specified 10 F (5.6 C) below the ambient temperature at which the engine is to be operated except where fuel oil heating facilities are provided.
[c] When pour point less than 0 F (−17.8 C) is specified, the minimum viscosity shall be 1.8 cSt (32.0 sec, Saybolt Universal) and the minimum 90% point shall be waived.
[d] For all products outside the U.S.A. the maximum sulfur limit shall be 1.0% by weight.
[e] Where cetane number by Method D 613, Test for Ignition Quality of Diesel Fuels by the Cetane Method, is not available, ASTM Method D 976, Calculated Cetane Index of Distillate Fuels Way be used as an approximation. Where there is disagreement, Method D 613 shall be the referee method.
[f] Low-atmospheric temperatures as well as engine operation at high altitudes may require use of fuels with higher cetane ratings.

TABLE 8.2. GAS TURBINE FUEL SPECIFICATIONS, ASTM D2880-71 [*1973 Annual Book of ASTM Standards*]

Designation[a]	Grade of Gas Turbine Fuel Oil	Flash Point, °F (°C)	Pour Point, °F (°C)	Water and sediment. Vol percent	Carbon Residue on 10% Residuum, percent	Ash, weight percent	Distillation Temperature, 90% Point °F (°C)		Saybolt Viscosity, s[b] Universal at 100 F (38 C)		Saybolt Viscosity, s[b] Furol at 122 F (50 C)
		Min	Max	Max	Max	Max	Min	Max	Min	Max	Max
No. 1-GT[c]	A volatile distillate for gas turbines requiring a fuel that burns cleaner than No. 2-GT.	100 (38) or legal	0 (−18)[d]	0.05	0.15	0.01	—	550 (288)	—	(34.4)	—
No. 2-GT	A distillate fuel of low ash and medium volatility suitable for gas turbines not requiring No. 1-GT.	100 (38) or legal	20 (−7)[d]	0.05	0.35	0.01	540 (282)	640 (338)	(32.6)	(40.2)	—
No. 3-GT[e]	A low-volatility, low-ash fuel that may contain residual components.	130 (54) or legal	—	1.0	—	0.03	—	—	45	—	300
No. 4-GT	A low-volatility fuel containing residual components and having higher vanadium content than No. 3-GT.	150 (66) or legal	—	1.0	—	—	—	—	45	—	300

Designation[a]	Grade of Gas Turbine Fuel Oil	Kinematic Viscosity, cSt[b] At 100 F (38 C)		At 122 F (50 C)	Gravity, deg API	Vanadium (V), ppm by weight	Sodium plus Potassium (Na + K), ppm by weight	Calcium (Ca), ppm by weight	Lead (Pb), ppm by weight	Magnesium to Vanadium Weight Ratio (Mg/V)		Sulfur, weight percent
		Min	Max	Max	Min	Max	Max	Max	Max	Min	Max	Max
No. 1-GT[c]	A volatile distillate for gas turbines requiring a fuel that burns cleaner than No. 2-GT.	1.4	2.5	—	35	2	5	5	5	—	—	0.5 or legal
No. 2-GT	A distillate fuel of low ash and medium volatility suitable for gas turbines not requiring No. 1-GT.	2.0	4.3	—	30	2	5	10	5	—	—	0.5[h] or legal
No. 3-GT[e]	A low-volatility, low-ash fuel that may contain residual components.	(5.8)	—	(638)	—	2[e]	5[e]	10[e]	5	—	—	
No. 4-GT	A low-volatility fuel containing residual components and having higher vanadium content than No. 3-GT.	(5.8)	—	(638)	—	500	10[f]	10[f]	5	3.0[g]	3.5[g]	

[a] No. 1-GT corresponds in general to Specification D 396 Grade No. 1 fuel and Classification D 975 Grade No. 1-D diesel fuel in physical properties. No. 2-GT corresponds in general to Specification D 396 Grade No. 2 fuel and Classification D 975 Grade No. 2-D diesel fuel in physical properties. No. 3-GT and No. 4-GT viscosity range brackets Specification D 396 Grades No. 4, No. 5 (light), No. 5 (heavy) and No. 6 and Classification D 975 Grade No. 4-D diesel fuel, which may be supplied provided metals composition requirements are met.

[b] Viscosity values in parentheses are for information only and are not limiting.

[c] Recognizing the necessity of additional requirements for certain types of gas turbines, the following may be specified for No. 1-GT fuel: luminometer number, min = 40. Thermal stability test for 5 hr at 250 F (121 C) preheater temperature, 350 F (177 C) filter temperature and at a flow rate of 6 lb (2.7 kg)/hr: filter pressure drop, max = 12 in. (300 mm) Hg; preheater deposit code, max = 2.

[d] For cold-weather operation, the pour point should be specified 10 F (5.6 C) below the ambient temperature at which the turbine is to be operated, except where fuel oil heating facilities are provided. When a pour point less than 0 F (−18 C) is specified for Grade No. 2-GT, the minimum viscosity shall be 1.8 cSt (32.0 sec, Saybolt Universal) and the minimum 90% point shall be waived.

[e] For gas turbines operating below 1200 F (649 C) maximum gas temperature, the limitation on vanadium, sodium plus potassium, and calcium may be waived, provided that a silicon-base additive, or equivalent, is employed. The special requirements covering the addition of and the type of additive shall be specified only by mutual agreement between purchaser and seller.

[f] Where water washing facilities are available at the point of use, these requirements may be waived by mutual agreement between the purchaser and seller.

[g] Special requirements covering the addition of and the type of magnesium-containing additive, or equivalent, to be used shall be specified only by mutual agreement between the purchaser and the seller.

[h] For products outside the U.S.A., the maximum sulfur limit shall be 1.0%.

[i] Legal requirements to be met.

Those of interest in gas turbine work are

MIL-T-5624C	Aviation fuel, JP-5
MIL-F-16884F	Navy diesel fuel
MIL-F-24397	Navy distillate fuel

A few selected items from these specifications are quoted in Table 8.3. Note that there is no coincidence with the ASTM limits of Tables 8.1 and 8.2, but that there is overlap, so that it is possible for a particular fuel to meet both ASTM and Navy requirements.

A navy distillate fuel specified by MIL-F-24376 is frequently mentioned in the literature, but this covers a developmental fuel not intended for general use. It is essentially the same as MIL-F-24397.

Some builders of gas turbines also issue specifications for the fuels to be used in their engines. Examples are the Turbo Power & Marine Systems specifications PWA 527 (distillate fuel, marine and industrial) and PWA 539 (heavy distillate fuel, marine and industrial). The former is similar to ASTM 2-GT, but metals contamination is more strictly limited; vanadium limit is 0.1 ppm, compared to 2 ppm for the ASTM fuel. The second specification is similar to

TABLE 8.3. SELECTED PROPERTIES FROM NAVY FUEL SPECIFICATIONS

	Aviation JP-5 MIL-T-5624G	Navy Diesel MIL-F-16884	Navy Distillate MIL-F-24397
Viscosity, cs	16.5 max at −30 F	2.1–6.0 at 100 F	10.0 max at 100 F
Distillation, max, °F			
10%	400	—	500
50%	—	—	644
90%	—	675	740
95%	—	—	765
end point	550	725	—
Flash point, min, °F	140	140	150
Pour point, max, °F	0	—	25
Carbon residue, max, %	—	0.2[a]	0.4[b]
Ash, max, %	—	0.0005	0.01
Sulfur, max, wt%	0.4	1.00	1.30
Vanadium, max, wt ppm	—	—	0.5

[a] On 10% bottoms.
[b] On whole fuel.

the navy distillate, but is also stricter on metals; vanadium limit is 0.2 ppm versus 0.5 ppm for the navy fuel.

The fuel specifications for the General Electric LM2500 engine is based on the simpler concept of listing acceptable fuels to ASTM or military specifications, plus certain additional restrictions. This specification is given here as Table 8.4 [*LM2500 Marine Gas Turbine Installation Design Manual* (1973)]. (The VV-F-800A in the table has not been mentioned above; it is a nonmarine military specification.) Note that the vanadium limit is the same as for navy distillate, but more restrictive than those of the ASTM fuels or of navy diesel (since MIL-F-16884 doesn't mention vanadium). Fuel meeting the ASTM and navy specifications, therefore, may be satisfactory, depending on whether it can pass the additional test.

Marine gas oil and marine diesel oil are generally commercial terms in world-wide use and are not names assigned by recognized specifications [Brubaker and Calvino (1972)]. The first of these, however, usually meets the ASTM 2-D specification and often meets the navy diesel specification. The other is a heavier distillate and varies more widely in quality, depending on the source of the crude and the practices of the refinery. It may in some cases meet the specifications for navy heavy distillate.

The intermediate diesel fuels likewise meet no recognized specification, save the generally understood viscosity level stated previously. Residual oil is also not defined by recognized specification, although many samples meet the requirements of No. 6 oil in the ASTM specification D396 for furnace oils, since the specification indeed intends that this oil be a residual. For this reason, residual oil is sometimes referred to as "No. 6."

8.4. FUEL SYSTEMS

A. General Requirements

In simplest terms, the fuel system is designed to deliver fuel to the engine in the quantity required, at a pressure sufficient to the engine fuel pump and in as clean a state as possible. The same may truthfully be said of the fuel systems for other types of engines, but the gas turbine requires greater emphasis on cleanliness of fuel than any other type. Such, at least, appears to be the lesson of experience from early gas-turbine vessels [for example, O'Hare and Holburn (1973); Zeien, Smith, and Hirst (1969)].

The number and type of features bearing on cleanliness of the fuel depends in an almost obvious way on the fuel used; an inherently dirty fuel, such as residual, requires a shipboard treatment system, while simple filtration may suf-

TABLE 8.4. GENERAL ELECTRIC FUEL SPECIFICATIONS FOR LM2500 ENGINE [6]

1. Fuels meeting the following specifications are acceptable for use in the 7LM2500 gas turbine provided they meet the additional criteria listed in paragraph 2.

Specification No.		Title
1.1	MIL-T-5624	Grades JP4, JP5
1.2	ASTM D975-68	Diesel fuel oil, Grades 1-D and 2-D
1.3	ASTM D1655-69	Turbine fuels
1.4	MIL-F-24397	Fuel: Navy distillate
1.5	MIL-F-16884	Fuel oil: diesel marine
1.6	VV-F-800A	Fuel oil: diesel, grades DF-A, DF-1, DF-2
1.7	ASTM D2880-71	Gas turbine fuel oils, grades 1-GT, 2-GT, 3-GT

2. *Additional requirements.* The following requirements supplement and supersede, where there is a conflict, the specifications listed above. However, if the specification requirement is more restrictive it applies.

a. The fuel shall consist of hydrocarbon compounds only and must be compatible between blends and batches.

b. The use of any additives requires the approval of the General Electric Company.

c. The viscosity of the fuel as supplied to the inlet connection on the gas turbine shall be 6.0 cSt or less for starting and 12.0 cSt or less during operation. The fuel may be heated to meet this requirement.

fice for a distillate. But in addition, complexity of the fuel system depends on the size and service of the vessel. In a small vessel, a single tank may suffice for the entire supply. If the tank is nonstructural (i.e., not part of the hull structure, as an inner bottom tank would be), its framing, if any, may be on the outside so that the interior is free of dirt-harboring pockets and corners, and it should be relatively free of condensation since its outer surface will not be in contact with cold sea water. Further, being small, it should be easy to clean to an acceptable degree following construction. And since a small vessel typically takes its fuel from a dockside pump, the fuel escapes the hazard of delivery via the tanks of a fueling barge. Figure 8.3 shows such a system, consisting in essence only of a single tank, a booster pump, and two stages of filtration.

The fuel system of a ship is more complex because the fuel is more subject to contamination in delivery and in storage aboard the ship. The reasons are simply the inverse of those put forward above for the small vessel: tanks are likely to be more subject to condensation, the probability of dirt within them is higher because of size and because there may be framing members inside, and

TABLE 8.4.—Continued

d.	With Marine Coatings	Without Marine Coatings
Ash, maximum percent	0.01 (100 ppm)	0.01 (100 ppm)
Sulfur, maximum percent	1.3	1.3
Vanadium, maximum, ppm	0.5	0.5
Sodium + potassium + lead, maximum ppm	2.0	1.0
Calcium, maximum ppm	2.0	1.0
Hydrogen Content, minimum percent	12.0	12.3
Demulsification, minutes, maximum	20.0	20.0

e. The fuel as delivered to the inlet connection on the gas turbine shall not contain more than 10 g of solid contaminates per 1000 gal of fuel. The contamination shall not exceed the following micron-size limits:

Particle Size, Microns	Percent of Total
0–5	Remainder
5–10	18% Maximum
10–20	16% Maximum

f. The fuel as delivered to the inlet connection on the gas turbine shall not contain more than 40 ppm of entrained water at 70 F.

the usual trip via the tanks of a fueling barge adds another step of possible contamination. The principle thus appears: what is done to the fuel in handling must be undone before it can be used. And that which is done is the addition of water and sediment (dirt, particulate matter).

Figure 8.4 contains block diagrams of two fuel systems—from the *Callaghan* and the *Euroliner*—that demonstrate the points made here, as well as serve for examples of actual systems. Both are designed to handle medium to heavy distillates, fuels that are inherently acceptable to the engines (in contrast to residual, which must be treated before becoming acceptable) and, therefore, have features intended only to remove contaminants from handling and storage. Both systems show the following significant features:

1. Fuel from the several storage tanks (only one appears in each diagram) is held in a settling tank to allow gravity separation of water and heavy solid matter.

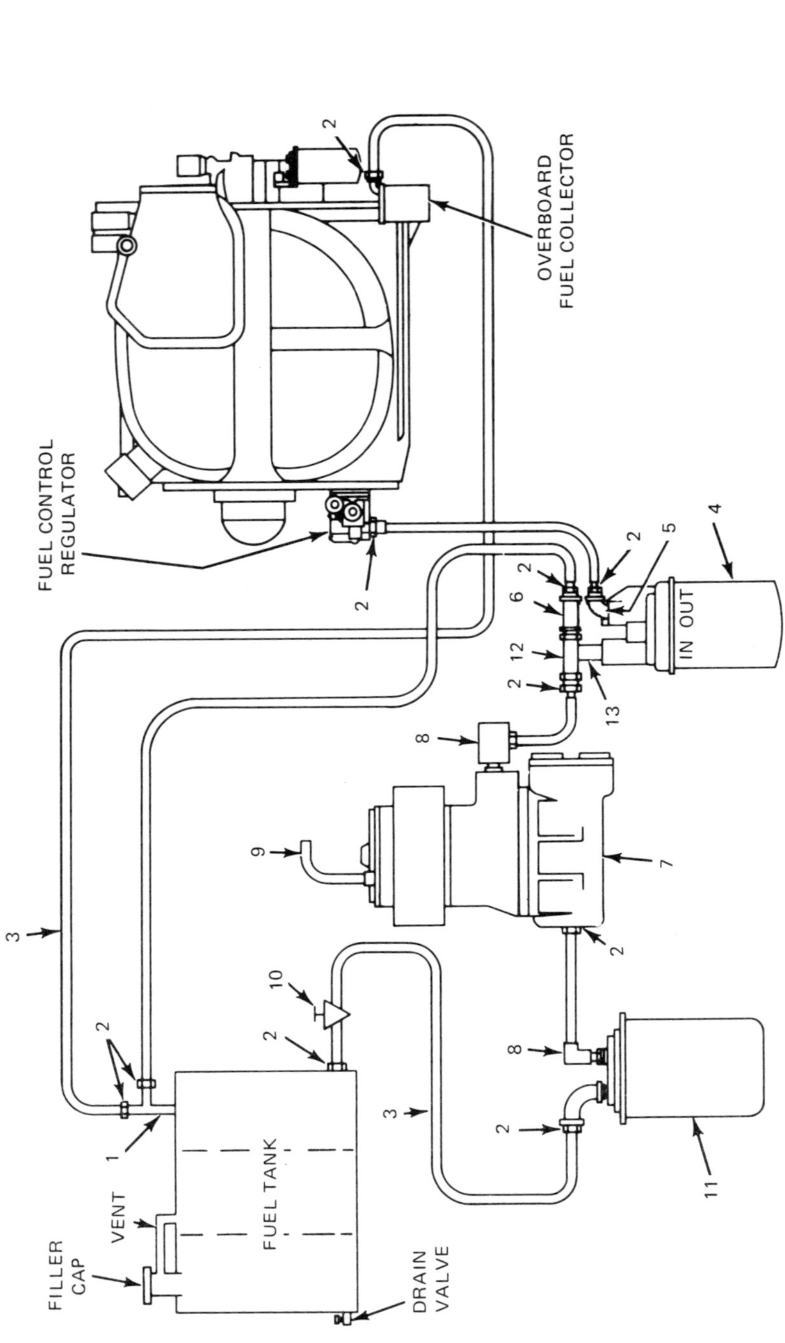

OVERBOARD
FUEL COLLECTOR
FUEL CONTROL
REGULATOR
FUEL TANK
FILLER
CAP
VENT
DRAIN
VALVE
IN OUT
1
2
3
4
5
6
7
8
9
10
11
12
13

Figure 8.3. Fuel system of Ford gas-turbine engine (from Ford literature).

1. T Pipe, 3/8′′ NPT 2 Female and 1 Male
2. 3/8 NPT, 45° Flare
*3. Stainless steel tube, 3/8 O.D., 028 wall, SAE 30316, single flare throughout
4. Secondary Filter, Fram P1101-PL, or equivalent; see specifications
5. 3/8 NPT 90° Street Elbow
6. Relief Value, 10 psig, Republic 451 3/8-D-2-10 (or equiv.)
7. Fuel Boost Pump; see specifications
8. 3/8 NPT to 3/8 45° Flare Tube, 90° Elbow
9. Boost Pump Electric Motor Leads (Refer to drawing MS-D19C-1399-A in Section 8).
10. Shut-off Valve, 3/8 NPT (min.)
11. Primary Fuel Filter, Fram P1109, or Equivalent
12. 3/8 NPT Straight Tee Female
13. 3/8 NPT × 1.00′′ Close Nipple

* Tube or hose material to be compatible with the fuel and the governing application specification.

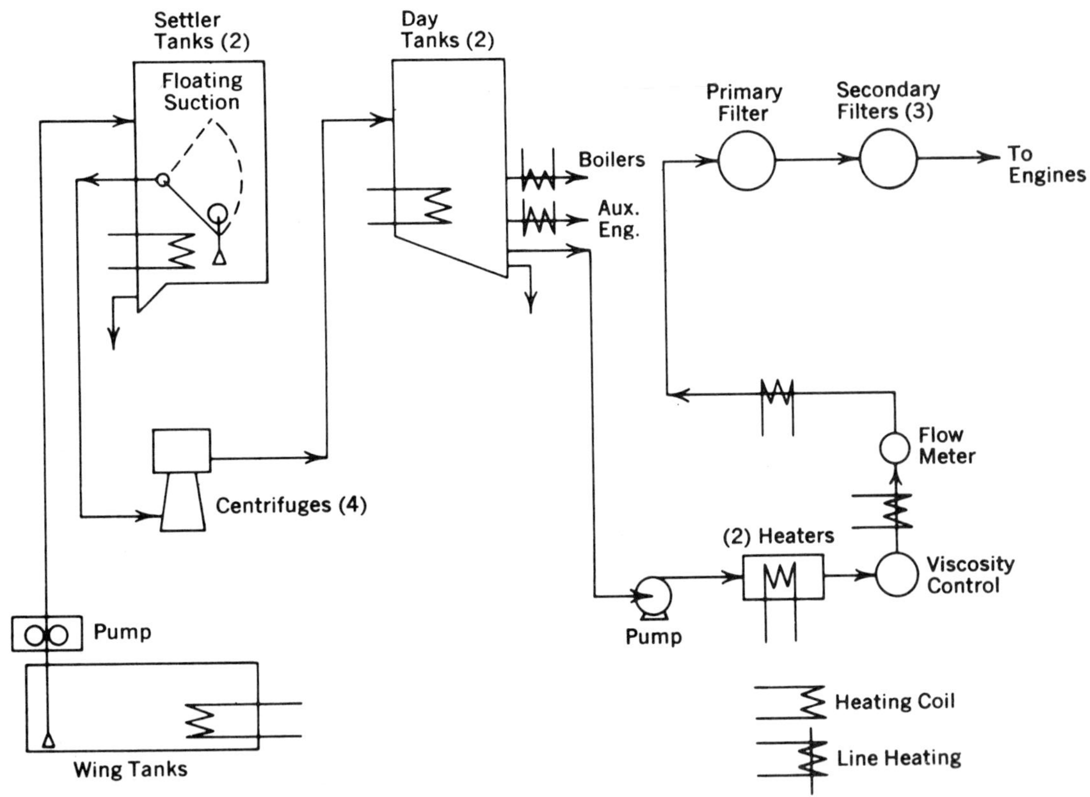

"Euroliner" schematic diagram of fuel system

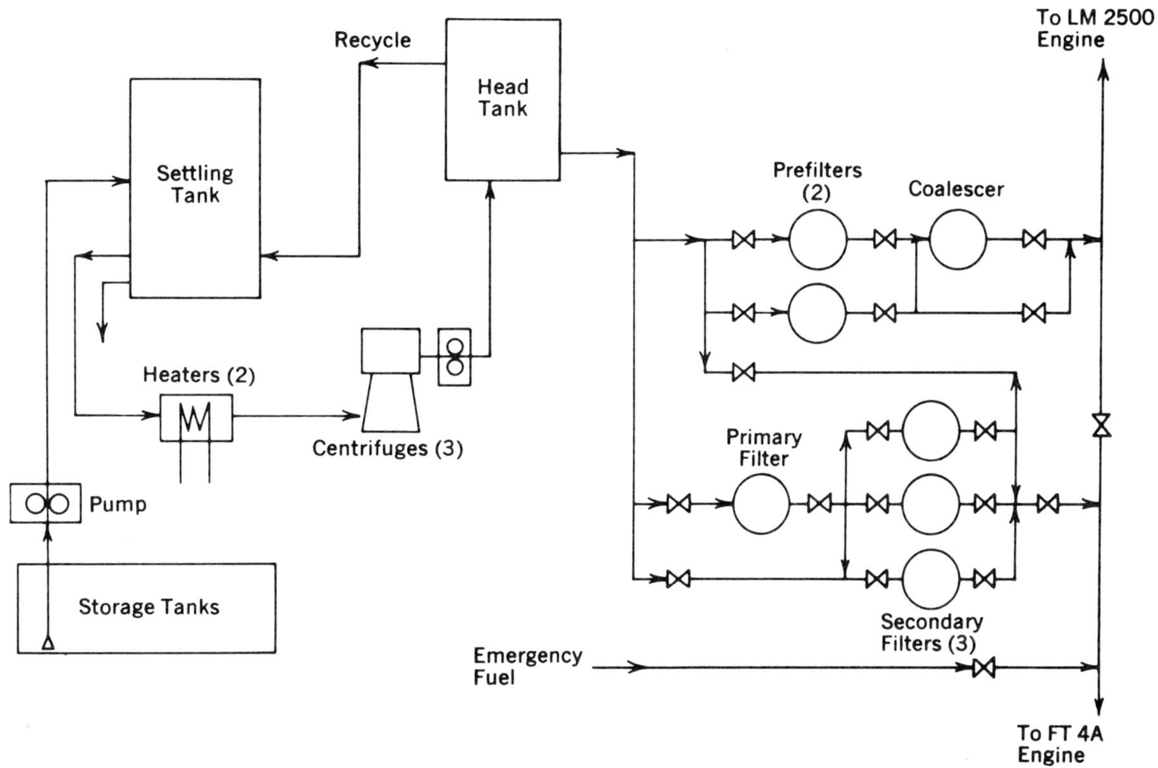

"Adm. Wm. M. Callaghan" schematic diagram of fuel system

Figure 8.4. Block diagrams of *Callaghan* and *Euroliner* fuel systems from Brubaker and Calvino (1972).

2. Fuel is centrifuged to continue the removal of water and solids.
3. Heating is provided to ease the pumping of the more viscous fuels, and to promote the separation process.
4. A final tank is provided (head tank or day tank) to hold cleansed fuel. This tank, and all piping from there to the engines, is stainless steel.
5. Two stages of filtering are provided between the final tank and the engines.

In the case of the *Callaghan* system, a *coalescer* (or coalescent filter) is indicated. This is a filter consisting of two elements in series. The first is an element of glass fibers or similar material that causes water droplets to coalesce into large drops on the fibers. When the drops become large enough, they flow down the element to a sump in the bottom, whence they are drained away. The second element consists of water-absorbing material (e.g., like blotting paper) that removes the moisture that may have escaped the first element. These devices are said to be capable of reducing water content of the fuel to essentially zero.

One may question why so many steps of water and solids removal are necessary. The essence of the answer is that each step protects the following ones from clogging by an excess of the removed matter. The centrifuge is a high-capacity device that does not clog, but which is unable to approach the near-perfection in water removal exhibited by the coalescer. The primary (or pre) filters are also relieved of part of the burden by the centrifuge and in turn protect the final filters or coalescer from clogging by the coarser particles.

B. Fuel-Systems Materials

A typical fuel system consists of carbon steel tanks and piping until the initial cleansing of the fuel, and stainless steel thereafter. The purpose of the stainless is to avoid formation of rust and scale that might be eroded into the fluid stream. Stainless steel throughout would be must too expensive, though in a small craft with a single tank it might be reasonable. An alternative is to coat the tank surfaces; epoxy compounds are a favorite. Coating is said to be a necessity when the jet fuels (JP-4, JP-5) are used, since they have strong scale-loosening properties.

The same situation applies with respect to the piping, but since coating of pipe interiors is less feasible than coating of tank interiors, a nonscaling material is more likely to be used throughout. Copper-nickel, especially 70–30, is a suggested alternative to stainless steel [Brubaker and Calvino (1972)]. Low-copper alloys are preferred (i.e., 70–30 Cu–Ni preferred to 90–10 Cu–Ni) because of the possibility of copper pickup by the fuel.

C. Fuel-Treatment Systems

Filtering, centrifuging, and heating as previously described are forms of treatment, to be sure, but this term is commonly reserved for the more elaborate shipboard handling given to residual fuels. The additional treatment includes features to remove or neutralize the metallic impurities that can cause corrosion and deposits of combustion products in the hot parts of the engine.

The removal is described as "washing," since it is a process in which clean fresh water is intimately mixed with the oil to dissolve the water-soluble contaminants. The removal of the water by centrifuging follows, usually assisted by the addition of a deemulsifying agent, such as tretolite. The significant contaminants removed in this way are the salts of sodium, potassium, and calcium (though some salts of calcium are not water soluble and hence are not removed in this way). They are not likely to be removed completely, but can be reduced to tolerable levels by washing.

Vanadium does not occur as a water-soluble contaminant and, therefore, cannot be removed by washing, nor by any other practicable shipboard process. Fortunately, though, it can be neutralized by the addition to the fuel of mangesium compounds, which are cheap and easy to handle. Magnesium sulfate (epsom salts) can be used and is widely available. Magnesium oxide or magnesium naphtenate may also be used. The latter compound has the advantage of being oil soluble and can be added to fuel as an oil-based slurry.

Quantitites of washing constitutents are based on the following [Gregg (1973)]:

1. Wash water is 5–10% of the oil being washed.
2. Tretolite (deemulsifier) is 0.025% of the oil.
3. Magnesium is 300–350% by weight of the vanadium.

Thus for a 20,000 hp engine with fuel rate of 0.5 lb/hp-hr burning a fuel of 50 ppm vanadium, quantities may be

1. $0.5 \times 20{,}000 \times 0.10 = 1000$ lb/hr or 2 gal/min fresh water to be provided and disposed of as oily waste after use.
2. $0.5 \times 20{,}000 \times 0.00025 = 2.5$ lb/hr tretolite.
3. $0.5 \times 20{,}000 \times 50 \times 10^{-6} \times 3.5 \times 10.25 = 18$ lb/hr epsom salts (10.25 lb $MgSO_4 \cdot 7H_2O$ per lb magnesium).

Figure 8.5 is a block diagram of a typical water-washing system. The system shown is that used aboard the *John Sergeant* in the early 1950s, but is essentially the same as those used aboard contemporary vessels, including diesel ships that burn residual oil.

An idea of the effectiveness of the treatment can be obtained from Table 8.5. This table, adapted from Gregg (1973), shows typical before-and-after

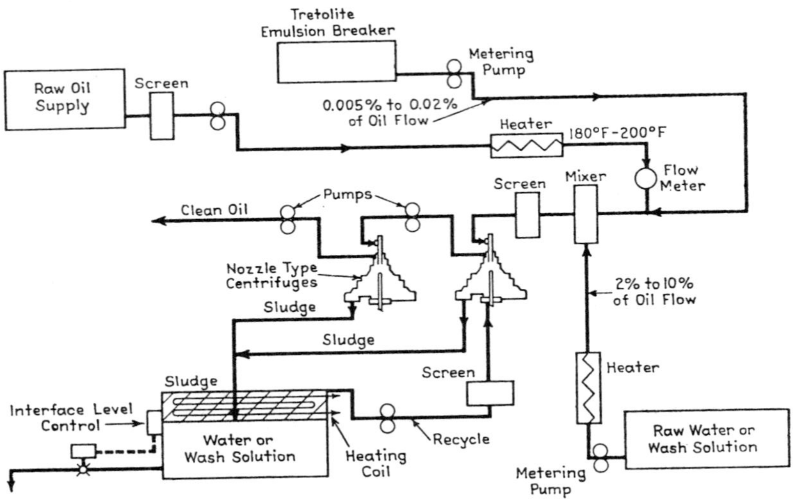

Figure 8.5. Block diagram of a fuel-washing [from McMullen (1957)]. From *Transactions,* 1957, published by the Society of Naval Architects and Marine Engineers.

concentrations of the metallic contaminants, as well as the magnesium that is added as the vanadium inhibitor. Note that the vanadium concentration is unaffected by washing, hence the need for the inhibitor.

8.5. USE OF GASEOUS FUEL

Natural gas is an excellent fuel for gas turbines. Its cleanliness, for example, is an outstanding characteristic (provided it is not contaminated in handling by rusty pipes, etc). It is widely used in stationary practice, where it may readily be supplied by pipline. A prominent example is the use of this fuel for engines driving gas-pipeline compressors; there, a portion of the gas being pumped is tapped off to run the engine.

Natural gas is similarly an excellent fuel for diesels and steam plants, but the difficulties in handling and storing it aboard ship have made it a rarely used fuel in any kind of engine at sea. An exception occurs in ships carrying liquified natural gas as cargo. This product is transported at low temperatures in insulated tanks. Since insulation is never perfect, some of the cargo evaporates, and in some instances this vapor ("boil-off") is used as fuel for the propulsion plant, whether it be steam, diesel, or gas turbine. Since the evaporation rate is unlikely to match the fuel demand of the engine, ability to burn gas of necessity includes the ability to burn gas *and* oil fuel simultaneously. The only such application for the gas turbines so far has used the General Electric heavy-duty engine. A description published by General Electric [Jones, Rexford, and Rowen (1972)] is epitomized here as a means of outlining how the dual-fuel capability can be achieved.

Figure 8.6, taken from Jones, Rexford, and Rowen (1972), illustrates the systems used. The figure depicts one of the fuel nozzles, the oil-fuel system (including the rudiments of the treatment components for residual fuel discussed earlier), an atomizing-air system, the gas-delivery system, and the means of simultaneously controlling gas and oil flows.

As the figure indicates, the nozzles are capable of handling both gas and oil simultaneously, or they can operate on either fuel alone. The *flow divider* on the oil side is a set of positive displacement pumps, one for each nozzle, mounted on a common drive shaft. Since the pumps are of identical displacement and turn at the same speed, this scheme ensures precisely equal flows of oil to all nozzles. Flow of oil to the engine is directly proportional to the speed of their common shaft, so that the speed sensing shown (MAG. P. U. = magnetic speed pickup) provides a signal proportional to oil flow. The *variable-control voltage signal* near the center of the figure is the power command to the engine, part of the General Electric SPEEDTRONIC control system described in Chapter 7. Were it not for complication of the gas fuel, the amplifier shown would simply compare this command to the signal of actual fuel flow from the flow divider, and adjust the solenoid (SV) bypass valve (BPV) to increase or decrease the flow going to the engine as required.

An important function of the gas-fuel system is removal of vapor from the cargo tanks without imposing an unacceptably low pressure on them. The delivery compressor is therefore fitted with a bypass (not shown in Figure 8.6)

TABLE 8.5. TYPICAL FUEL WASHING EFFECTIVENESS [GREGG (1973)]

Element	Before Washing	After Washing
Sodium	100 ppm max	2 ppm max
Sodium + potassium	100 ppm max	5 ppm max
Calcium (water soluble)	20 ppm max	10 ppm max
Calcium (Non-water soluble)	10 ppm max	10 ppm max
Lead	5 ppm max	5 ppm max
Magnesium	—	$3 \times V$ concentration (below)
Vanadium	400 ppm max	400 ppm max

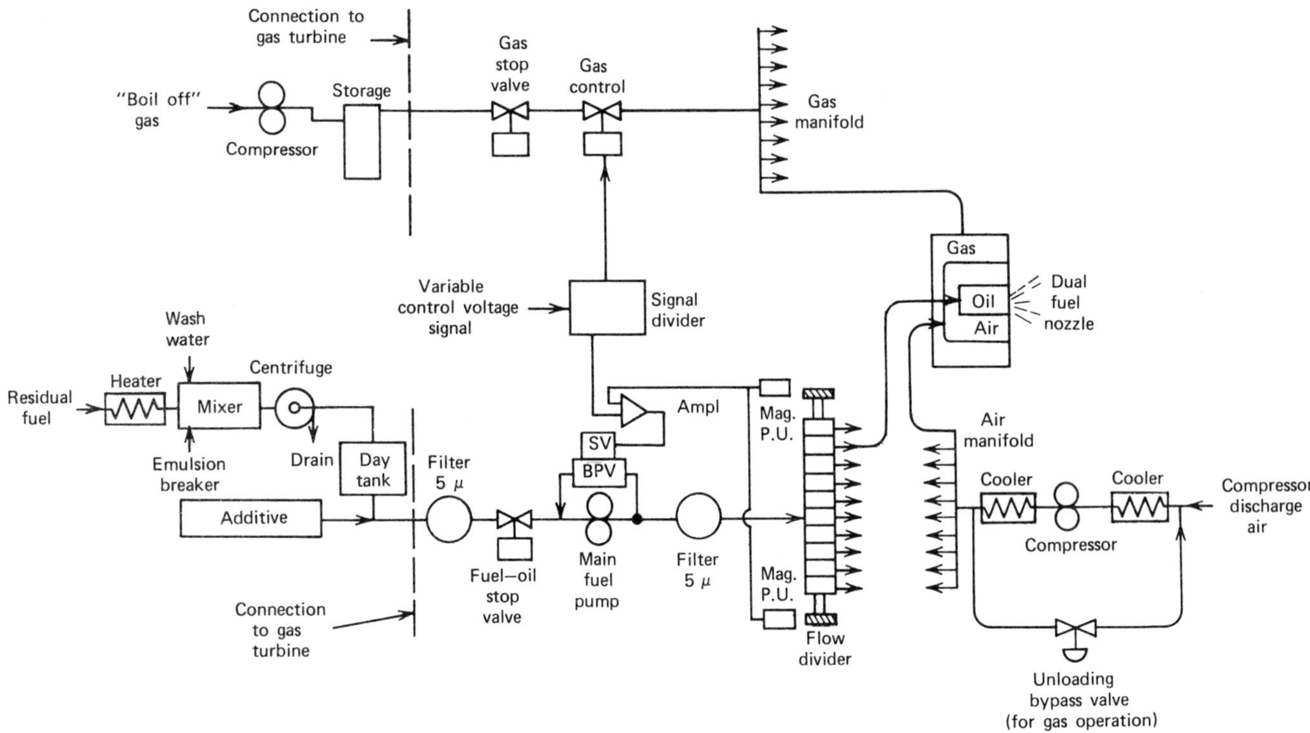

Figure 8.6. General Electric dual-fuel system [from Jones, Rexford, and Rowen (1972)].

that opens to return gas to the tank should the pressure there tend to fall as a consequence of a decrease in the boil-off rate. A second function is maintenance of constant pressure at the accumulator (*storage* in Figure 8.6) from which the engine draws fuel. Thus excess pressure at the accumulator also causes the bypass to open. However, the latter action occurs only if the gas fuel control is unable to keep pressure below the set point by forwarding the gas into the engine, that is, only if the engine demand for fuel is less than the boil-off rate.

Under normal at-sea conditions—conditions in which boil-off can supply a large fraction of the engine fuel requirements—flow of gas to the engine is regulated by the gas-control valve in response to accumulator pressure, and not at all in response to engine demands. The deficiency in fuel is met by oil; the *signal divider,* in calling for more or less fuel as power demand or gas supply fluctuates, sends its commands only to the oil side, adjusting the fuel-pump bypass as required.

Should gas supply rise toward the total engine demand, oil flow is correspondingly reduced until it reaches the minimum level at which oil combustion can take place. Reduction past this limit is not permitted. If increase in gas supply continues, a cutback signal from the SPEEDTRONIC circuits is generated, and the signal divider now reduces the gas valve opening to balance fuel flow and demand. As pointed out previously, a bypass around the delivery compressor opens to return excess gas to the cargo tanks, whence it may be removed by means other than engine consumption.

On the other hand, a deficiency of gas causes the gas accumulator pressure to fall, causing the gas control valve to close and the oil flow to increase in compensation, as described above. Since a minimum gas flow is required for accurate metering to the burners, the signal divider acts to close the gas valve entirely if the minimum is passed.

In all cases, feedback to the power controller (recall Figures 7.5 and 7.11) is constructed from the sum of signals from the oil-fuel-flow-divider speed and the gas-control-valve position.

Significant fluctuations can occur in the gas-heating value, as well as in quantity. The cause is a gradual change in cargo composition due to boil-off as the voyage progresses. As a result of the change, the components evaporating change, and consequently the heating value changes. For example, boil-off gas may be 18–20% nitrogen at the start of a voyage, but may be almost 100% methane at the end. An increase in heating value causes an increase in propulsion power at any gas-control-valve position and hence at any position of the ship operator's control handle on the bridge. This is sensed as an increase in propeller rev/min, a signal that is used to introduce a bias into the power control. The bias adjusts the handle position-power relationship so that propeller rev/min remains constant for a given handle position. This feature is needed only to ease the navigational problems of the operator. Protection of the engine against overheating due to the increase in heating value is provided by the normal functions of the SPEEDTRONIC control system.

8.6. APPENDIX—SOME THINGS ABOUT VISCOSITY

The shear force per unit area (τ) between layers of fluid in a viscous flow field, Figure 8.7, is proportional to the velocity gradient, $du(y)/dy$. The constant of porportionality (μ) in the resulting equation

$$\tau = \mu \frac{du(y)}{dy}$$

is the *viscosity* of the fluid. Dimensions of viscosity are, therefore, force × time/area, or the equivalent mass/time × length. The common scientific unit is the *poise* (1 gram/centimeter-second), or for more manageable numerical magnitudes, the *centipoise,* 0.01 poise.

Viscosity divided by density of the fluid in question is called *kinematic viscosity.* The common scientific unit is the *stoke,* or the *centistoke,* 0.01 stoke (density in grams/cubic centimeter).

In practice, viscosity is measured by an *efflux viscosimeter,* a device in which a standard amount of the fluid being tested is allowed to drain through a standard hole. The time required is a function of kinematic viscosity and so is used to express this quantity directly in the time dimension, usually in seconds. There are several different viscosimeters in general use. The most common in liquid fuel technology are the Saybolt Universal (SSU = Saybolt

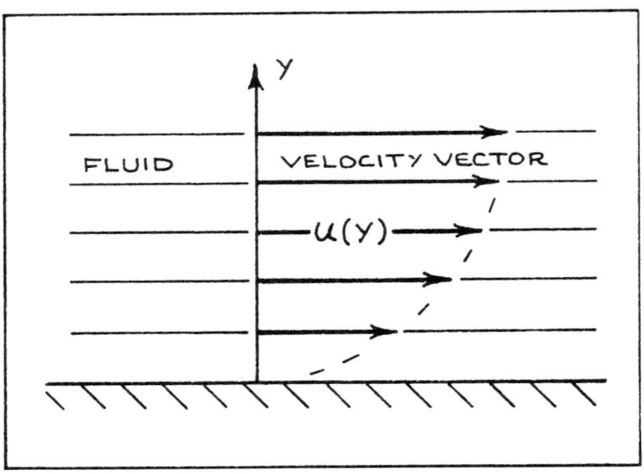

Figure 8.7. Defining viscosity.

Universal Seconds) and the Redwood No. 1. The SSU and Redwood units appear in the specifications that have been quoted in this chapter and in Figure 8.1.

Conversion equations from the Redwood and Saybolt units (designated t) to centistokes (C_s) are

$$\text{Redwood 1:} \qquad C_s = 0.26t - \frac{172}{t} \tag{8.1}$$

$$\text{Saybolt:} \qquad C_s = 0.22t - \frac{180}{t} \tag{8.2}$$

8.7. REFERENCES

1973 Annual Book of ASTM Standards, Part 17, American Society of Testing and Materials.

LM2500 Marine Gas Turbine Installation Design Manual (1973), General Electric Report MID-IDM-2500-2, Marine and Industrial Projects Department.

Brubaker, N. H., and Calvino, L. N. (1972), "Fuels for Marine Gas Turbines," American Society of Mechanical Engineers, International Gas Turbine Conference, San Francisco.

Gregg, S. W. (1973), "Modern Fuel Processing Plant," General Electric 18th Gas Turbine State of the Art Engineering Seminar.

Installation Design Criteria for Gas Turbine Application in Naval Vessels, NAVSHIPS 0941-038-7010 United States Navy Technical Manual.

Jones, R. M., Rexford, D. L., and Rowen, W. I. (1972), "The Control of Heavy Duty Gas Turbines Burning LNG Cargo Boiloff," presented at the third annual ship control systems symposium, Bath, England.

McMullen, J. J. (1957), "Trials and Maiden Voyage of GTS *John Sargeant,*" *Society of Naval Architects and Marine Engineers, Transactions,* **65,** pages 25–48.

Morrow, A. S. (1972), "Fuels for Marine Gas Turbines—Requirements and Availability in the 1970s," American Society of Mechanical Engineers, Paper 72-GT-13.

O'Hare, T. L. R., and Holburn, J. G. (1973), "Operating Experience with Gas Turbine Container Ships," *Transactions, Institute of Marine Engineers,* (Series A, Part I), **85,** pages 1–23.

Sigal, E. B. (1971), "The Navy Distillate Fuel Conversion Program," *Naval Engineers Journal,* **83,** pages 51–58.

Zeien, C., Smith, H. F., and Hirst, F. W. (1969), "The Gas Turbine Ship *Callaghan*'s First Two Year of Operation," *Transactions, Society of Naval Architects and Marine Engineers,* **77,** pages 344–371.

The following references are not cited in this chapter, but are nonetheless worthy of mention. Several are interesting in that they relate some of the steps in the development of present fuels, and in the development of gas turbines to enhance their capability for burning these fuels.

Cabal, A. V., and Warren, J. H. (1967), "Fuels for Marine Gas Turbines," American Society of Mechanical Engineers, Paper 67-GT-50.

Carlton, G. A. (1970), "New Fuel for Naval Gas Turbine Engines," American Society of Mechanical Engineers, Paper 70-GT-29.

Cullen, P. J., and Urbas, T. A. (1972), "The Treatment of Marine Heavy Fuels for Gas Turbine Combustion," American Society of Mechanical Engineers, Paper 72-GT-97.

Ekstedt, E. E., and Macaulay, R. W. (1972), "Combustion of Heavy Distillate Fuel for the LM2500 Gas Turbine," American Society of Mechanical Engineers, Paper 72-GT-24.

Schiefer, R. B. (1971), "The Combustion of Heavy Distillate Fuels in Heavy Duty Gas Turbines," American Society of Mechanical Engineers, Paper 71-GT-56.

Siemietkowshi, J. S. (1962), "Gas Turbine Testing on Naval Distillate Fuel," American Society of Mechanical Engineers, Paper 71-GT-62.

Taylor, W. G. (1967), "Residual Oil Combustion in Gas Turbines—State of the Art," American Society of Mechanical Engineers, Paper 67-GT-19.

Chapter

NINE

INLET AND EXHAUST SYSTEMS

9.1. INTRODUCTION

Let us begin the chapter with three figures (9.1, 9.2, and 9.3) that together typify the problem of intake and exhaust for the gas-turbine marine engine. The first of these is a reproduction of a figure that has appeared in several papers on the *Callaghan*; it shows the two propulsion engines and their associated inlet and exhaust ducting. The significant thing is the magnitude of the ducts. Note that the volume of the engines is almost trivial by comparison. The second figure is a set of correction factor curves showing the loss in power and increase in fuel rate that is caused by pressure losses in the ducting. The third figure shows an example of the major loss in performance that can be experienced by a marine gas turbine when its compressor becomes fouled with sea salt entering with the combustion air.

The fundamental factor in the technology of gas-turbine air supply and exhaust is the high rate of air usage by this type of engine. Even the most efficient gas turbine uses about 5 ft^3/hp-min, which is roughly twice that of the biggest air consumers among the diesels, and about $2\frac{1}{2}$ that of a steam-turbine propulsion plant. Ducting, therefore, tends to be large, and naturally there is incentive to make it smaller by moving the air at high velocity. But high velocity means high pressure drop with the consequences shown. High velocity also tends to increase the transport of salt-laden droplets of sea spray, and efforts to remove the droplets by filters or traps increases the pressure loss. Design compromise is consequently very much in order, with the objective of usurping the minimum of ship volume while providing adequately low pressure drops and thwarting the ingestion of salt via the inlet duct.

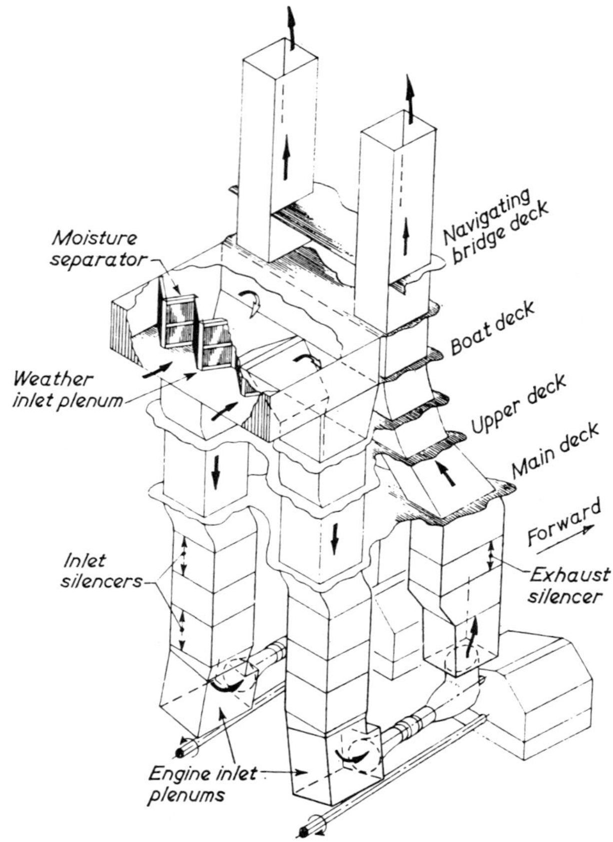

Figure 9.1. Inlet and exhaust ducting of the *Admiral William Callaghan*.

There are other features of inlet and exhaust ducting that the designer must provide. For a first example, both ducts are pathways for sound escape to the surroundings, and, therefore, silencing features usually must be incorporated. Exhaust gas should not be reingested by the intake; avoiding this defect is complicated by the several strengths and relative directions of winds that the vessel may encounter. Likewise, oil vapors, and indeed all other foreign substances and objects, must be kept out of the intake. The exhaust gas may be dangerously hot and so must be kept well clear of wooden structures (such as could be found in the proximity of a small vessel alongside a pier). And there are other points mentioned in this chapter, all requiring careful attention of the designer for a satisfactory propulsion plant.

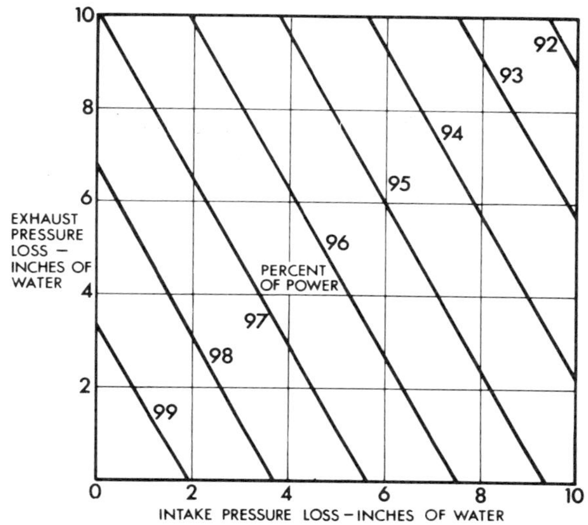

Figure 9.2. Typical loss of performance as consequence of pressure drops in inlet and exhaust ducts [Lindsay (1968)]. From *Transactions of the Diamond Jubilee Meeting*, 1968, published by the Society of Naval Architects and Marine Engineers.

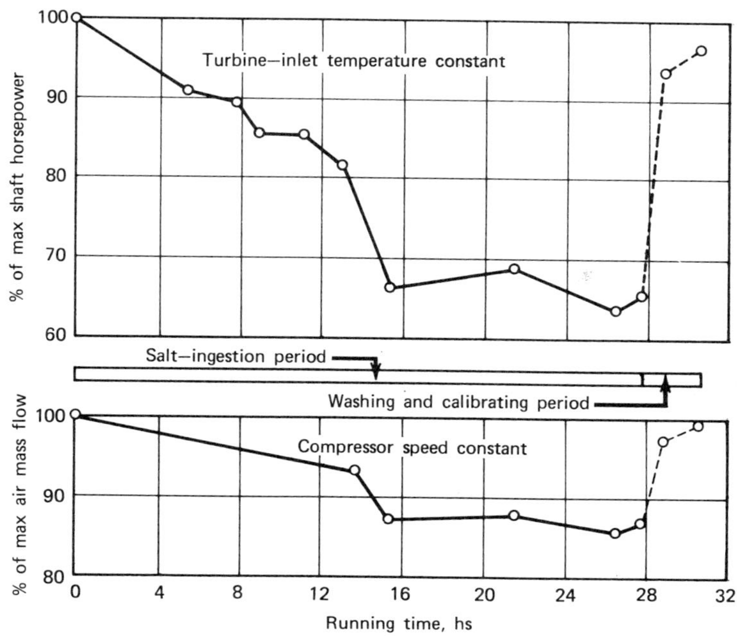

Figure 9.3. Performance degradation caused by salt ingestion [Hardy (1966)].

9.2. DUCT PRESSURE DROP CALCULATIONS

A. General Considerations

Although it is perhaps impossible to label any one aspect of inlet and exhaust design as "most important," the calculation of pressure drops in the ducting should be the first consideration, since it establishes the cross-sectional area required. Most of the other aspects of duct design follow from that.

Consider the fundamental situation at the intake of a gas-turbine engine, without intervening ducting and without motion relative to the ambient air (see Figure 9.4, left sketch). The compressor accelerates air from the surroundings toward the inlet, as indicated by the arrows in the figure. The mechanical energy per unit mass of the air, or its *head,* is largely unaffected by the process, though static pressure declines as the air accelerates toward the inlet. Head is the sum of three components, as shown by the defining relation

$$H = \frac{P}{\rho} + \frac{V^2}{2} + gZ \tag{9.1}$$

where P is pressure, ρ is density, V is velocity, g is acceleration of gravity, and Z is height above datum (note that unitary constants are omitted). Head is reduced only by extraction of mechanical work, heat flow from the fluid, or by viscous degradation of mechanical energy to heat within the fluid. The first two processes are absent here, and the third is negligible. The decrease in static pressure comes about from interchange between components of head, P/ρ decreasing as V increases. (Change in gZ is usually negligible.) However, this interchange is reversible within the compressor and so is not a loss.

Now consider the slightly modified situation brought about by the addition of a straight duct to the engine inlet, Figure 9.4, right sketch. The acceleration of air again takes place at the entrance to the system, and the static pressure component of the head becomes less than the ambient pressure because of the interchange with the velocity component. Again, this is not a loss, since it is a

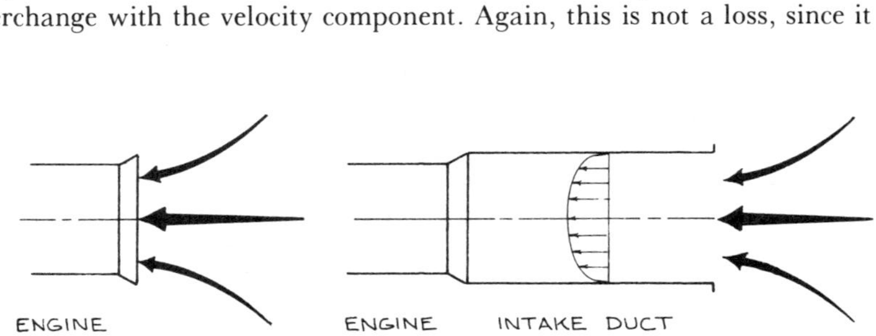

Figure 9.4. Elementary concept of air flow to engine, with and without inlet duct.

reversible process. But the solid boundary (duct walls) of the column of air moving toward the engine causes steep velocity gradients within the flow, especially near the boundaries. At any point where such a gradient exists, a viscous frictional shear force is developed according to

$$\tau = \mu \nabla u \tag{9.2}$$

where μ is viscosity of the air and ∇u is the velocity gradient. Such forces also exist in the accelerating region outside the duct entrance, and near the engine inlet in the first case, but are neglibly small compared to those existing in duct flow. (The flow is doubtless turbulent, which adds to local velocity gradients, but consideration of this detail is not essential to this discussion.)

A friction force between moving bodies or fluid elements represents a degradation of energy to heat, and this process is not reversible by the compressor. This loss of head, which is typically measured by a pressure-sensing instrument, is the phenomenon commonly called the *pressure drop*.

A practical duct is unlikely to be straight and unobstructed. For instance, see Figure 9.5, which shows the intake duct for the DD963-class destroyers of the U.S. Navy [Rains and d'Arcy (1972)]. A sharp 180° reversal of direction at the inlet, obstruction by screens and moisture separators, diverging cross sections, converging cross sections, narrow passages, and a 90° elbow are evident and are typical. Each of these increases the pressure drop, basically because they increase velocity gradients, and hence viscous degradation of head. A screen, moisture separator, or other restriction can do this simply by raising the average local value of velocity. A diverging section or change of direction can cause separation of flow from the walls, with resulting high turbulence between the wall and the main stream of the flow.

Events in the exhaust duct are the same, although the velocity component of head represents energy that is generally not recoverable. A diffuser, or section of gradually diverging flow area, might be expected to convert this component back to pressure, thereby reducing static pressure within the duct, but it is rarely practical to build such a device into the limited volume available for shipboard ducting. In addition, it is usually necessary to expend this velocity in ejecting the exhaust gases a goodly distance from the ship.

B. Calculational Principles

Air, like all gases, is a highly compressible fluid. In flow through a constant area duct under steady, subsonic, adiabatic conditions, the degradation of head in a compressible fluid causes its density, pressure, and temperature to decrease and its velocity to increase. A precise analysis to determine, say,

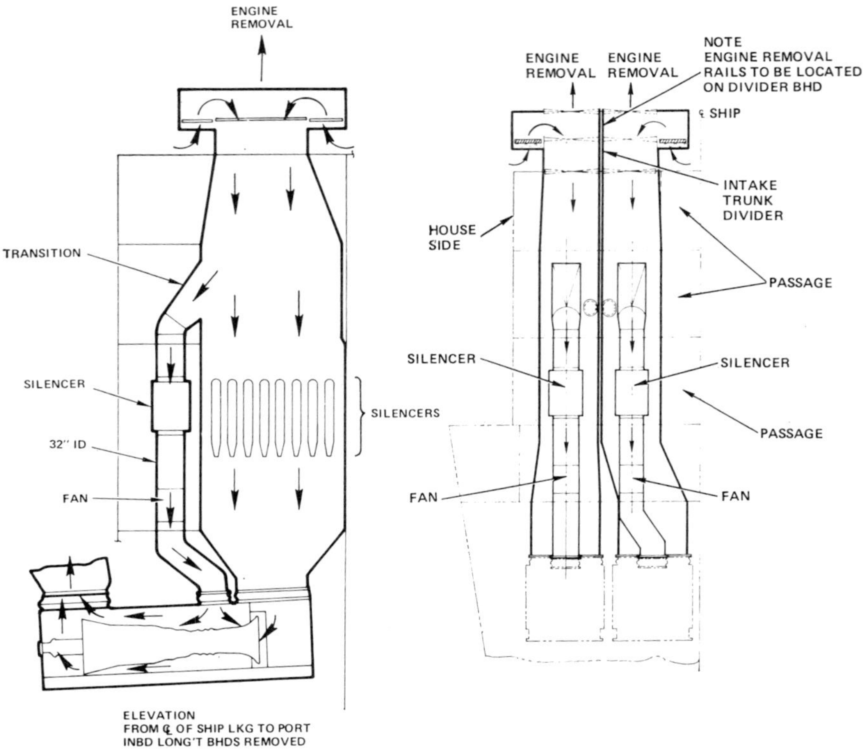

Figure 9.5. Inlet duct, U.S. Navy DD963 class destroyers [Rains and d'Arcy (1972)].

pressure drop, must take into account all of these changes due to compressibility. On the other hand, the change in properties of the fluid is often so small in duct flow that practical calculations can ignore them, that is, the fluid can be treated as if it were incompressible. Such is the universal approach to ventilation-duct calculations: air density is regarded as constant since it is not likely to change by more than 1% over the length of the duct. Although the latter approach is usually preferred for gas-turbine ducting also, suggestions that compressibility is significant have been published [Sawyer, Kelnhofer, and Smith (1966)]. Therefore, let's look at both viewpoints.

The fundamental equation relating pressure changes to flow rate, duct dimensions, and shear forces (represented by a *friction factor*) is

$$\int_1^2 \rho\, dP + \left(f\frac{l}{2d} + \ln\frac{\rho_1}{\rho_2}\right)\left(\frac{\dot{m}}{A}\right)^2 = 0 \tag{9.3}$$

where ρ is the fluid density; P is pressure; $\dot{m}/A$ is the mass flow rate per unit area; l is the length of duct between points 1 and 2; f is the friction factor, a nondimensional relation of shear forces to velocity head; d is the duct hydraulic diameter, which is 4 × cross-sectional area/wetted perimeter (note that unitary constants are omitted). The friction factor is generally a function of Reynolds number and of surface roughness, and must be evaluated from experimental data. Data by Moody (1944) is commonly used and is widely found in handbooks, for example the *Heating, Ventilating, and Air Conditioning Guide* (1974) and the *Flow of Fluids Through Valves, Fittings, and Pipe* (1965).

Since ρ is a function of P, the first-term integral cannot be evaluated without aid of the functional relationship. This relationship is found by simultaneous application of three equations: (1) an energy equation, (2) the one-dimensional continuity equation, and (3) an equation of state. Handy short-cut methods of making the evaluation with the aid of tabulated solutions are well known [see, for example, Shapiro (1953)]. Alternatively, the individual equations just mentioned can be evaluated, these being, respectively

$$T + \frac{V^2}{2g_oJc_p} = T_0 \tag{9.4}$$

$$\frac{\dot{m}}{A} = \rho V \tag{9.5}$$

$$P = \rho RT \tag{9.6}$$

where T is temperature; T_0 is stagnation temperature, that is, temperature of air at points where $V = 0$, usually the ambient temperature; V is the average fluid velocity in the duct; g_o is the unitary conversion constant (32.17 lbm-ft/lbf sec^2); J is the mechanical equivalent of thermal energy (778 ft lbf/degree); c_p is constant-pressure specific heat (0.24 Btu/lbm degree); and R is the gas constant (53.3 lbt-ft/lbm degree). Other terms are as in equations above. To be consistent with the units given here, P must be in lbf/ft^2, T in degrees Rankine, $\dot{m}/A$ in lbm/ft^2 sec, V in ft/sec, and ρ in lbm/ft^3. Note that air is here treated as a perfect gas, quite an acceptable procedure for duct calculations.

A reasonable application procedure for these equations in finding a pressure drop over a length, l, involves plotting a pressure drop (ΔP) versus l curve by the following steps:

1. Choose a T slightly less (typically, one or two degrees) than T at the beginning of length, l, and use Eq. 9.4 to solve for the corresponding V.
2. Use Eq. 9.5 to solve for ρ ($\dot{m}/A$ is a constant whose value is presumably known).

3. Use Eq. 9.6 to solve for P.
4. Evaluate the third term of Eq. 9.4, using the ρ just found as ρ_2.
5. Also use this ρ to evaluate $\int \rho \, dP$ by simple trapezoidal rule, that is, $\int \rho \, dP = (\rho_2 + \rho_1)(P_2 - P_1)g_o/2$.
6. Solve Eq. 9.4 for l.

Figure 9.6 demonstrates the results of making such calculations for several initial velocities and several nondimensional lengths ($f\, l/d$).

The incompressible approximation follows simply from treating density as a constant. Equation 9.3 immediately becomes

$$\Delta P = f \frac{l}{d} \frac{(\dot{m}/A)^2}{2g_o\rho} \qquad (\text{lbf/ft}^2) \qquad (9.7)$$

The other equations are superfluous, since properties of the air remain constant. Figure 9.6 also demonstrates curves calculated by this equation. In practice, it is unlikely that velocities greater than 100 ft/sec in the intake duct, or 200 ft/sec in the exhaust duct, will be used, and $f\, l/d$ is unlikely to exceed 5.0. The small differences evident in Figure 9.6 suggest that the simpler procedures of Eq. 9.7 be used in practice. The error so introduced is probably less than the uncertainty in evaluating the friction factor or other empirical data used.

Equation 9.7 serves for straight ducts and also for the expansions, elbows, and so forth, previously mentioned, provided that data on their equivalent lengths is available (equivalent length = length of straight duct that causes the same pressure drop as the feature in question). Equivalent length data may be given in the form, l/d, in which case the friction factor used with straight duct is to be applied, or it may be given in the form, $f\, l/d$, often symbolized by K. Therefore, Eq. 9.7 might be written as

$$\Delta P = \sum_i K_i \frac{(\dot{m}/A)^2}{2g_o\rho} (0.192) \qquad (\text{in. water}) \qquad (9.8)$$

where K is $f\, l/d$ (f read from sources mentioned earlier) for a straight duct, or is a factor read directly from experimental data for features that contribute more than just length to the pressure drop. The latter data can be found in many handbooks dealing with ventilation; see, for example, Moody (1944) and Shapiro (1953). The factor, 0.192, converts lbf/ft^2 to the more commonly used units, inches of water.

C. An Example Calculation

Let us attempt a calculation of pressure drop through the intake duct shown in Figure 9.5. Since the duct is part of a combat naval vessel, its dimensions are

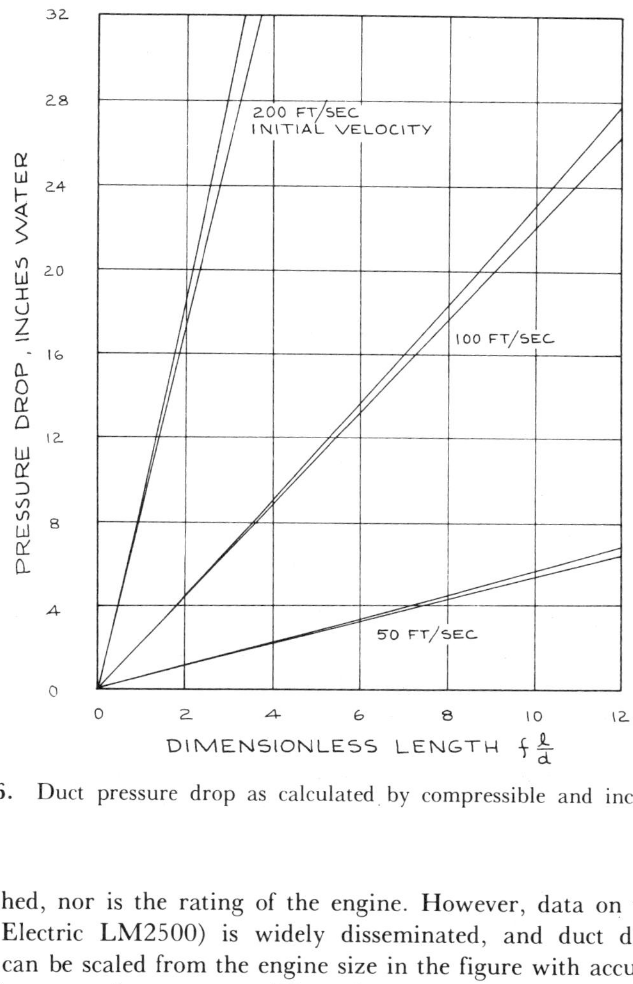

Figure 9.6. Duct pressure drop as calculated by compressible and incompressible techniques.

not published, nor is the rating of the engine. However, data on the engine (General Electric LM2500) is widely disseminated, and duct dimensions, therefore, can be scaled from the engine size in the figure with accuracy sufficient for demonstration purposes. These dimensions are found in Table 9.1; other basic numbers are

Engine power	21,000 hp
Air flow	1,740 ft^3/sec
Air temperature	80 F
Air density	0.0735 lb/ft^3
Air viscosity	0.0185 centipoise

The air flow listed is to the engine intake only. The cooling air that uses the same intake, but is tapped off by the small parallel duct, is ignored in this calculation. The calculation uses Eq. 9.8, with friction factors and K factors being taken from handbooks previously cited [*Heating, Ventilating, and Air Conditioning Guide* (1974); *Flow of Fluids Through Valves, Fittings, and Pipe* (1965); and *Pressure Loss Ventilation Fittings* (1950)]. Details are given in Table 9.1. Pressure drop is found to be 1.12 in. water.

The calculation does not include the pressure drop caused by inlet screens and moisture separators, since the figures depend so much on the details of the particular components. Since the Navy standard allowance for inlet duct loss is 4.0 in. water, it might be said that the present calculation shows that 2.88 in. remain to be expended in screening and drying the air.

Other than the effects of the omissions mentioned, how accurate is the calculation likely to be? It is hopefully a good approximation. The principal doubt in precision comes from the uncertainty in the use of the empirical data for friction factor and for the *K* factors for elbows, expansions, and so forth. Whatever the precision with which this data was obtained, it is often difficult to match the job at hand with the experimental conditions, hence judgement—which may be imperfect—is required in reading an *f* or a *K*. Designers of equal probity and experience are likely to differ noticeably. Another point, not accounted for in the available design data, is the possible interaction between components of the ducting. For example, the effect of a bend or elbow depends to some degree on the length of duct following before the next deviation from constant-area straightness. Although this point has been discussed by Sawyer, Kelnhofer, and Smith (1966), it is not usually accounted for in handbook data. The calculated result should consequently be regarded with some skepticism, and margin allowed in any use of the results, such as correcting engine rating for duct losses. An example may be cited: *Euroliner* intake-duct loss was calculated in design to be 1.87 in., but was found to be 0.57 in. higher during sea trials [O'Hare and Holburn (1973)].

D. Velocity and Pressure Distribution Within the Inlet Duct

The calculational techniques discussed here—both compressible and incompressible—are one dimensional, that is, they are based on an average velocity, namely, the volume flow rate divided by the flow area. Attempts to account for the actual velocity distribution at any cross section would in fact probably add nothing but complexity to the calculations, so that the one-dimensional approach is not to be criticized from the standpoint of accuracy in pressure-drop predictions. The velocity or pressure distribution is of no interest, except at the inlet face of the compressor. A nonuniform distribution at the compressor inlet can cause poor performance [Reid (1969)], or even fatigue failure of compressor blading. A widely reported case of such failure occurred aboard the British frigate, *Exmouth* [Sawyer, Kelnhofer, and Smith

TABLE 9.1. DETAILS OF EXAMPLE CALCULATION

Description	Units	En-trance	Straight	Ex-pansion	Straight[a]	Con-traction	Straight (Silencer)	Ex-pansion	Con-traction	90° Turn	
1. Duct width	in.	—	122	188[b]	188	—	10[c]	—	94[d]	85	
2. Duct depth	in.	—	80	80	80	80	80	80	80	80	Note[e]
3. Flow area	sq ft	—	68	—	104	—	50	—	52	—	Note[f]
4. Length	in.	—	40	130	290	—	110	—	56	—	
5. Velocity	ft/sec	—	25.6	—	16.7	—	34.8	—	33.5	37.0	Note[g]
6. Velocity head	in. H_2O	0.144	0.144	0.144	0.061	0.061	0.265	0.265	0.246	0.301	Note[h]
7. Hydraulic diameter	ft	—	8.1	—	7.9	—	1.3	—	6.4	6.8	Note[i]
8. Reynolds number	—	—	1.25	—	0.80	—	0.27	—	1.30	1.53	Note[j]
9. Friction factor	—	—	0.012	—	0.013	—	0.018	—	0.012	0.012	Note[k]
10. *K* or fl/d	—	1.5	0.059	0.12	0.477	0.20	1.52	0.24	0.105	1.15	
11. Pressure drop	in. H_2O	0.216	0.009	0.017	0.029	0.012	0.404	0.063	0.026	0.346	

[a] Sum of length before and after silencer.
[b] Maximum.
[c] Per channel.
[d] Minimum. Calculations in this column treat the expansion as a straight duct of this width.
[e] Perpendicular to plane of figure.
[f] **1** × **2**/144.
[g] 1740/**3**
[h] 0.192 ρ $\mathbf{5}^2/2g_0$
[i] 4 × **3** × 12/2 (**1** + **2**)
[j] 6060 **5** × **7**
[k] From *Flow of Fluids Through Valves, Fittings, and Pipe.*

(1966); Mathews and Hodgson (1971); Miller (1969)]. This vessel experienced compressor-blading failure caused by high excitation as the blades passed through a standing vortex that developed close to the engine intake. The vortex was discovered by smoke tracing tests, after the failure had led to a thorough investigation into its cause. Approximate location of the vortex is indicated by a scale sketch in Figure 9.7, and the same figure also shows a correction that was developed by model testing.

To avoid the difficulty experienced by the *Exmouth,* it has been recommended that, within the main body of the flow (i.e., outside the boundary layer at duct walls), the total pressure at no point on the compressor intake plane should differ by more than 5% from the average value [Miller (1969)]. Since the pressure variation at any cross section cannot be reliably found by calculation, it is generally recommended that models of the intake ducts be tested in the design stage so that predictions can be based on measurements. Since the models are easily and cheaply constructed and tested, such a precaution is well justified. Literature on recent ship designs shows that modeling is indeed widely used to check the flow distributions.

9.3. SEA-SALT INGESTION VIA THE INLET DUCT

Air above the sea surface contains significant concentration of the salts of that sea. The salt may be present in the gross drops of spray that are so often

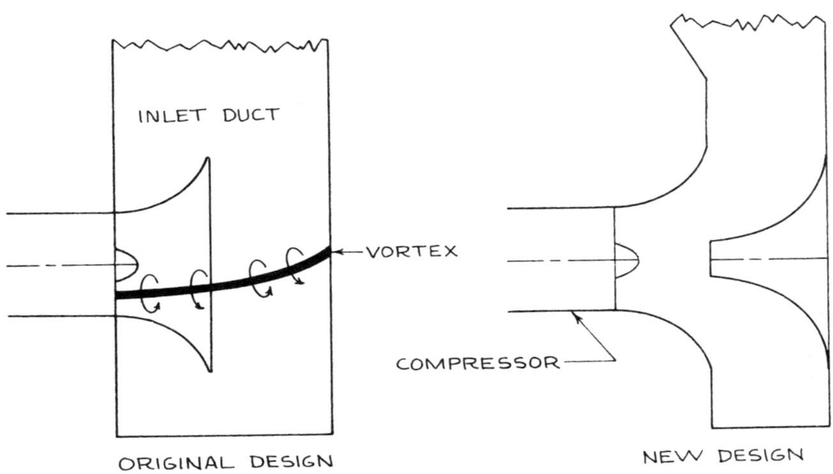

Figure 9.7. Original *Exmouth* inlet design with vortex, and corrected design [sketch by Miller (1969)].

present, or in small droplets forming an aerosol. The former source is relatively easy to isolate from the engine air supply, since duct intakes can be placed high (assuming a large vessel) and in a protected location, low duct velocity or settling in a plenum chamber readily removes large particles, and the addition of baffles and modest filters can make it virtually impossible for sea spray to reach the engine. The aerosol component is more difficult to avoid and remove, and it may be impossible to remove completely with any reasonable combination of measures. Generally speaking, "reasonable" in this context would imply avoidance of the excessive pressure drop that complete removal might require.

Figure 9.8 gives an idea of the aerosol concentration to be expected. It represents data taken by the Naval Ship Engineering Center (Philadelphia Division) [*Installation Design Criteria for Gas Turbine Application in Naval Vessels* (1968)] aboard destroyer-type vessels and is for inlets about 35 ft above the waterline in locations protected from direct spray.

The most direct manifestation of salt reaching an engine is loss of performance due to the salt coating compressor blades. The coating distorts the blade profile and narrows flow passages. The degradation is largely reversible by means of water washing of the compressor, a necessary process that has been followed with many marine gas turbines. Figure 9.3 demonstrates one such case of drastic performance loss, followed by recovery after washing.

A possible longer-term consequence of salt deposition in the cold end is corrosion, although the problem is trivial in contemporary engines since resistant alloys are used. Early marine gas-turbine engines adapted from aircraft service tended to suffer severe pitting of compressor parts because of their common use of magnesium alloys, which are highly subject to this type of corrosion. Replacements have typically been aluminum or titanium alloys.

More serious consequences stem from corrosion of the hot section—especially first-stage turbine nozzles and blades—in the process known as sulfidation. Sulfur is not present in sea water, of course, but is contributed by several of the fuels used in marine service. If present, it reacts in the combustion process with sea salt contributed by either water in the fuel or in the air to produce sodium sulfate. This substance, upon coating the hot surfaces, reacts with the chromium in the usual nickel-chromium alloy. The chromium-depleted metal is then subject to accelerated corrosion [Weinert and Carlton (1965)].

Since some sulfur in fuel apparently must be accepted, great care has to be taken to remove water/salt from both fuel and air.

Protected inlets and low inlet duct velocities are universally recognized as necessary for reducing the airborne salt. Deliberate reduction measures by means of separators have also been used on some vessels, and from the lessons

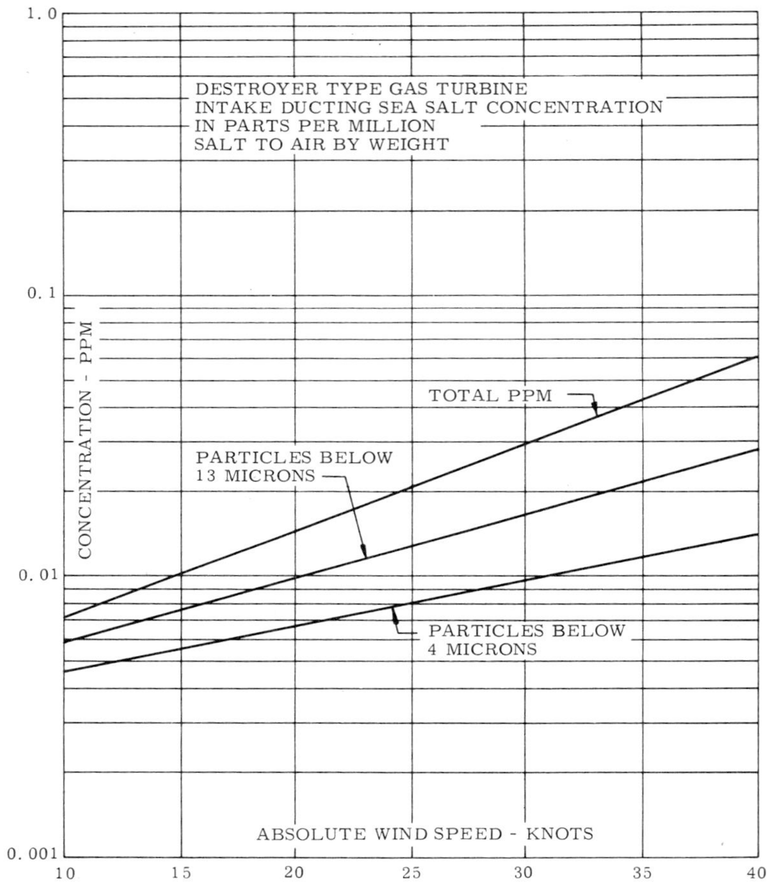

Figure 9.8. Sea-salt aerosol concentrations [*Installation Design Criteria for Gas Turbine Application in Naval Vessels* (1968)].

of experience this appears to be becoming universal practice. Two general types of separators (also called demisters) are favored, the inertial type and the filter type.

Inertial separators typically take the form of baffles that force the air to make abrupt changes in direction so that the entrained particles tend to move to the outside of the turn. Figure 9.9 illustrates the baffle elements of two different configurations; the (*b*) type is used aboard the *Euroliner* [O'Hare and Holburn (1973)]. Figure 9.10 shows the removal capability under actual service for the *Euroliner* demisters. It demonstrates what appears to be an inherent characteristic of inertial separation: high removal efficiency for large particles, but poor efficiency in removing the smaller. The *Euroliner* experienced the sulfidation problem in its early voyages, in part because of defects in the separator installation, but also attributed to the almost unimpeded passage for the smaller particles [O'Hare and Holburn (1973)].

The filter type usually takes the form of a knitted mesh, of either monel or polypropylene, through which the air is forced to pass. Although small particles also penetrate more freely than the larger, the difference in separation efficiency for the two is less abrupt than with the inertial separator. Also, the efficiency of separation does not decline as badly when the velocity goes down, so that the device works well at low engine power. There are disadvantages, however. The pressure loss tends to be much higher than that of the inertial type; a case is cited [Bogus and Weinert (1972)] of the two types used in series with 0.57 in. pressure drop through the inertial separator, and 2.0 in. through the knitted mesh. The mesh is subject to icing, to blast damage when used in military service, and to deterioration with age (corrosion if metal, embrittlement if plastic) that might result in pieces coming adrift to be sucked into the engine. Also, the mesh is likely to become encrusted with salt, requiring frequent washing.

Figure 9.11 depicts an inlet configuration with mesh demisters as used in a British naval installation [Morgan, Smith, and Lamport (1971)]. The washing spray is evident. The large valve that would bypass the demisters is for emergency use in the event of clogging by ice.

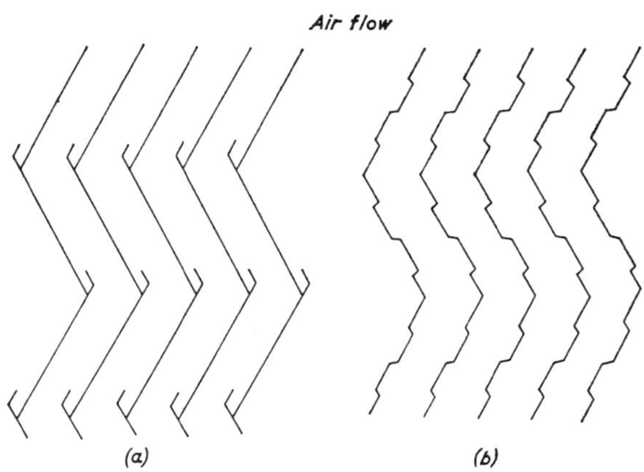

Figure 9.9. Typical inertial separator elements [O'Hare and Holburn (1973)].

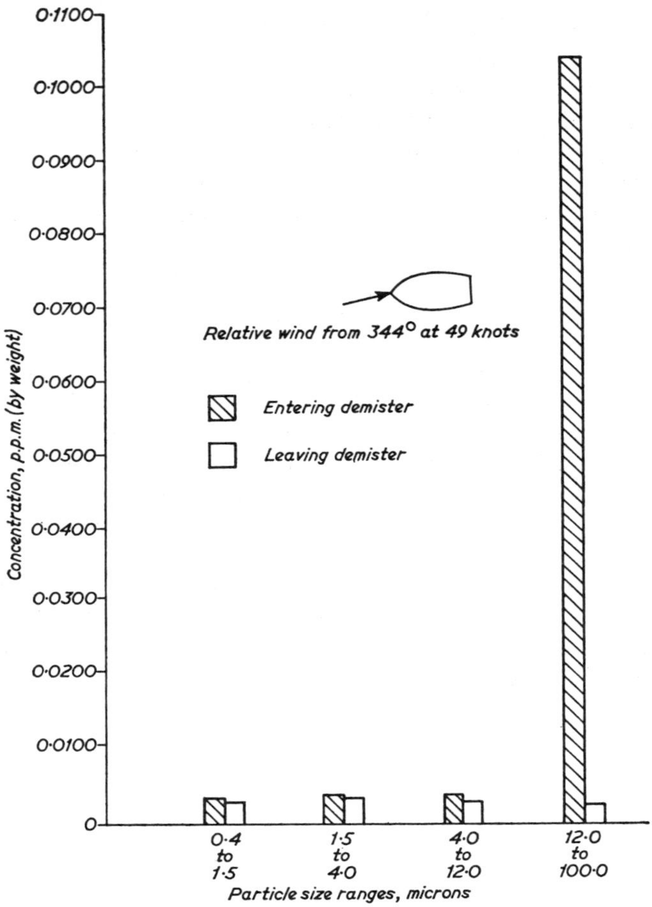

Figure 9.10 Inertial separator performance during early voyage of *Euroliner* [O'Hare and Holbrun (1973)].

9.4. NOISE ATTENUATION IN THE INLET AND EXHAUST DUCTS

Figure 9.12 shows as typical examples the sound power levels at inlet and exhaust openings of the General Electric LM2500 and the Avco-Lycoming TF35 gas-turbine engines: their sound power levels in decibels (dB = 10 log (W/W_0) where W is the power in watts and W_0 is the base power of 10^{-12} W) as a function of frequency over the audible range. Since these levels are intolerably high from the standpoint of nearby human speech, comfort, or even safety, and since the intake and exhaust ducts can act as channels for the almost unimpeded passage of sound to the decks of the vessel, attenuation measures are essential in most installations.

A significant measure of sound attenuation is provided merely by

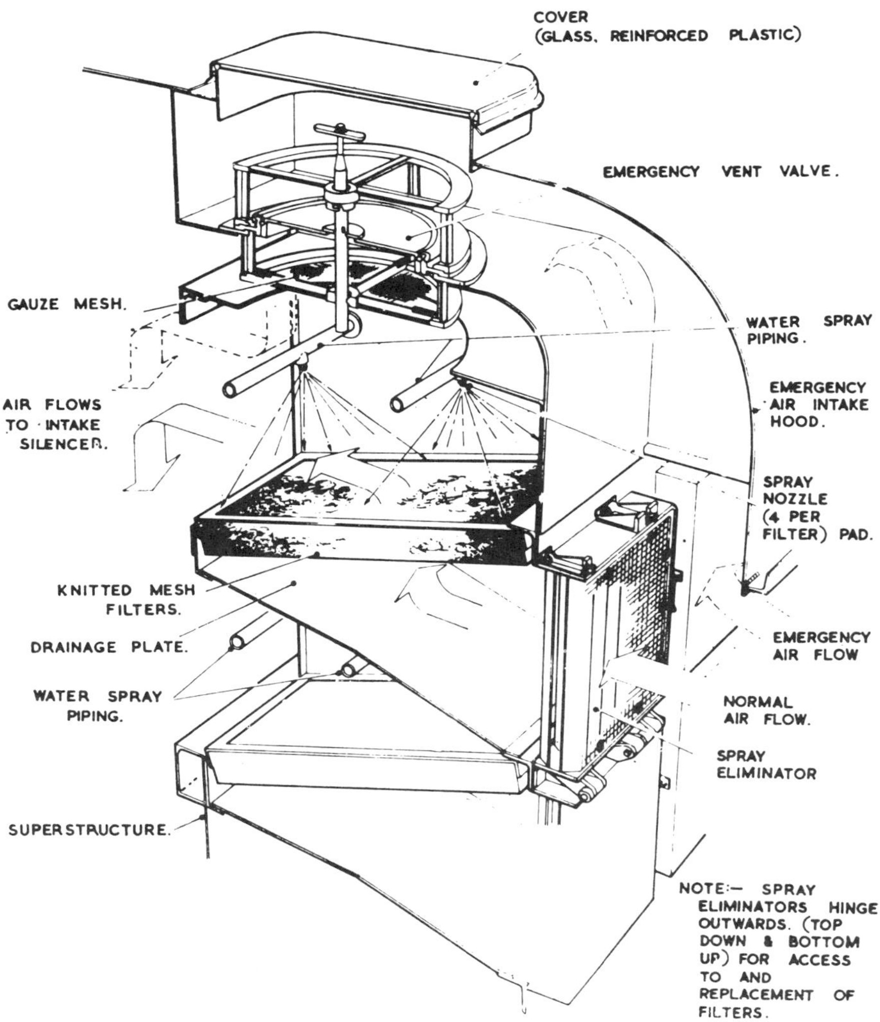

Figure 9.11. Mesh demister installation, British frigate [Morgan, Smith, and Lamport (1971)].

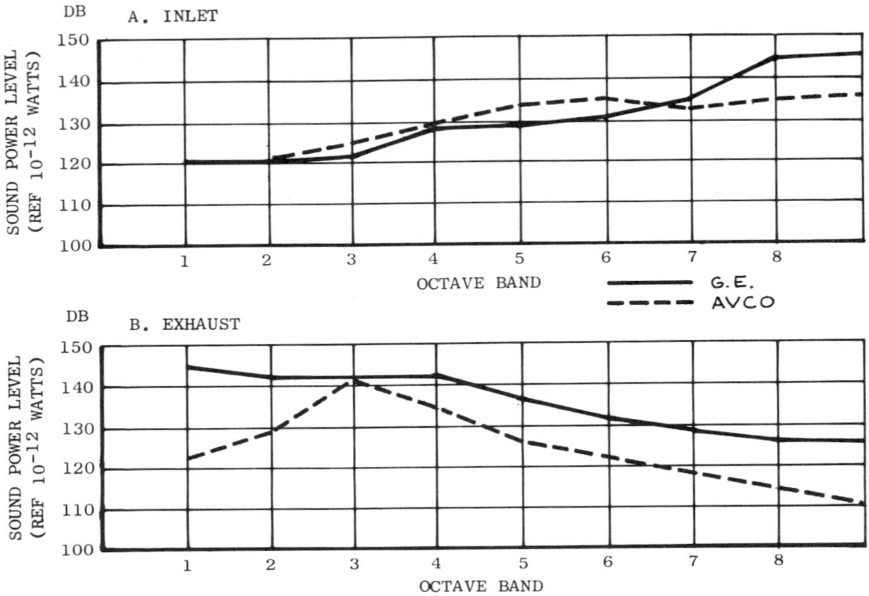

Figure 9.12. Typical sound-power measurements at inlet and exhaust for General Electric LM2500 and Avco TF35 engines (from manufacturers' published data sheets).

obstructing the line of sight via the ducting from engine to the exterior. Any change of direction in a duct, a feature that is often necessary anyhow, consequently contributes to attenuation. Usually, however, some measure of absorption by surfaces within the ducting is required. The basic relation for attenuation of sound by absorption within a straight duct is [*Industrial Noise Manual* (1958); Savine (1940)]

$$R = 12.6\alpha^{1/4}\frac{P}{A} \qquad \text{(dB/ft)} \tag{9.9}$$

where α is the absorption coefficient of the duct surfaces, P is the perimeter of the duct cross section, and A is the area of the duct cross section. The absorption coefficient is the fraction of the sound impinging that is absorbed, and as intuition based on everyday experience indicates, it is high for thick, fibrous materials and low for materials with hard, impervious surfaces. It is also a function of frequency, generally increasing in value for the good absorbers as the frequency increases, thereby making high-frequency sounds easier to attenuate through absorption than low-frequency sounds. Figure 9.13, as an example, shows the absorption coefficient for fiberglas blanket of

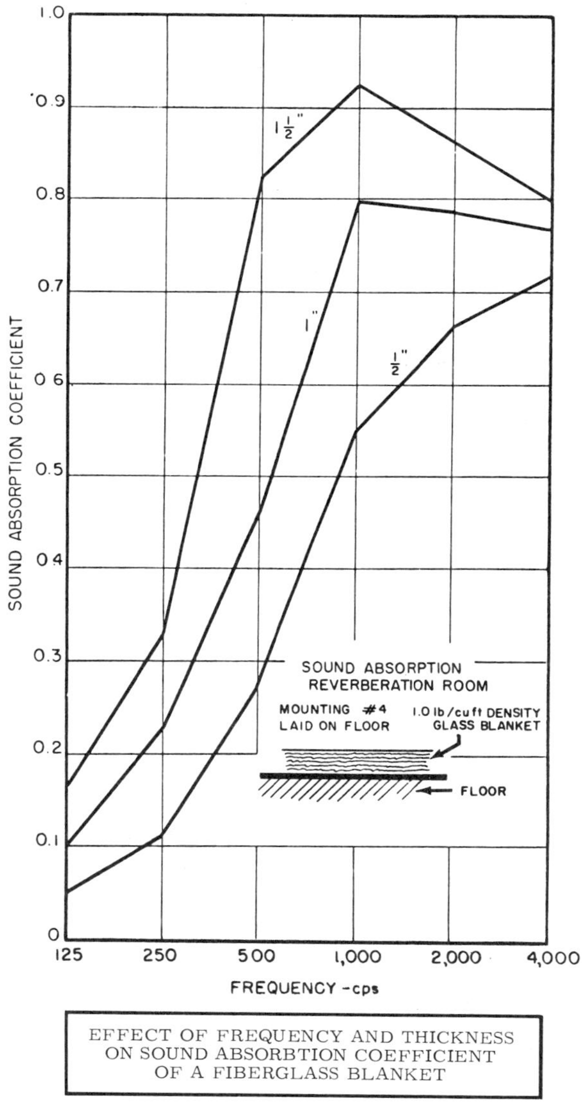

Figure 9.13. Typical sound-absorption coefficient data [*Installation Criteria for Gas Turbine Application in Naval Vessels* (1968)].

several thicknesses. In contrast, ordinary glass is one of the poorest absorbers, having a coefficient ranging of from 0.04 at low frequencies to 0.02 at high frequencies (above about 2000 Hz) [*Industrial Noise Manual* (1958)].

The absorption coefficient of steel is close to that of glass, so that the typical intake and exhaust systems must be lined with a better absorber for adequate attenuation. The *P/A* ratio of the duct also may not be adequate, that is, there may not be enough surface area to effect the desired absorption. Often, therefore, the surface is deliberately increased. Figure 9.14 is a sketch of the *Euroliner* ducts showing the added silencers in both intake and exhaust ducts. The intake silencer consists of a set of absorbent splitters in the flow path, the basic intent being to place more absorbing area in close contact with the sound path. Similar silencing is also seen in Figure 9.5.

The exhaust silencers in Figure 9.14 are of a different configuration, being two torpedo-shaped absorbing bodies placed in the gas stream. Note from Figure 9.12 that exhaust noises are of lower frequency than intake noises, and from Figure 9.13 that absorption is poor for low frequencies, although enhanced by added thickness. Attenuation is, therefore, more difficult in the exhaust duct, although there is the ameliorating fact that the low-frequency component of sound is less bothersome to the human ear. (Also, the high temperature of exhaust gas dictates that the outlet be more removed from human activity than the inlet.) The lower frequency means longer wavelength, so that the closely spaced absorbers common to intake passages are less effective than widely spaced absorbers. The exhaust silencers shown in Figure 9.14 are consistent with the points here mentioned; they provide additional, thick, widely spaced absorbing surface and compensate somewhat for poor absorption at low frequency by obstructing the line-of-sight along the duct axis.

Figure 9.15 depicts an exhaust silencing device developed by the Navy [*Installation Design Criteria for Gas Turbine Application in Naval Vessels* (1968)] for a particular application, in principle quite similar to the *Euroliner* silencer. Note how it obstructs the line-of-sight down the duct without reducing flow area.

Absorbing materials in the exhaust must, of course, be able to withstand the high temperature of the gas and must be protected against mechanical erosion. The latter point is also important—perhaps vital—in the inlet ducting, since any material being swept away passes into the engine. The fibrous blankets that may be used as the absorbing material are consequently protected by perforated sheet metal, with wire screening, fiberglas cloth, or plastic sheeting between the metal and the absorber. These coverings usually reduce the absorption coefficient only slightly [*Installation Design Criteria for Gas Turbine Application in Naval Vessels* (1968)].

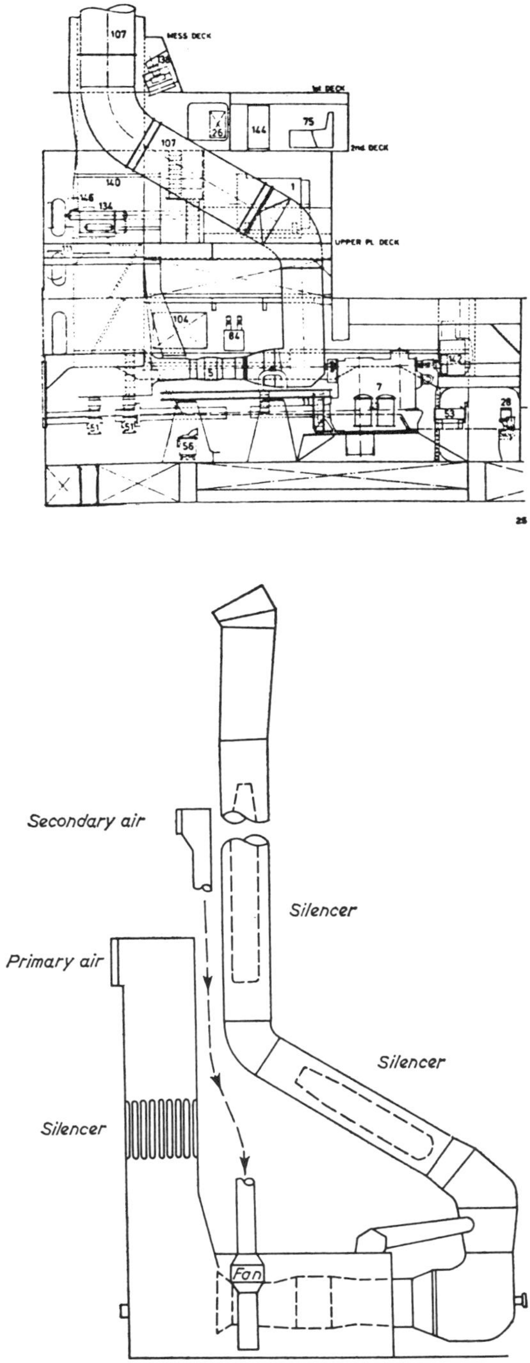

Figure 9.14. *Euroliner* exhaust ducts [O'Hare and Holbrun (1973)].

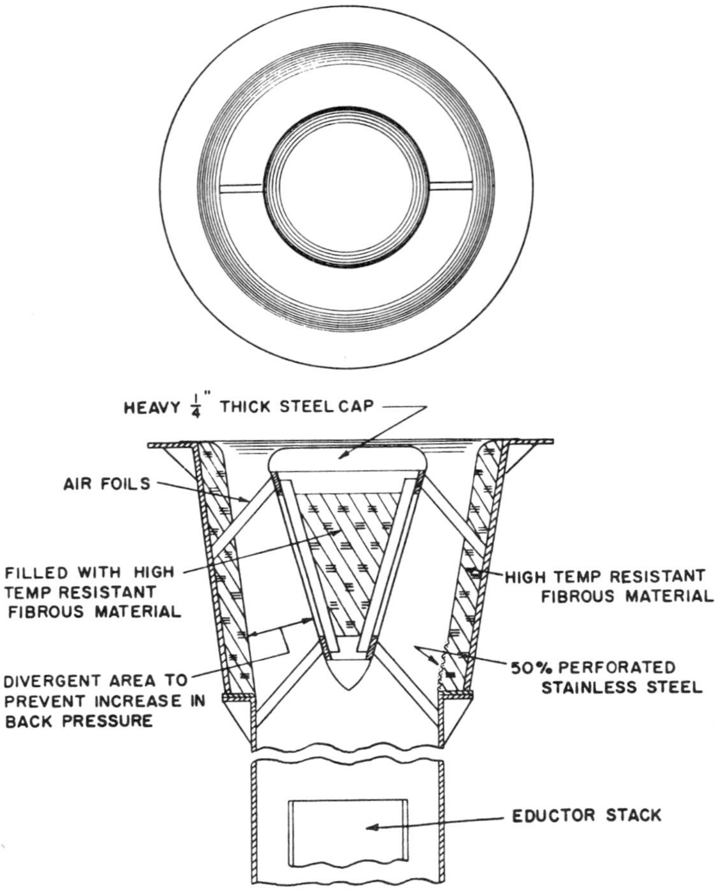

Figure 9.15. Exhaust silencer from small naval vessel [*Installation Design Criteria for Gas Turbine Application in Naval Vessels* (1968).]

A key bit of information that the duct designer must have in providing attenuation measures is the desired sound level at the duct exterior terminal. It is not practicable in a general treatment such as this to state what that level must be, for it depends on the distance and direction to the nearest human activity, and what that activity is to be (e.g., must conversation at normal voice levels and distances be possible?). The sound level perhaps may be stated in the specifications for the vessel. If the designer must provide his own specification, he can find guidance from acoustical handbooks such as the *Industrial Noise Manual* (1958) and the *Handbook of Noise Control* [Harris (1957)].

The sound attenuation treatment increases pressure drop, because of the obstructions in the duct, or simply because of reduced flow area. As with the inlet-salt attenuation measures, the designer must compromise needs for attenuation against the necessity for low head loss.

9.5. ADDITIONAL POINTS IN DUCT DESIGN

A. Inlet Problems

Ideally, the duct inlet should be located so that air can be inducted without hazard of contamination. In Section 9.3 it is noted that the pervasiveness of sea-salt aerosols makes this ideal an impossible one for sea-going vessels. On the other hand, salt ingestion can be greatly reduced by locating inlets to protect them from direct spray, a principle that seems to be universally recognized. Location of intakes in well-sheltered locations, especially the aft side of deckhouses, therefore, is a common practice.

Heat can also be a contaminant, since high-temperature ambient air reduces engine output. Inlets are consequently to be placed in locations remote from any heat source that could warm the air as it enters (or at any point along the duct, for that matter). A likely culprit is the engine exhaust, which might be reingested by the intake. Generally, the problem is avoided by placing the exhaust outlet well above the intake (see Figure 9.14, for example), although this feature may be difficult in a small vessel. Under normal circumstances, intakes forward of exhaust will be safe from reingestion, but operation astern or ahead with a following relative wind are abnormal circumstances of almost certain occurrence. Wind-tunnel tests to determine exhaust behavior may be a necessary design step when there is doubt about an acceptable location relative to the inlet.

Because ice is likely to form on inlet screens and demisters during the appropriate weather conditions, deliberate heating at the inlet is sometimes essential. Since the ambient air is obviously quite cold when icing occurs, this heating is not a negative factor in engine performance, though the heat used is, to be sure, an energy loss from the propulsion plant. A suitable source of heat is air bled from the propulsion engine at its compressor discharge.

Water and salt removed from the air by demisters must in turn be removed from the vicinity, lest they be reentrained into the air stream. Salt caking the vanes of an inertial separator may, for example, break off and be swept into the engine. Washing at frequent intervals is likely to be necessary for both inertial and mesh separators. Removed sea water draining away may be reentrained if it is exposed to the air stream; such a problem was reported in early *Euroliner* experience [O'Hare and Holburn (1973)] and necessitated alteration of inertial separator drains to remove water immediately, rather than allowing it to cascade from element to element of the separator.

Oily vapors, as might emanate from a fuel-tank vent, are also a possible contaminant, since oil in the air coats compressor blades, and subsequently acts as a binder for dirt. A problem of this nature was also reported from the *Euroliner* [O'Hare and Holburn (1973)], caused by a tank vent terminating on deck near the engine air inlet.

And it is perhaps a trite point, but let's also note that *any* foreign object ingested by the engine is likely to be deleterious. Screens at inlet are therefore always employed. Less obvious but also essential is care in avoiding the generation of foreign objects within the duct: a leaking pipe joint, a shred of accoustical lining, a nut worked off a stud by vibration—these are common hazards that come readily to mind.

B. Individual Ducts

When more than one propulsion engine is used, as in the case of twin-screw, or cruising and boost engines on the same shaft, a designer may expect some saving in duct volume and weight by serving two (or more) engines with a single duct. Although this practice is reasonable at first glance, it has such a potential for damage to an engine that the preponderance of advice is strongly in favor of individual ducts.

The hazard exists when an engine sharing a duct is not operating. If the inlet and exhaust ducts are both common, a backflow of exhaust gas from the operating engine passes through the idle engine, since exhaust pressure is always higher than inlet pressure. A likely consequence is destruction by overheating of the lube oil in the idle bearings, since this oil does not circulate through its cooler as in normal operation. If only the inlet duct is common, an operating engine draws a backflow of ambient air through the idle engine, via its exhaust duct. The likely consequence is salt deposition, since exhaust ducts are not protected against this phenomenon.

A possible solution that would allow common ducts is the use of dampers to isolate idle engines, but this, too, has strongly unfavorable aspects. For one thing, dampers are difficult to build airtight and difficult to maintain in their original degree of tightness. The most probable effect of a damper, therefore, is reduction of backflows, rather than the complete elimination that would be necessary to avoid consequences mentioned.

In sum, the elements of good design heavily favor the use of individual ducts.

C. Use of Machinery Rooms As Inlet Plenums

A large plenum for engine inlet air is beneficial, since it allows the low approach velocities that encourage the settling of salt and other contaminants. The engine compartment itself is a natural choice for this function, especially since it has been common practice for diesel and spark ignition engines to take suction directly from the space in which they are located. There is a negative factor, however, in gas-turbine practice: oily vapors are likely to be present in an engine room atmosphere, and these coat its blades on being drawn into the gas-turbine compressor. Dust and other fine matter in the air then adhere to the oil, and eventually enough fouling can be accumulated to degrade engine performance significantly. The problem apparently can be minimized if all oil sources, such as vents from tanks, gear cases, and crankcases, are piped to the exterior.

Use of engine rooms as plenums is most often seen in smaller craft, since they may badly need the maximum plenum effect to compensate for the lack of high and sheltered locations for duct intakes.

D. Engine Removal Via the Inlet Duct

The gasifier section of an aircraft-derivative gas turbine, being light and compact, is usually removed from the vessel and replaced when maintenance is necessary. In some instances, the inlet duct provides the most convenient route by which this may be done, if the duct is big enough and the necessary guide rails and lifting gear are furnished. Figure 9.16 illustrates this feature of inlet duct design by views of duct arrangements aboard two naval vessels.

E. Thermal Problems of the Exhaust

The temperature of a gas-turbine exhaust may be as high as 1200 F at rated power; this is quite high enough to lead to problems of several kinds. High temperature is, of course, a fire hazard and a safety hazard to personnel, hot exhaust in the vicinity of antennas can interfere with electromagnetic communication through ionization of the air, and heating of duct structure causes significant thermal expansion.

The most obvious and direct response is heavy insulation of the duct within the vessel and use of expansion joints adequate to absorb the expansion without putting excessive thrusts or moments on either engine or structure. In these things the gas turbine is much like all other marine engines, since they

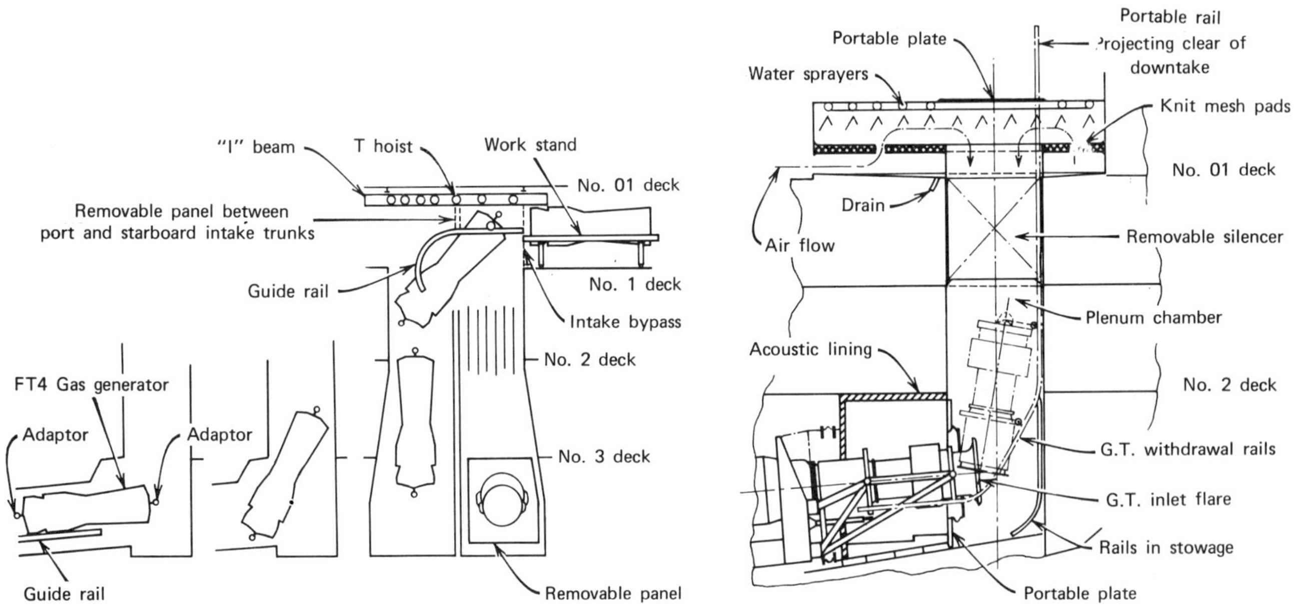

Figure 9.16. Use of inlet ducts for engine removal and installation: Canadian DDH-280 destroyer on left [Sachs (1969)], British frigate on right [Good (1967)].

must also cope with the hot exhaust problem. The most noticeable difference in the practice of other engine types is in the use of wet exhaust by small-craft diesel and spark-ignition engines. (In the common wet exhaust, part of engine cooling water is injected into the exhaust line, which consequently must slope downward toward the exterior to preclude flow toward the engine.) The large volume of gas-turbine exhaust makes the wet-exhaust scheme impracticable as a means of keeping an entire duct line cool. It should be noted, though, that water sprays have been used in gas-turbine exhausts to produce a cooler product leaving the vessel. This technique may be necessary in small vessels with side or transom exhaust that would otherwise be a grave hazard to piers and other vessels close alongside.

A more commonly use method of cooling marine gas-turbine exhaust is the *eductor* stack, widely adopted for smaller craft and also used on some larger vessels. An exhaust system using the eductor principle is shown by Figure 9.17. Note that the essence of the system is in two elements: an inner stack through which flows the hot gas and an outer stack into which the primary flow induces a secondary flow of cool air. The exhaust is, therefore, isolated from outer-duct structure by an annulus of the cool air. Although this annulus may be disrupted by turbulent mixing, the mixing also cools the hot gas. In the figure the cooling air is being drawn from the outside; the same technique is also used in some installations to exhaust air from the engine compartment. An example of the latter scheme is shown in Figure 9.18.

Figure 9.19 depicts an eductor installation on a larger vessel, the U.S. Navy DD963 destroyer class. Here the problem is mainly cooling the gas at exit to protect nearby antennas, and not generally cooling or protecting the exhaust system itself.

As mentioned earlier, the thermal-expansion phenomenon dictates the use of expansion joints in exhaust ducting. In some installations the need for these devices is bypassed by use of the eductor. Such is the case in the arrangement shown in Figure 9.17. Note there that the inner exhaust stack is supported by the engine and the outer stack by the vessel structure. The air gap is, in effect, the expansion joint.

A designer may need to estimate the flow of cooling air produced by an eductor, or may need to design to obtain a particular flow. Unfortunately, this topic has not been adequately covered in the published literature. Aside from model tests of a proposed arrangement, the most directly applicable source of design information appears to be in research reports contracted for by the U.S. Navy. Several are referenced in Section 9.8 of *Installation Design Criteria for Gas Turbine Application in Naval Vessels* (1968).

Mention should be made of the exhaust from regenerative turbine engines. The regenerative heat exchanger reduces both the outlet temperature and the sound level, so that problems from both sources are of lesser magnitude than

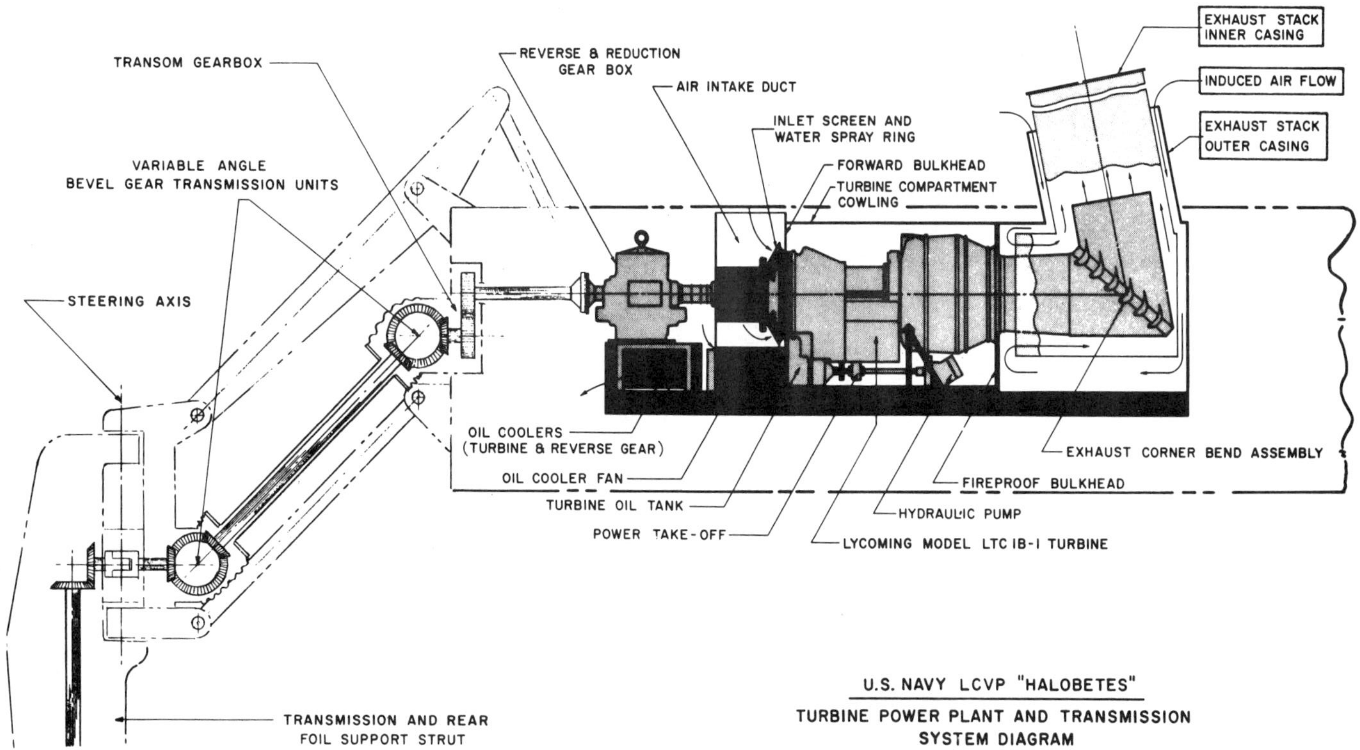

Figure 9.17. Exhaust duct illustrating the eductor concept [Austin (1959)].

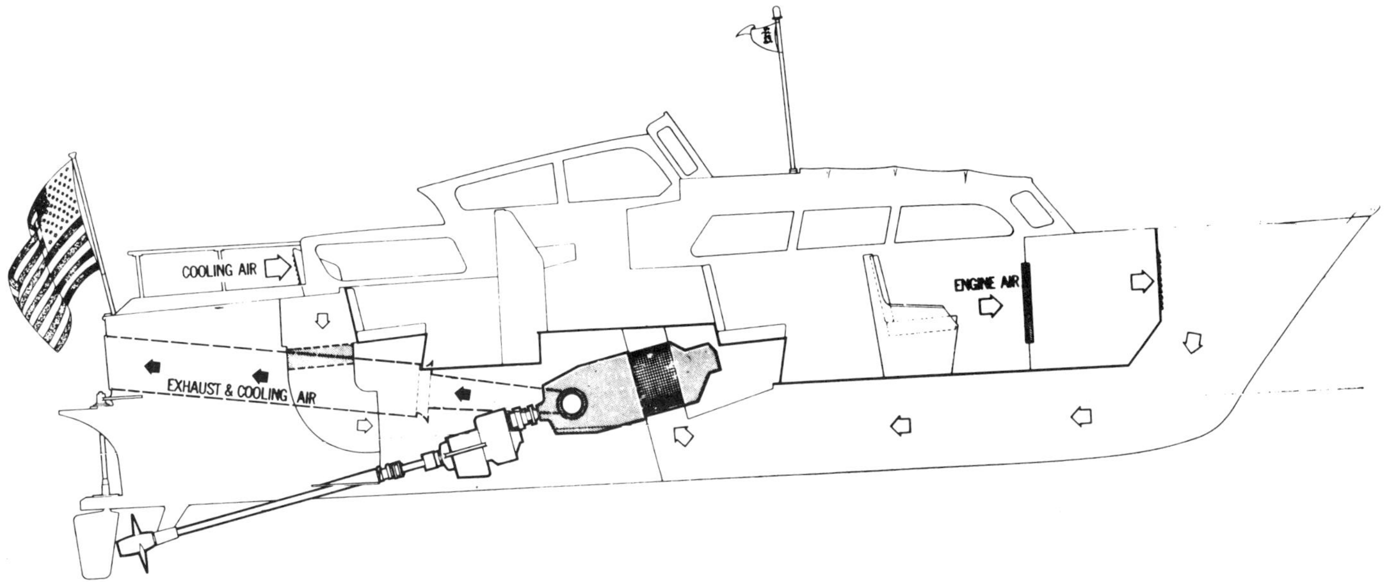

Figure 9.18. Experimental gas-turbine motor boat [Hardy (1966)].

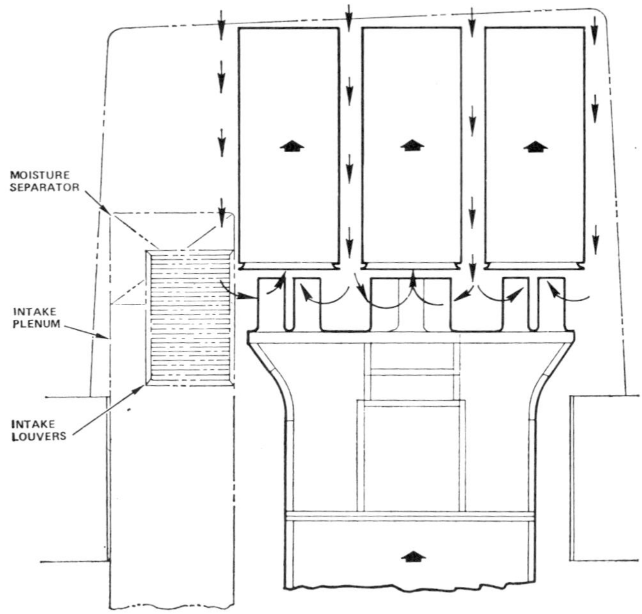

Figure 9.19. Exhaust eductors, U.S. Navy DD963 class destroyer.

with the simple-cycle engine. In one contemporary example, powered by General Electric regenerative heavy-duty engines (Chevron tankers), a waste-heat boiler in the exhaust line further attenuates temperature and sound. The ultimate exhaust line is much the same as the uptake from any modern steam boiler, though the magnitude of flow per horsepower is naturally much greater.

9.6. INLET AND EXHAUST FOR SMALL CRAFT

Inlet and exhaust problems for small craft may be both lesser and greater than for large vessels. Since the engine is closer to the exterior, the duct paths are usually short, and in consequence, pressure-drop problems are likely to be minimized. On the other hand, arrangement of the vessel may deny the designer the use of the direct path upward to the atmosphere, forcing him to use an excessive amount of internal volume for intake and exhaust space. Such

Figure 9.20. Pacemaker gas-turbine boat, showing inlet louvers (Pacemaker Corp.).

Side and top views of turbine installation on Pacemaker show porting for intake and exhausts, positioning of turbine for direct drive.

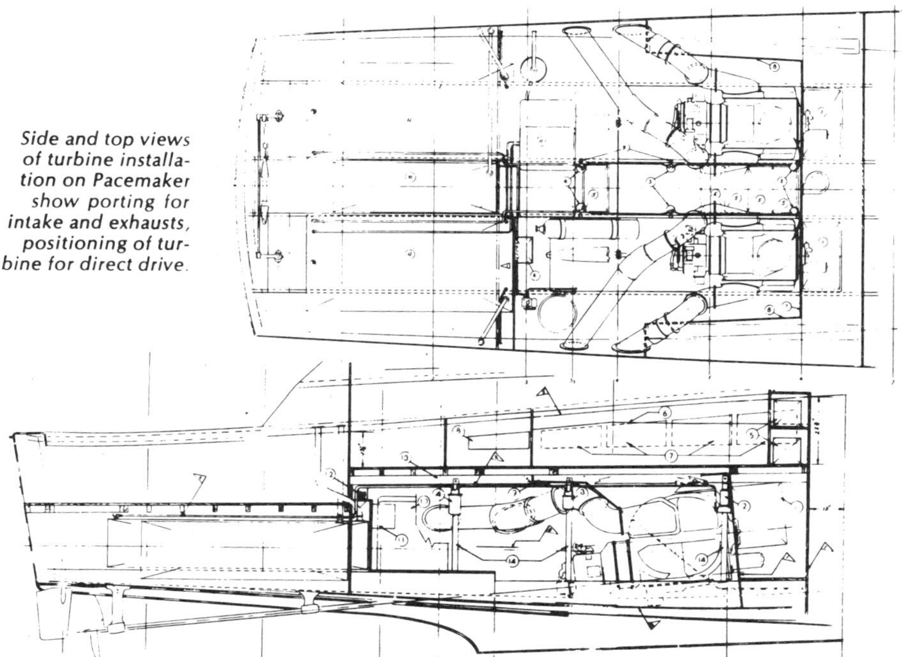

Figure 9.21. Machinery arrangement of boat shown in Figure 9.20, illustrating inlet and exhaust configurations (Pacemaker Corp.).

appears to be the case in several pleasure-boat applications that have been attempted. And the inevitable closeness of the intake to the water surface seems likely to aggravate the salt-removal problem, requiring even greater volume for air stilling.

Figure 9.18 illustrates how inlet and exhaust were fitted into a traditional motor cruiser arrangement and shows the rather large volume devoted to these functions in a boat of this type. Figures 9.20 and 9.21 together show a later attempt in a production boat to reduce the volume used. The first is a picture of the boat at high speed. The intake louvers show clearly along the topsides amidships; proximity to heavy spray is also evident. Figure 9.21 gives two views of the machinery arrangement of this vessel, powered by twin Ford engines. The louvers are seen to lead into a plenum space forward of the engines. Engine-inlet air boxes (symbol 1) are located in the plenum so that the engine can take suction directly from them through the bulkhead. These boxes contain inertial demister elements and dirt filters, and also act as inlet silencers. Two exhausts per engine lead as directly as possible through the sides of the boat just above the waterline (outlets are hidden by spray in Figure 9.20). Cooling by injection of water into the exhaust is used, but no silencing is possible in such short duct lengths. The engines, however, are of the regenerative type, so that the exhaust is both silenced and cooled by the engine itself. It is likely that a through-hull exhaust such as this would not be feasible with a simple-cycle engine unless some cooling feature, such as a water spray, were included in the design.

9.7. REFERENCES

Austin, K. A. (1959), "Gas Turbine Propulsion for Hydrofoil Craft," American Society of Mechanical Engineers, Paper 59-A-324.

Bogus, B. J., and Weinert, E. P. (1972), "Generalized Outlook on Marine Gas Turbine Hot Section Deterioration," American Society of Mechanical Engineering, Paper 72-GT-95.

Flows of Fluids Through Valves, Fittings, and Pipe (1965), Crane Company Technical Report 410.

Good, E. B. (1967), "Gas Turbine Installation Design for Naval Ships," *Journal of Engineering for Power, Transactions of the American Society of Mechanical Engineers,* Series A, **89,** pages 150–164.

Hardy, G. (1966), "The Development of a Marine Version of the United Aircraft of Canada PT6 Engine," American Society of Mechanical Engineering, Paper 66-GT/M-31.

Harris, C. M. (1957), *Handbook of Noise Control,* McGraw-Hill.

Heating, Ventilating, and Air Conditioning Guide (1974), American Society of Heating, Refrigerating, and Air Conditioning Engineers, (published annually).

Industrial Noise Manual (1958), American Industrial Hygiene Association.

Installation Design Criteria for Gas Turbine Application in Naval Vessels (1968), NAVSHIPS 0941-038-7010 United States Navy Technical Manual.

Lindsey, W. H. (1968), "The Marine Gas Turbine," Society of Naval Architects and Marine Engineers, Proceedings of Diamond Jubilee Meeting.

Mathews, D. B. M., and Hodgson, J. R. F. (1971), "Sea Trials of Gas Turbine Machinery in HMS *Exmouth*," *Proceedings, Institute of Marine Engineers International Conference.*

Miller, C. L., (1969), unpublished notes on marine gas turbines, University of Michigan, short course on "Small Craft Engineering."

Moody, L. F., (1944), "Friction Factors for Pipe Flow," *Transactions, American Society of Mechanical Engineers,* **66,** pages 671–696.

Morgan, S. G., Smith, A. J. R., and Lamport, A. W. (1971), "Gas Turbines in the Royal Navy 1965–1969," *Naval Engineers Journal,* **83,** pages 35–58.

O'Hare, T. L. R., and Holburn, J. G. (1973), "Operating Experience with Gas Turbine Container and Ships," *Transactions, Institute of Marine Engineers,* Vol. 85, Series A, Part I, .

Pressure Loss of Ventilation Fittings (1950), U.S. Navy Design Data Sheet, Section DDS3801-2.

Rains, D. A., and d'Arcy, R. J. (1972), "Considerations in the DD963 Propulsion System Design," *Naval Engineers Journal,* **84,** pages 65–77.

Reid, C. (1969), "The Response of Axial Flow Compressors to Intake Flow Distortion," American Society of Mechanical Engineers, Paper 69-GT-29.

Sachs, R. M. (1969), "Description of Propulsion Systems for DDH-280 Class Gas Turbine Destroyers," American Society of Mechanical Engineers, Paper 69-GT-26.

Savine, H. J. (1940), "The Absorption of Noise in Ventilating Ducts," *Journal of the Acoustical Society of America,* **12,** pages 53–69.

Sawyer, W. T., Kelnhofer, W. J., and Smith, R. A. (1966), "Design Considerations for Marine Gas Turbine Ducting," American Society of Mechanical Engineers, Paper 66-GT/M-27.

Shapiro, A. H. (1953), *The Dynamics and Thermodynamics of Compressible Flow,* Ronald Press.

Weinert, E. P., and Carlton, G. A. (1965), "Salt-Water Problems in Marine Gas Turbines," American Society of Mechanical Engineers, Paper 65-GTP-18.

Chapter TEN

MACHINERY ARRANGEMENT AND INSTALLATION

10.1. INTRODUCTION

This chapter discusses the physical interfaces of a gas-turbine engine with the vessel it propels. There are several aspects of this, perhaps only loosely related. Most general is the question of *location* within the vessel, or its arrangement with respect to its propulsor, to other machinery items, and to the spaces for payload, crew accommodations, stores, fuel, ballast, weapons, communications, and so forth. The type of propulsor and the type of transmission from engine to propulsor have a strong influence on location; for example, electric transmission usually allows a much greater latitude in engine placement than does the usual mechanical transmission. but another aspect of the transmission is the *alignment* of the engine with whatever device it directly drives. This word has the general implication that connected devices must be arranged with their shafts concentric (usually), but also implies the fine points of obtaining concentricity, or whatever modification is appropriate to minimize the stresses that one unit imposes on the other.

Related to the alignment question is that of *mounting* of the engine. The engine is a weight that must be supported by vessel structure, but to the obvious weight must be added the contributions of thrust and torque, live loads (e.g., gyroscopic interaction between the rotating turbine and compressor and the pitching hull). Above all, the inevitable thermal expansion must be allowed for, both from the standpoint of its influence on engine-load alignment and from the standpoint of resulting loads on the engine structure.

The gas turbine is frequently used as a propulsion engine in combination with a diesel, or with another gas turbine. Such combinations always require means of connecting and disconnecting alternative engines and their load, so that *clutches* are a prominent consideration in the design of the propulsion train of many gas-turbine installations.

The engine is a source of high temperature and noise that usually demand ameliorating measures in design. These factors bear on the choice of location and to some extent on engine mounting, but also have additional aspects. A typical response to noise and high temperature is to place the engine within an individual enclosure that isolates these undesired outputs from the balance of the machinery space.

Whether trapped within an enclosure or not, the stray heat from the engine must be removed from the vessel, usually by ventilation. If there is an enclosure, especially, the ventilation is likely to be integrated with the combustion-air-inlet and exhaust systems.

The inlet and exhaust systems should also be mentioned as forming an important part of the engine-vessel interfaces. They are, of course, treated as a separate topic in Chapter 9.

10.2. ENGINE ARRANGEMENT

Arrangement implies the location of the engine within the vessel, its orientation, and its positioning with respect to other components.

The location of propulsion engines of all types is affected by several general considerations. One is alignment with the propulsion shafting, avoidable only if a nonmechanical (most always electrical) drive is used. For example, a directly connected engine—typified in modern practice only by low-speed diesel engines—must be located with its axis coincident with the propulsion-shaft centerline, and there is usually only minor latitude in where that shaft can be run. With a reduction gear, the engine is offset from the shaft by the center-to-center dimension of the gear set; often there is some choice as to how the offset may be oriented, that is, either vertically or horizontally. In smaller vessels, where outdrives and V-drives are fairly common practice, the orientation of engines with respect to propulsion shafts takes on much wider scope. And if electric drive is used, almost complete freedom from the dictates of the shaft is available, though the propulsion motor must be aligned to the shaft. Occasionally, this freedom for the engine is obtained with mechanical drive through the agency of angled gear boxes. Such has been the necessity on several hydrofoil boats in which the engine is mounted within the hull, while the propeller is mounted on the foils. Figure 10.1 gives an example showing the drive line used by the Canadian hydrofoil vessel, *Bras d'Or* [Fischer and Callow (1970)].

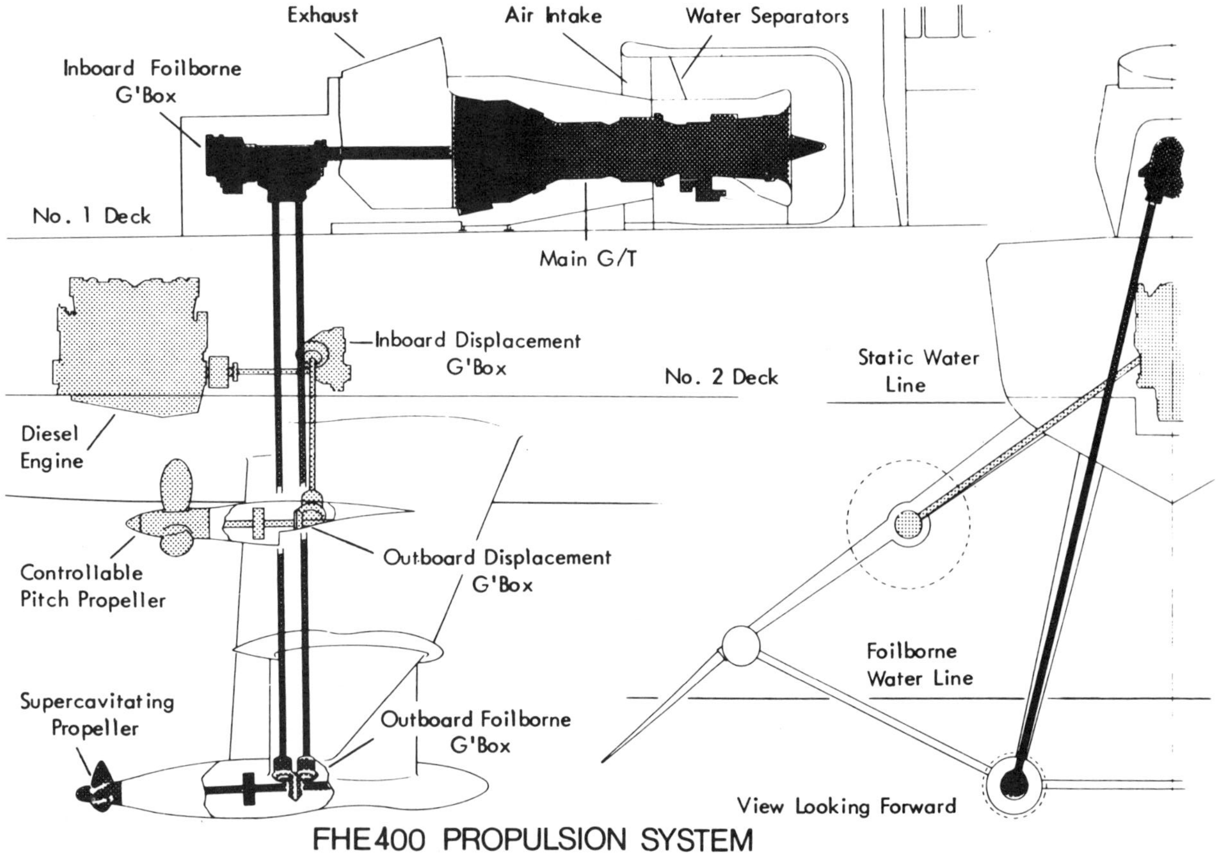

Figure 10.1. Shafting arrangement, H.M.C.S. *Bras d'Or* [Fischer and Callow (1970)].

The fore-and-aft location of propulsion machinery is less affected by the propulsion shaft. The influence may be significant, but indirect. For instance, the fact that shafting is shorter, hence lighter, and uses less internal volume when the engine is close to the propulsor, is a strong incentive for placing the machinery in the stern. The overall arrangement of the vessel, however, also has a major influence on fore-and-aft location. It is usually desirable to integrate accommodation space with machinery space, and the needs of the payload space often have the greatest influence of all. General rules are difficult to state, however, because of the great variations among types of vessels and among individual designs.

Certain prominent problems of naval architecture may override other considerations. The engine does represent a concentrated weight, a weight of greater density than the general internal volume of the hull. Its location has a significant influence on trim and on transverse stability. For this reason, propulsion machinery has traditionally been located low and near amidships, although as machinery of all types has become lighter, permissable locations have tended to diffuse toward the ends and upward.

The gas turbine has prominent characteristics that modify some of the general considerations. Its lightness gives it greater freedom than its competitors from stability strictures, so that location above the main deck is feasible for both large and small craft, even for the heavier industrial-type engine, provided the previously noted demands on the shaft are satisfied. Figures 10.2 and 10.3 show two tanker machinery arrangements with heavy-duty industrial engines mounted on the main deck, and driving through electrical transmissions. The first is a General Electric design and represents the arrangement used for the Chevron Oil Company tankers built in the early 1970s. In both cases, however, the weight of the propulsion motor and shaft doubtless causes the center of gravity of the overall propulsion machine (engine-generator-motor-shaft) to be below the main deck.

The requirements of inlet and exhaust ducts may have a strong influence on gas-turbine arrangement within the hull, because of their size and because of the need for the intake to be in a sheltered location. The size of the ducting is apparent in several of the figures of Chapter 9. A high location for the engine has an advantage in that it shortens duct length and thereby reduces weight and volume devoted to air and exhaust handling. The very definite need for a sheltered intake is probably a stronger (though not necessarily conflicting) requirement and usually dictates the arrangement of an engine with respect to deckhouses. Figure 10.4, showing the inboard profile of the *Euroliner,* is an example. Note the location of the compressor end of the engine directly below the aft end of the deck house, allowing the shortest run of air duct. The engine is in consequence oriented with its output shaft forward; the reduction gear is forward of the engine with the propulsion shafting running aft at a lower level. The *Callaghan* propulsion machinery is oriented in the same way.

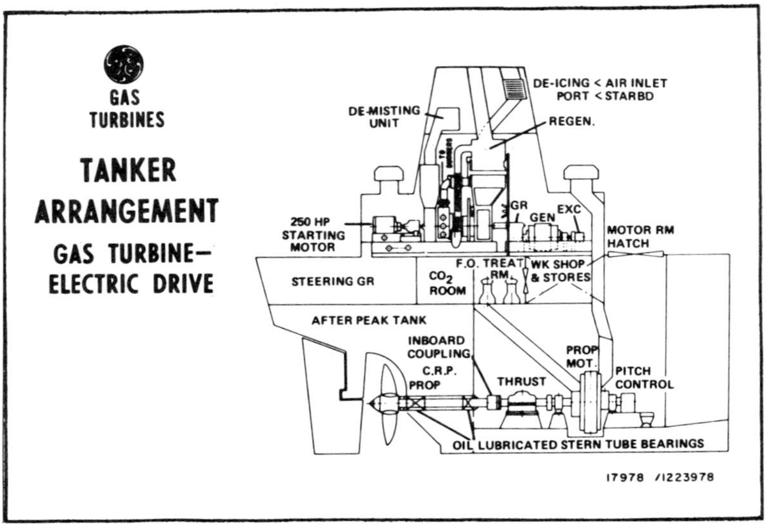

Figure 10.2. Machinery arrangement, General Electric heavy-duty industrial engine with electric drive.

Figures 10.5, 10.6, 10.7, and 10.8 show several examples of small vessels powered by aircraft-derivative gas turbines located in the stern. The first is a British patrol boat powered by the Bristol Proteus engine driving a propeller via a V-drive. Note that the output shaft of this engine is at the compressor (inlet) end, fitting in conveniently with the V-drive and transom-exhaust arrangement. The other three are of vessels driven by Avco Lycoming engines; two use waterjet propulsors, while the other powers propellers via right-angle strut drives. These engines also have output shafts at the compressor end, reflecting their origin as turboprop aircraft engines (likewise with the Proteus). In all four examples the engine compartment serves as the inlet plenum. In Figure 10.7 and 10.8, the engine-inlet filter boxes are evident as rectangular shapes on either side of the aft end of each engine. Exhaust expansion joints are quite evident, and in Figure 10.8 an array of mesh-type demisters is indicated in the aft-facing air opening into the engine compartment.

All types of propulsion machinery are subject to certain detailed rules of arrangement, many of them simple common-sense precepts. As obvious examples, adequate access to operating and maintenance points is necessary, and interconnecting piping and wiring should be as short as practicable. More special rules may apply to particular types or to particular items of machinery. This seems especially to be the case with steam propulsion. There, for example, gravitational head is usually enlisted to suppress cavitation in the

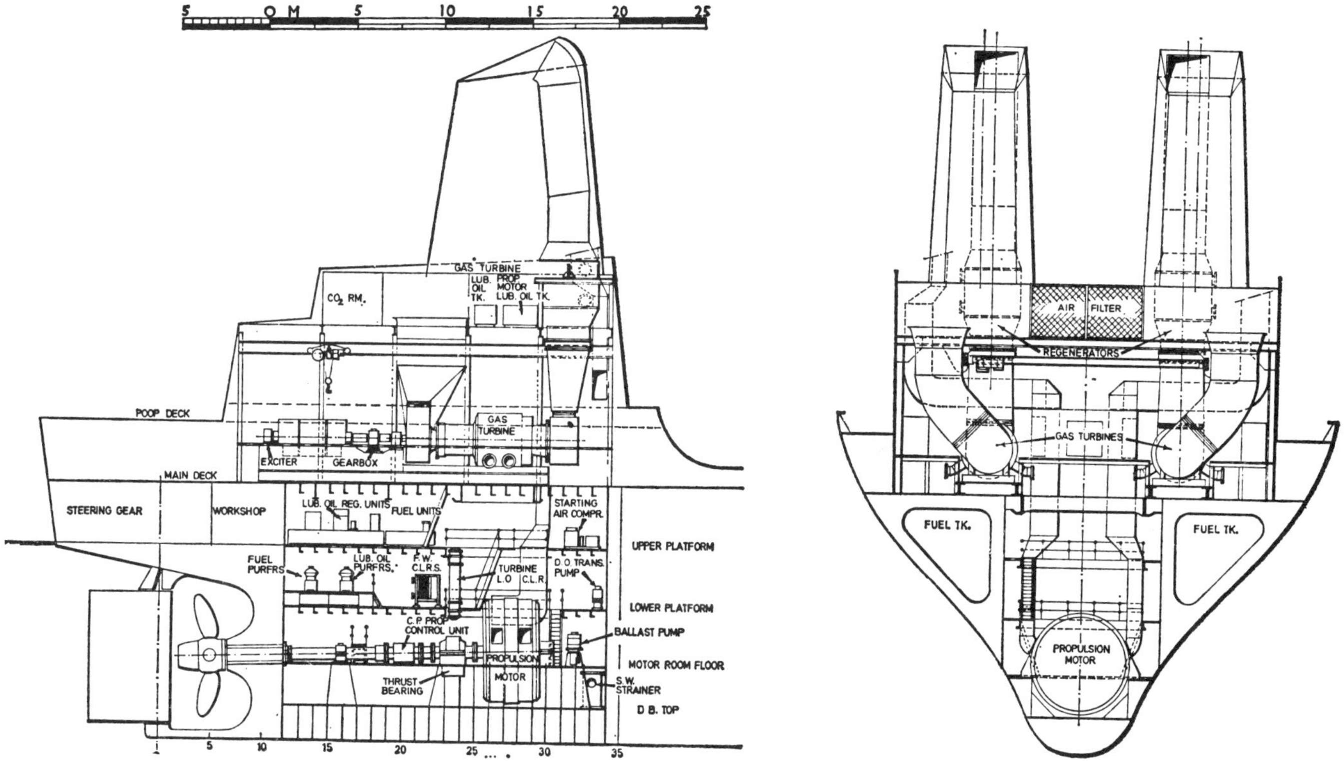

Figure 10.3. Gas-turbine machinery arrangement with electric drive as proposed by Sulzer Brothers [*The Motorship* (1973)].

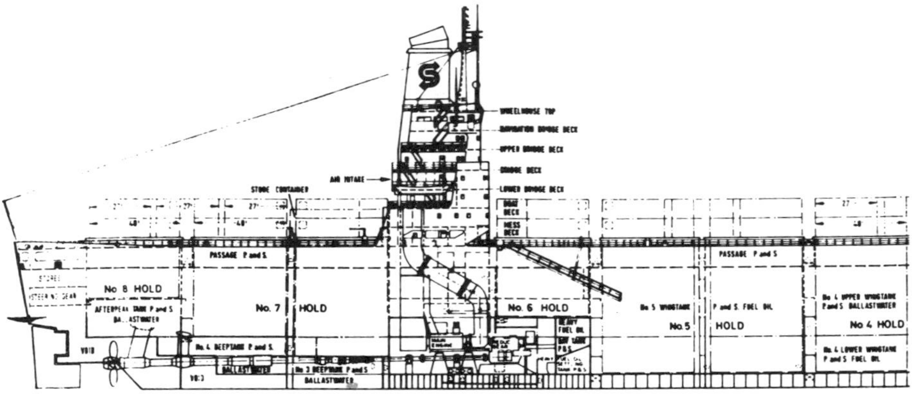

Above: General arrangement of G.T.S. Euroliner

G.T.S. EUROLINER

world's first gas turbine container ship

Principal particulars:

Length, o.a.	798 ft 6 in
Length, b.p.	734 ft 11 in
Breadth, moulded	98 ft 5 in
Depth to upper deck	63 ft 0 in
Depth to freeboard deck	54 ft 2 in
Draught, loaded	32 ft 6 in
Deadweight	23 100 tons
Service speed	25 knots

Classification ABS ✠ AI(E) ✠ AMS ✠ ACCU

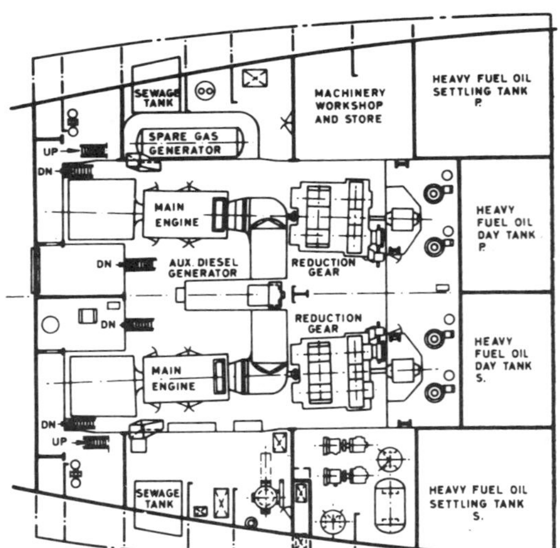

Lower platform deck/III deck

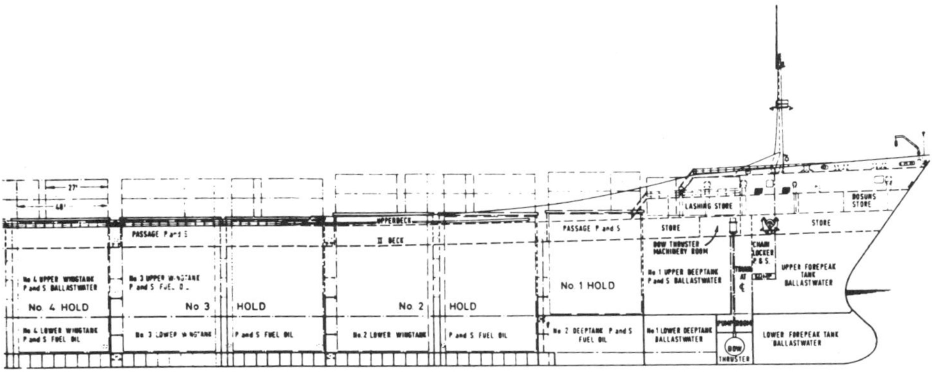

Right: Port elevation

1) Hot water boiler
5) Main engine
7) Reduction gear
11) Shaft brake
26) Feed water ballast tank
28) Main S.W. circulating pump
43) F.O. filter unit for main engine
51) Main gear L.O. pump
53) L.O. coalescer system
56) Main gear L.O. suction filter
75) Main engine control console
84) Pump priming system
107) Main engine exhaust gas silencer
111) C.p.p. hydraulic oil gravity tank
134) Air conditioning system
138) Engine room exhaust fan
140) Main engine inlet silencer
142) Shaft generator
144) Main switchboard
146) Rotary condenser

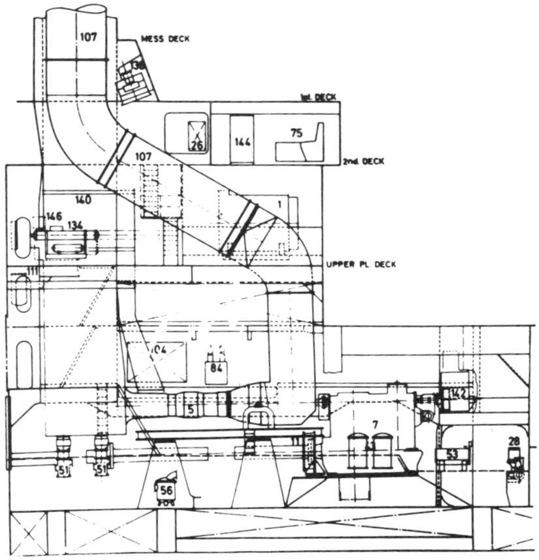

Figure 10.4. *Euroliner* machinery arrangement [*The Motorship* (1970)].

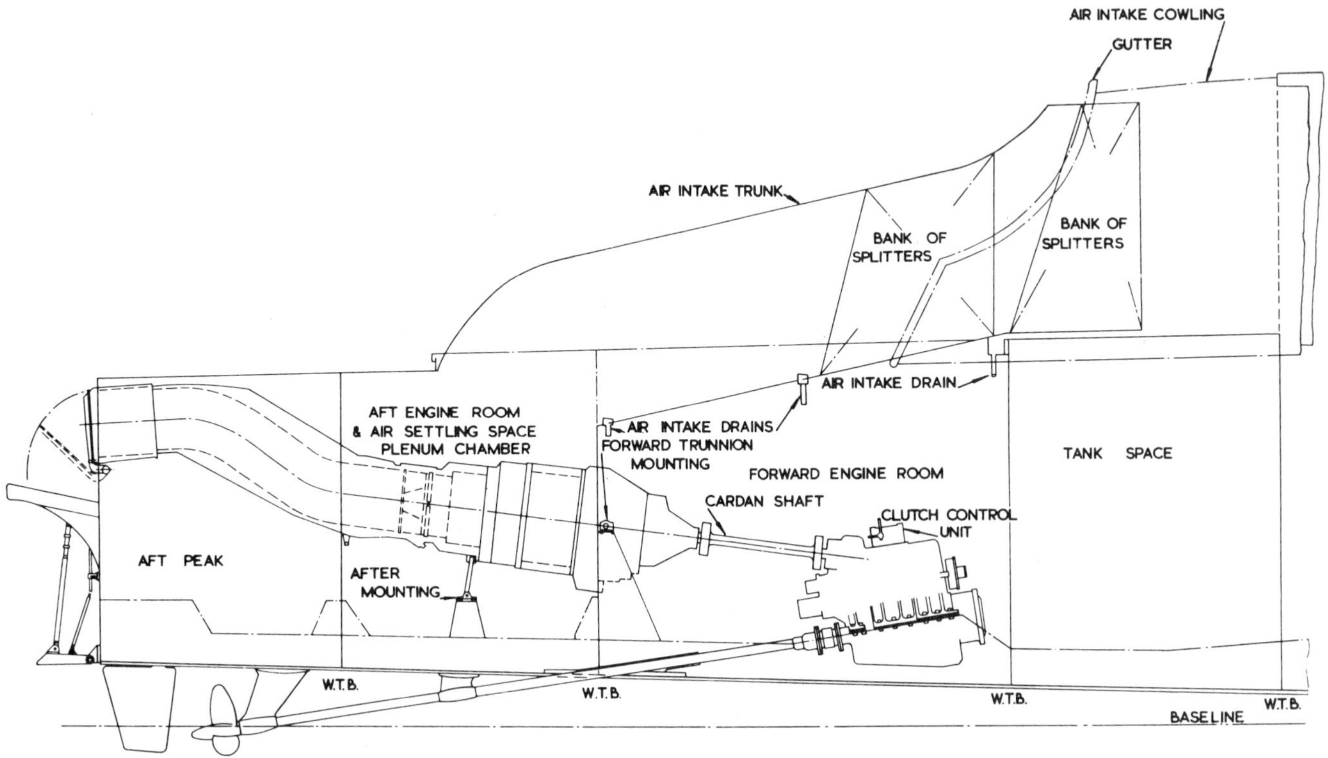

Figure 10.5. Machinery arrangement, British patrol boat with Proteus engine [Trewby (1963)].

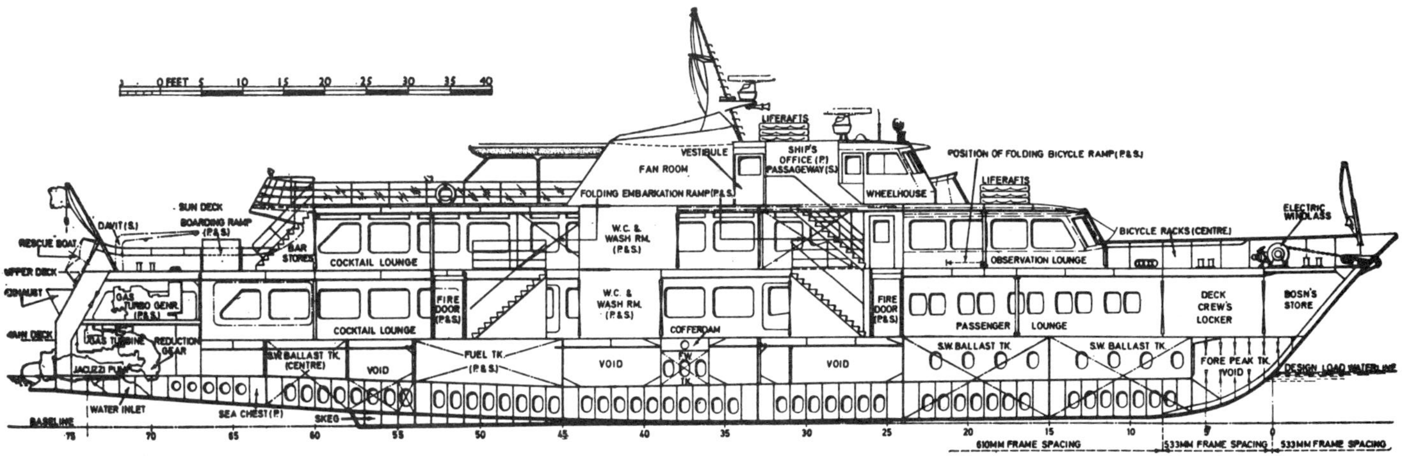

Figure 10.6. Inboard profile, showing machinery, San Francisco Bay Ferry with Avco-Lycoming engines, Jacuzzi jet drives [*The Motorship* (1973)].

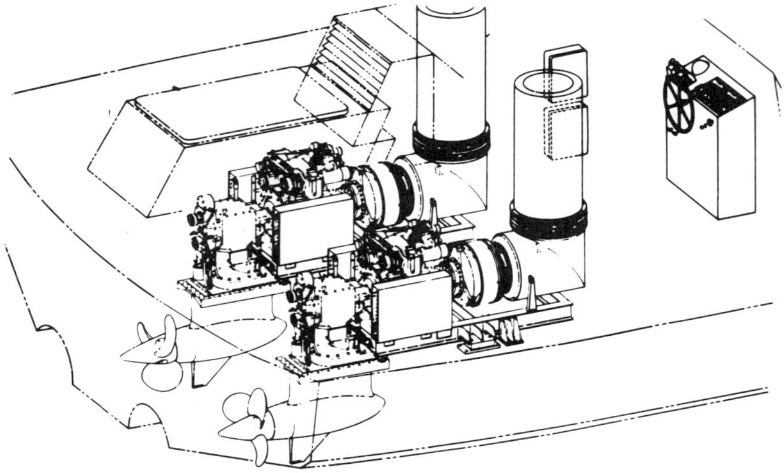

Figure 10.7. Machinery arrangement, river patrol boat, Avco-Lycoming engines with strut drive [Lauriat (1972)].

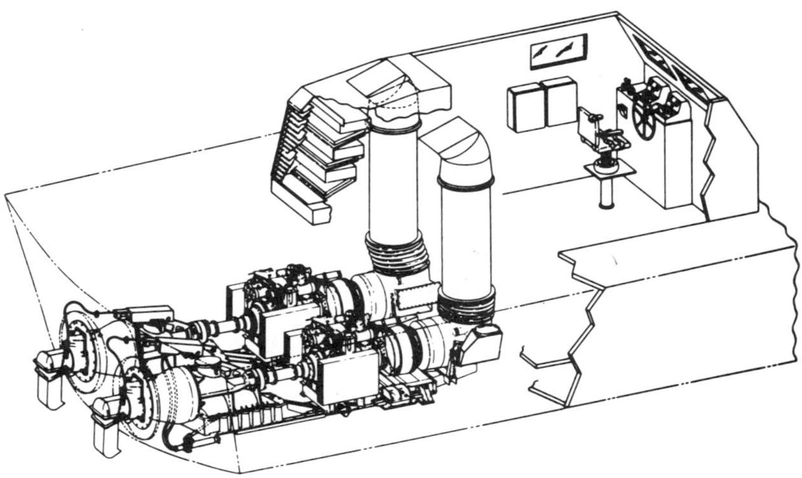

Figure 10.8. Machinery arrangement, river patrol boat, Avco-Lycoming engines with jet drives [Lauriat (1972)].

condensate and feed systems, so that definite elevation differences must be maintained among condensers, pumps, and feedwater heaters. No analogous stipulations are to be found for the gas turbine; auxiliary component locations with respect to one another, and to each other, are subject only to the generalities suggested above. Exceptions nonetheless may be necessary for particular engines. For example, engines that are derived from aircraft service often have accessories mounted low on the engine, since this is where servicing would be most convenient for an aircraft mechanic. In such cases, the marine designer needs to mount the engine so that service can be performed in his environment also. And the arrangements shown in Figures 10.7 and 10.8 suggest another consideration that may face the particular engine. Note the inlet filter boxes mentioned in the preceding paragraph; where pairs of engines are close together in a side-by-side configuration, the inlets may in effect compete for air, with consequent distortions in flow pattern to the inlets. Such engines, therefore, must be placed an adequate distance apart, or baffles must be provided to isolate the facing inlets.

Engines of all types are subject to restrictions on attitude, mostly to prevent malfunction of lubrication systems, and gas turbines are in no way unique in this respect. For instance, General Electric quotes [*LM2500 Marine Gas Turbine Installation Design Manual* (1973)] for the LM2500 a maximum permanent trim angle of 5° in either direction, added to a maximum installation angle of 8° of the engine centerline from horizontal, and a permanent list angle of 15°. Momentary trim and list of 10° and 45°, respectively, for durations of up to 10 sec, are quoted. These stipulations are typical for marine machinery.

It is traditional to avoid aligning the axes of marine rotating machinery athwartship, since the resulting gyroscopic moment on bearings and foundations would thereby be a maximum because of rolling acceleration. This stricture is perhaps a bit stronger for gas turbines because their rotational speeds are typically higher than other types of machines.

Combined propulsion plants (CODAG or CODOG) present no unique arrangement requirements beyond the need to contend with the inevitable complications of two or more engines per shaft. Several strikingly different arrangements have been used for diesel and gas turbine engines powering common propeller shafts. In the PG 84 gunboats of the U.S. Navy, each of the two shafts is driven by a diesel engine located aft of the reduction gear, while a single gas turbine located forward of these gears drives both shafts via a combining gear. The *Hamilton* class of U.S. Coast Guard cutters employs what might be called a more traditional arrangement: each shaft is driven by a diesel and a gas turbine located side-by-side aft of their common reduction gear sets. Figure 10.9 sketches these arrangements.

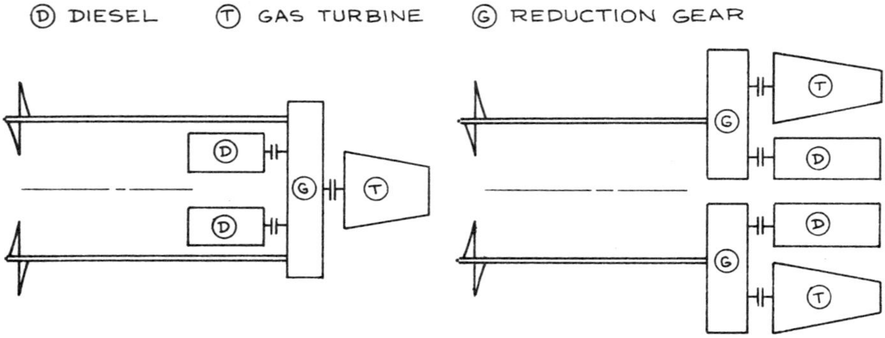

Figure 10.9. Two arrangements of combined-engine propulsion plants.

10.3. FOUNDATIONS AND MOUNTINGS

A. Requirements

The *foundation* is a rigid—and perhaps massive—component of ship structure upon which the engine rests. The *mounting* is the means by which the engine is attached to the foundation. In its simplest form, say, for a small craft reciprocating (diesel or spark-ignition) engine, the mounting may be a few hold-down bolts, plus liners or shims by which alignment to the propulsion shaft is obtained. More complexity is often found and especially in a gas-turbine installation.

Several functional requirements are to be met by the combined foundation and mounting. First and most obvious is the support of the machinery dead weight. The torque produced by the engine and the thrust produced by the propeller both must be transmitted to the hull. (Thrust is usually received by a thrust bearing in the reduction gear, so that the foundation of that unit, if separate from the engine foundation, is the one transmitting the thrust.) Motion of the vessel produces accelerative forces. As an example, rolling motion of θ_R degrees (to one side of vertical) with a complete cycle period of T_R seconds produces a maximum angular acceleration of $4\pi^3\theta_R/180T_R^2$, and the horizontal and vertical components of the resulting forces are [Cashman (1962)]

$$H = \frac{0.0214W\theta_R y}{T_R^2} \quad \text{(lb)} \qquad (10.1)$$

$$V = \frac{0.0214W\theta_R x}{T_R^2} \quad \text{(lb)} \qquad (10.2)$$

where W is the weight of engine (lb); x is the horizontal distance, engine center of gravity to centerline of vessel (ft); and Y is the vertical distance, engine center of gravity to center of the roll motion (ft).

Angular motions (pitching, rolling, yawing) also produce gyroscopic moments on mountings and foundations as they change the plane of rotation of the turbines and compressor. The moment acts in a plane perpendicular to the plane of the rotating member and in the plane of motion. The magnitude of the moment is [Cashman (1962)]

$$M = \frac{WR^2}{9}\frac{2\pi N}{60}\dot{\theta} = 0.00325WR^2N\dot{\theta} \tag{10.3}$$

where M is the moment (lb-ft), W is the weight of the rotating element (lb), N is the rev/min of the rotating element, R is the radius of gyration of the rotating element (ft), and $\dot{\theta}$ is the angular velocity of the imposed motion (rdn/sec). The magnitude of $\dot{\theta}$ is greatest for the rolling component of motion; therefore, rotating members are usually placed in the plane of rolling, that is, with shafts fore-and-aft. With this orientation, pitching produces a moment in the vertical fore-and-aft plane, and consequently forces act alternatingly up and down at the forward and aft ends of the engine. Yawing in similar fashion produces horizontal forces, but they are usually negligible.

The forces resulting from vessel motions are, of course, dynamic, but their frequencies are typically so low that the forces can be regarded as static. In the design process they are simply added to the deadweight forces.

All engines experience thermal expansion as they heat from cold to operating conditions. Restraint of thermal expansion produces stress proportional (usually) to the amount of unrealized expansion, and such stress can be high enough to cause damaging distortions in the engine. This is a special concern to turbine engines, since they are essentially uncooled, and especially a concern to the aircraft-derivative gas turbine, since its light-weight structure is not intended to resist heavy distorting forces and moments. The mounting must therefore be carefully designed to allow thermal expansions; it must also minimize distortions that might be imposed externally, as from distortions of the hull.

Changes in momentum of air and exhaust gas produces forces that must be resisted by mounting and foundation. As an example of possible magnitudes, such loads for the General Electric LM2500 engine [*LM2500 Marine Gas Turbine Installation Design Manual* (1973)] are 1500 lb acting axially forward on the engine, 6400 lb acting axially aft (aft = direction of engine output shaft and of exhaust) on the exhaust duct, and 4800 lb vertically downward on the exhaust duct.

The mounting may be required to isolate structure-borne noise originating within the engine, this being a requirement especially likely aboard naval vessels. This requirement implies resilient members that offer a high impedance to sound transmission. Naval service also usually requires shock protection, and this feature affects mounting design, especially if resilient.

Resilient mountings and mountings that allow thermal movements must perform their functions without allowing excessive misalignment of the engine to reduction gear, or of reduction gear to propulsion shafting. Indeed, it might be said that the principal function of foundations and mountings is to maintain the engine, gear, and shafting in the proper alignment—or more completely said, to maintain it while simultaneously meeting the demands of thermal expansion and possible hull distortions, and of noise isolation and shock specifications.

B. Engine Foundations and Mountings

Figure 10.10 illustrates the basic elements of a typical marine gas turbine mounting: a subbase and three points of suspension for the engine. (This particular mounting is for the Turbo Power & Marine Systems FT-4 engine [McAllister (1972)]). The two U-supports mounted on the subbase siderails support the engine by trunnions at its horizontal midplane near the power turbine (output) end. These supports restrain the unit only in the vertical for-and-aft plane on one side and only vertically on the other side. The trunnions slide to allow radial expansion. A guideway on the centerline at the same location restrains side-to-side motion to maintain transverse alignment. At the compressor end the engine hangs from an overhead frame by a pivoted ball joint that restrains it only in the vertical transverse plane (see Figure 10.11 for

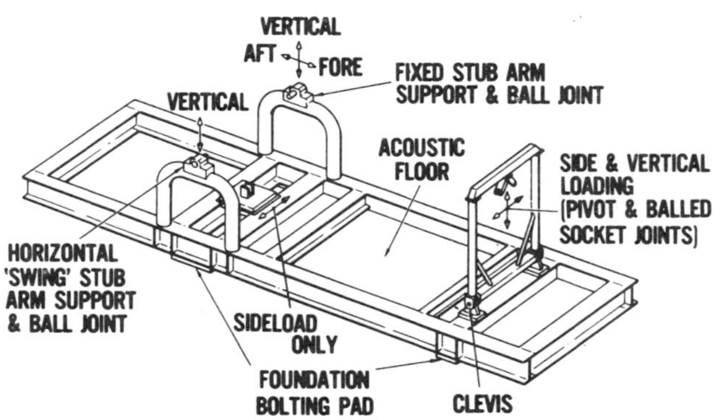

Figure 10.10. Subbase for FT4 engine [McAllister (1972)].

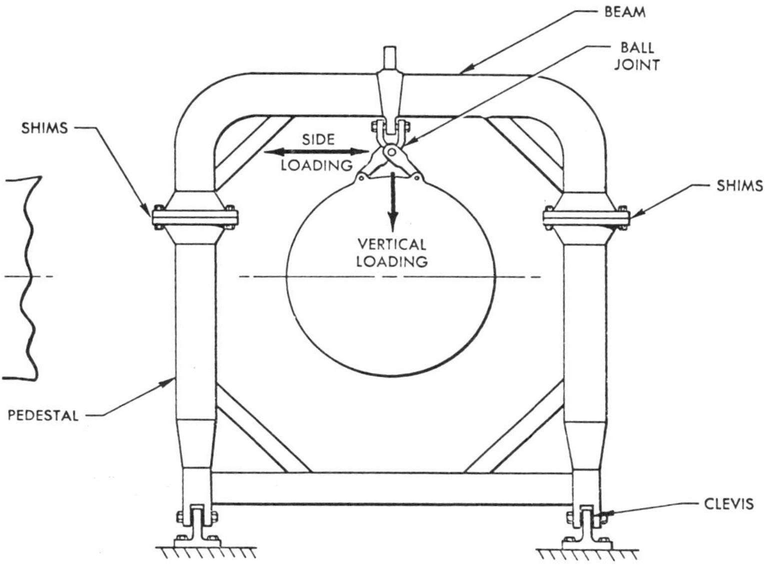

Figure 10.11. Details of FT4 mounting [*Installation Design Criteria for Gas Turbine Application in Naval Vessels* (1968)].

a detailed view of this last). The frame mounts in a clevis joint so that it may swing forward, allowing the fore-and-aft thermal expansion as well as radial expansion.

Figure 10.12 shows a pair of FT4 engines in a twin-screw arrangement like that of the *Euroliner* (engines themselves are hidden within enclosures). The subbase and the U-supports of Figure 10.11 are evident, as are four foundation pedestals by which the subbase is supported from the ship tank top.

Figure 10.13 is a cut-away view of the General Electric LM2500 engine enclosure, showing subbase and engine supports similar to those of the preceding figures. The aft (right) support is not of the U-configuration of Figure 10.10; a support of this type is used on one side only by this engine, and it happens to be the other side here. Figure 10.14 shows a pair of LM2500 engines with reduction gear as installed in the U.S. Navy DD963 class destroyers [Rains and d'Arcy (1972)] (engines actually hidden by enclosures). A more complicated mounting arrangement than the *Euroliner* (Figure 10.12) is evident, made necessary by the sound-isolation and shock requirements of military service. Turbines, subbase, and gears are supported by resilient mounts from a common bedplate, which in turn is supported without intermediate resilient elements from three foundation pedestals.

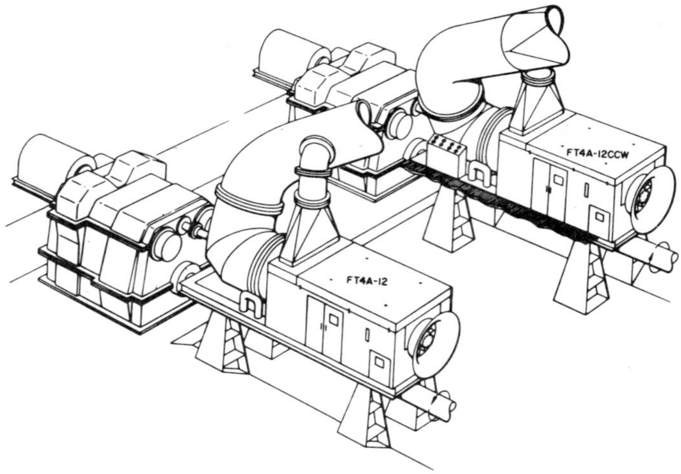

Figure 10.12. FT4 engines in twin-screw installation, showing foundation pedestals, U-supports, and enclosures [Daresta and Smith (1971)].

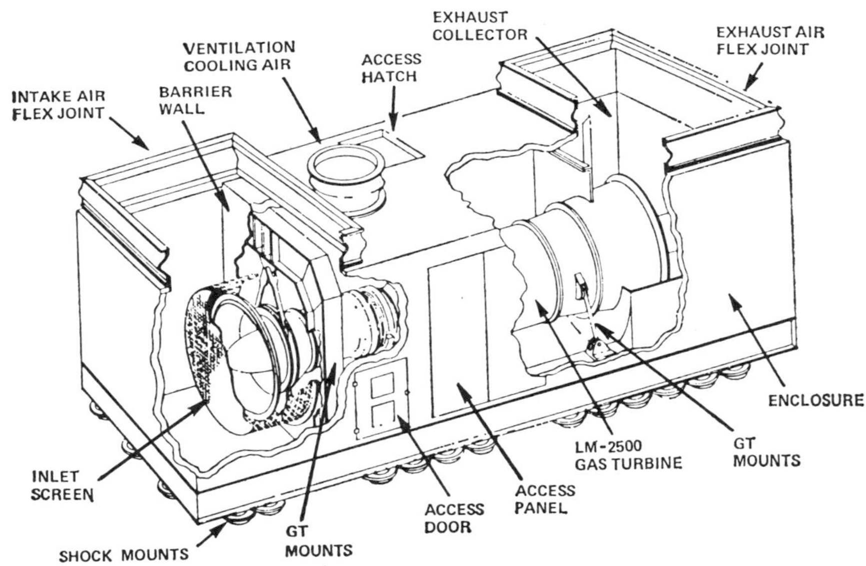

Figure 10.13. LM2500 engine showing mounting and enclosures [Rains and d'Arcy (1972)].

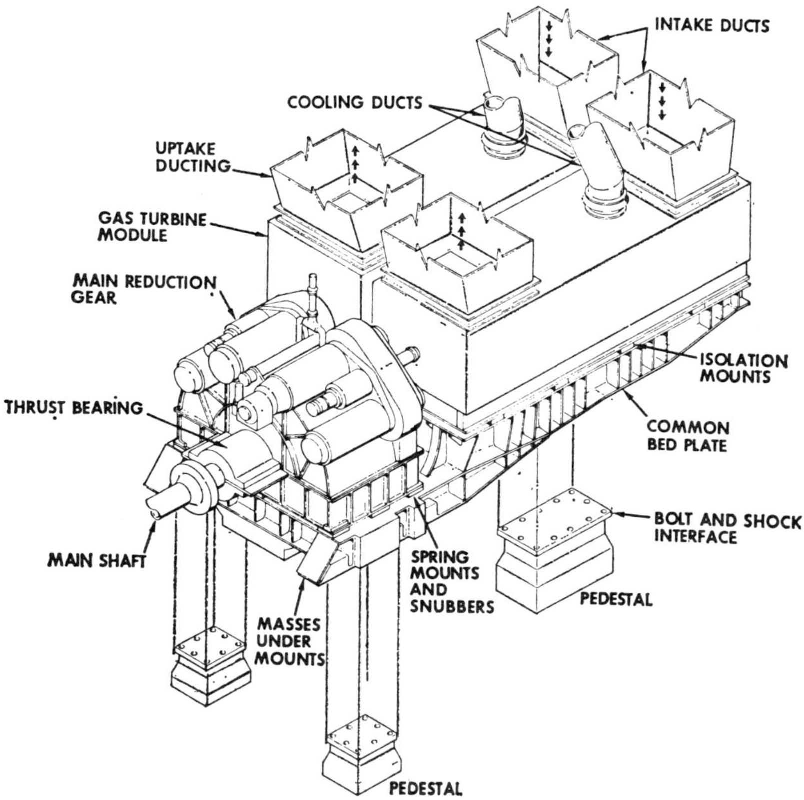

Figure 10.14. LM2500 engine, twin arrangement in DD963 destroyers [Rains and d'Arcy (1972)].

A somewhat different concept is pictured in Figure 10.15. This is the British Olympus engine as it is mounted in a naval vessel. The subbase is a rather short structure at the output end of the engine, from which the engine is firmly supported at the horizontal center plane. The gas generator end is supported cantilever-fashion from the subbase by substantial tubular arms.

The heavy-duty industrial-type engine is also mounted on a subbase and must be supported from it in such a way that thermal expansion can take place. The supports may differ somewhat in detail from the aero engine, however, since the engine casing is capable of bearing reasonable loads. The technique used resembles more that used with steam turbines in that expansion is allowed by a simple, flexing support plate, rather than the clevis and ball joints of Figure 10.10. In a typical configuration, the hot end is rigidly supported by a vertical plate and the cold end by a thin plate that flexes

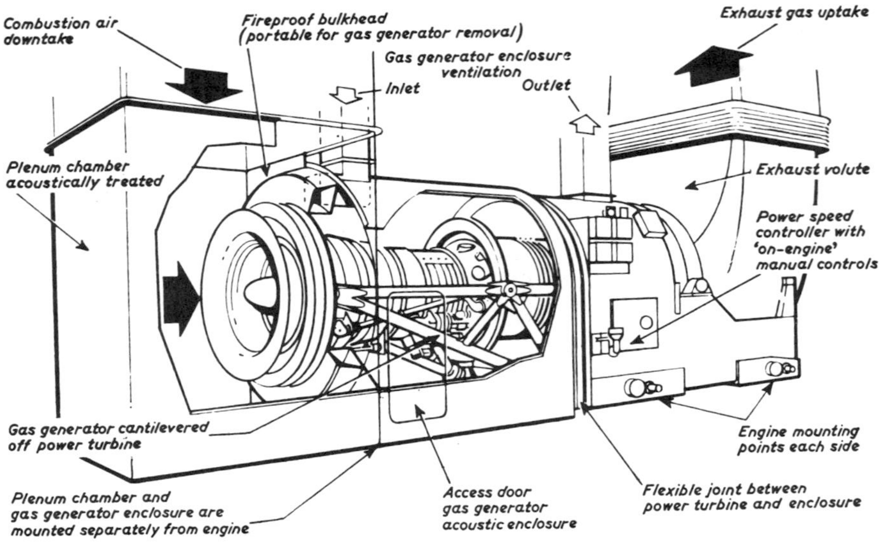

Figure 10.15. Olympus engine, showing cantilever mountings and enclosure.

in response to axial thermal growth without imposing an excessive load on the engine casing. An intermediate center post supports the engine near the hot end, approximately under the power turbine. It is constructed to allow free vertical movement. Figure 10.16 is a sketch depicting this type of support.

The smaller gas turbine engines, such as those of Ford, Avco Lycoming, and United Aircraft of Canada, are subject to the same mounting problems, and the solution in principle is usually the same, although simpler in practical details. The engines are typically provided with two mounting pads on or near the horizontal center plane near the output end and a third pad on the vertical

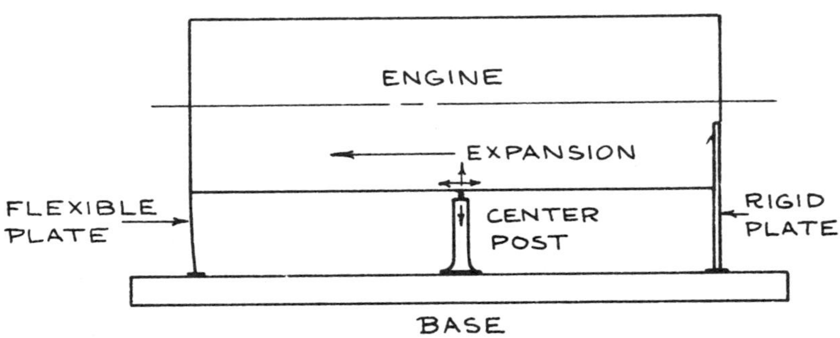

Figure 10.16. Typical mounting for industrial-type engine.

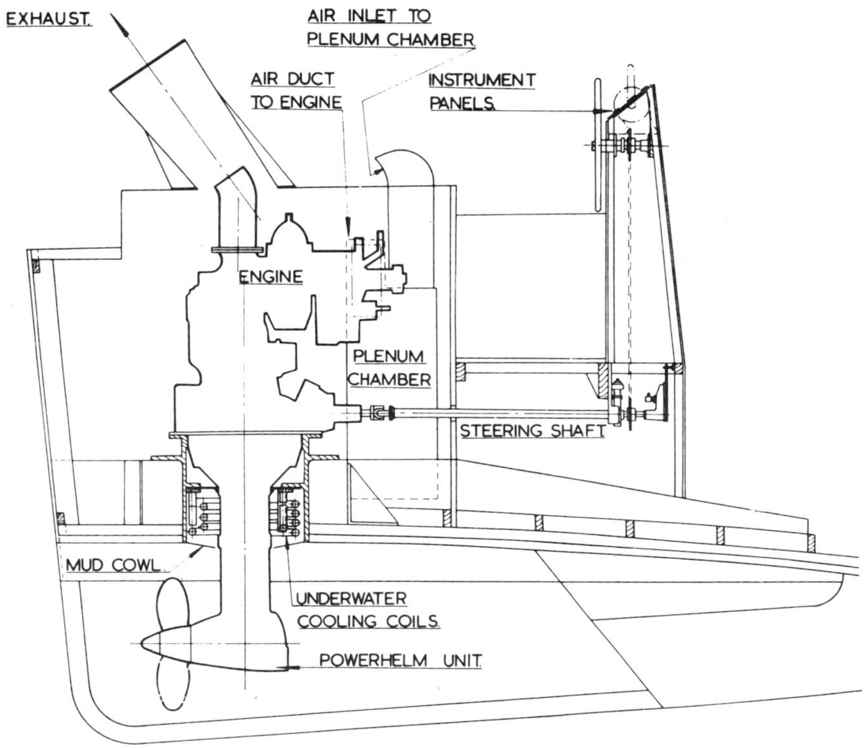

Figure 10.17. "Outboard" mounting of gas-turbine engine (Rover Aurora) [Trewby (1963)].

centerplane near the other end. The latter must be left free to slide axially to allow thermal expansion. Figure 10.17 outlines the mounting of an even smaller engine (less than 100 hp) in what closely resembles an outboard-motor configuration. The engine is supported solely by the propulsion unit.

10.4. MOUNTING FOR SHOCK, VIBRATION, AND SOUND ATTENUATION

A. General Considerations

Shock, vibration, and sound are dynamic phenomena that share important characteristics; their principal differences are in the many magnitudes that may separate their respective intensities and durations. All may pass to or from an engine via its mounting, and all are deleterious in some circumstances, so that ameliorating features are required in mounting design. These commonly take the form of resilient pads or "feet" of several possible designs between engine subbase and foundation (though resilience along may be insufficient in the case of shock design). These units can be seen in small scale in Figures 10.13 and 10.14. Figure 10.18 shows the details of a typical resilient pad.

Shock is, of course, the most likely of the three phenomena to produce immediate damage—perhaps catastrophic damage—to a machinery unit. Fortunately, it is not a normal occurrence and is something that can be ignored in commercial marine foundations and mountings. On the other hand, naval vessels must expect shock from enemy action in combat, and their machinery must be designed to survive shocks that reasonable design can attenuate or isolate. Such is the expectation with noncontact underwater explosions that do not directly impinge on machinery, but that produce large accelerations transmitted via hull structure, foundations, and mountings. These accelerations in turn can produce excessive displacements and rupturing stresses in connecting members. Naval machinery components, therefore, are designed for resistance to shock and are qualified for service by being subjected to specified test shocks. Although the naval propulsion gas turbine may consequently be highly resistant to shock, its protection by isolating features in its mounting is usually necessary.

Sound in its usual occurrences in a machinery plant originates within the components themselves, and its displacements and accelerations are too insignificant to be damaging to inanimate objects. Protection is often necessary for human ear—this is certainly true in the case of a gas-turbine source; thus attenuation of sound escaping via the ducting has been described in Chapter 9, and enclosure of the engine to protect the crew in its vicinity is sketched in a following section. Sound also flows from a machinery source via structural pathways, whence it may emerge as airborne or waterborne (i.e., outside the hull) sound at points remote from the engine. Sound traveling this way is usually not of consequence for human comfort and, therefore, does not usually require attention in commercial-ship design. But for the Navy, however, sound

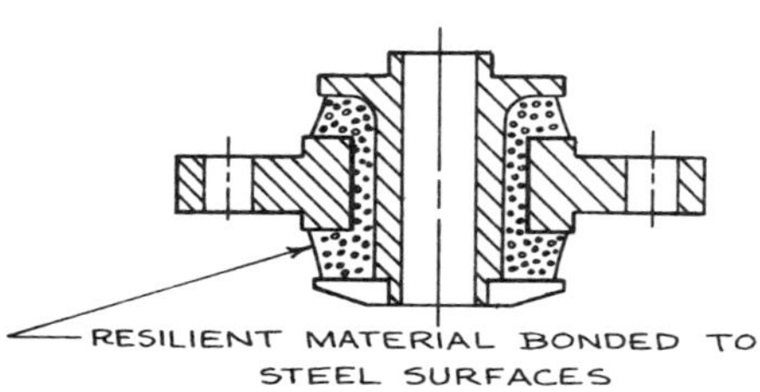

Figure 10.18. Typical resilient mounting.

radiating into the water is a phenomenon of great consequence, especially in its antisubmarine vessels. Resilient mountings are thus a necessity for the gas turbines or other types of engines in such vessels. Likewise, resilient sections in piping and ducting attached to the engines are necessary.

Vibration is intermediate in the shock-vibration-sound triumvirate. Like sound, it is often annoying to humans, but unlike sound sometimes involves such magnitudes of displacements that damage to machinery can result from the accompanying cyclic stresses. In this way it resembles shock, though usually differing radically in the time required for damage to develop. Turbine machinery is nominally free of vibration (especially in comparison to reciprocating machinery), though this condition is usually attained only through careful design. For instance, destructive vibration of turbine blading can occur if the natural vibrational frequency of the blades coincides with the frequency at which they pass nozzle jets.

Although the well-designed and well-built turbine may be free of inherent vibration hazards, it may nonetheless be subject to dangerous vibrations excited by external sources. A rare (hopefully) instance of excitation of gas-turbine compressor blades by a vortex formed within its inlet duct is cited in Chapter 9. A more common phenomenon is torsional vibration of the rotating system consisting of turbine, reduction gears, shafting, and propeller. The excitation is produced by the passage of the propeller blades through regions of differing wake behind the hull. This problem is shared by all types of engine (with a reciprocating engine, its varying torque is usually the major excitation), and vibration analysis of the rotating system is a traditional feature of propulsion-system design. The purpose of the analysis is usually to ensure that no natural frequency of the system coincides (resonates) with a significant excitation frequency. Longitudinal and lateral vibrations of the propulsion system are also possible, and its design is a check for these hazards also. The gas turbine propulsion engine is not unique in these respects; the analyses and their techniques are well established in marine design, so that it is not necessary or appropriate to give a dissertation upon them here.

But in addition to the standard phenomena associated with shaft vibrations, an engine mounted on resilient pads forms a spring-mass combination that can vibrate when excited by the shaft vibration, or perhaps by hull vibration transmitted through the foundation. In its simplest form, this vibration might be described as a bouncing up and down on the resilient elements. This misfortune is to be avoided by choosing these elements so that the natural frequency of the spring-mass combination does not coincide with any excitation frequency of significant magnitude. The possible vibration is nominally single-degree-of-freedom, for which analysis is elementary, but the single degree may lie in or about any of the three coordinate directions (e.g., the engine may vibrate fore and aft, or transversely, as well as vertically, and it may vibrate in any of the three corresponding rotational modes). The marine designer, therefore, must determine the center of gravity and principal moments of inertia of the mounted unit. Necessary data should be furnished by the engine builder since it is not available in the public literature. Naval literature is likely to be a usable source, though, since it is for naval applications the resilient mountings are most often used; [*A Guide for the Selection and Application of Resilient Mountings to Shipboard Equipment* (1958)] is an example.

B. Principles of Resilient Mountings

An engine on resilient mountings is represented in elementary fashion by a mass, m, on springs of spring constant, k, in Figure 10.19. Translational motion in the vertical direction only is considered and is perhaps caused by a vibratory excitation that can be represented by a periodic force of frequency, ω_f, represented by $F_0 \cos \omega_f t$. (I am borrowing here from any one of the many textbooks on mechanical vibrations; all treat this problem of single-degree-of-freedom forced vibration in essentially the same way.) The steady-state response of the mass is a periodic motion of the same frequency, given by

$$x = \frac{F_0/k}{1 - (\omega_f/\omega)^2} \cos \omega_f t \tag{10.4}$$

The maximum displacement of the mass during the vibratory cycle is, therefore,

$$x = \frac{F_0/k}{1 - (\omega_f/\omega)^2} \tag{10.5}$$

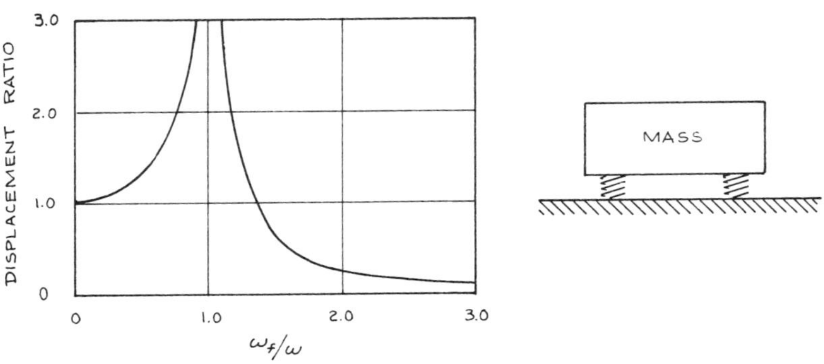

Figure 10.19. Simple spring-mass system and its response to periodic exciting function.

where $\omega = (k/m)^{1/2}$ is the natural frequency of the spring-mass combination. The displacement is quite apparently a function of the ratio, ω_f/ω, of exciting frequency to natural frequency. A steady force ($\omega_f = 0$) produces the displacement, F_0/k; the ratio of displacement to this base displacement is plotted in Figure 10.19 as a function of the frequency ratio. Obviously, the condition, $\omega_f/\omega = 1.0$ (resonance), is to be avoided, even though in practice the peak would not be so extreme because some damping is always present in real systems. The preferred condition is $\omega_f \gg \omega$, for which the resulting displacement is very small compared to the static displacement. Resilient mountings, therefore, are always made "soft" to obtain this state of essential isolation from the exciting forces of expected frequency, ω_f.

C. Mounting for Shock Attenuation

If the force exciting the spring-mass system is a shock, this force is likely to appear as a brief and irregular function of time. Figure 10.20 gives an example. The shock pulse can be resolved by Fourier analysis into a sum of harmonic components and the displacement of the mass caused by each component can be calculated by elementary vibration formulas (but not by Eq. 10.5, since it does not include the transient part of x). The total displacement, x, is then obtained by adding the individual components of response. If all of the harmonic components of the shock were of high frequency (i.e., $\omega_f \gg \omega$), the total displacement would be small and the resilient element would serve well as a shock isolator. But some of the shock harmonics are of low frequency, and some are likely to fall near the resonant condition ($\omega_f = \omega$). Unless it is only a mild shock, a large displacement results, perhaps exceeding the elastic range of the resilient element and rupturing it.

The essence of shock mounting is consequently the provision of a snubbing element in addition to the elastic resilience of the sound-vibration isolator. In its simplest form, this could be a hard stop that would limit the displacement of the subbase. More usually, the snubber incorporates resilience also, but at a higher spring constant than the normal elements. The snubber is not engaged during normal circumstances and so does not interfere with the isolating function. It is expected to act only in compression and hence is not at hazard from a stress exceeding its elastic range.

Figure 10.21 depicts a British (Yarrow and Company) design of a resilient mounting, including the shock snubber. Note that a gap of 0.04 in. vertically isolates the snubber from normal contact with the engine subbase ("main machinery raft," implying that the reduction gear is mounted on the same base), and that horizontal as well as vertical snubbing is provided.

The normal resilient part of the mounting shown in the figure is more complex than our previous discussion has implied for these things (cf. Figure

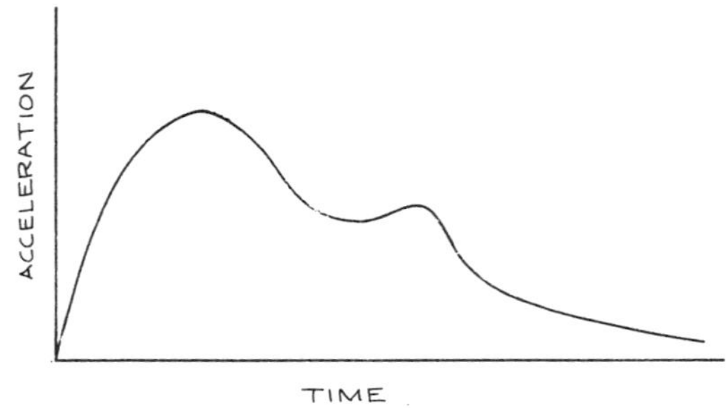

Figure 10.20. Example of a shock input.

10.18). The resilient element is compressed air confined by a diaphragm. The main feature of complexity, however, is the hydraulic device that maintains the static position of the engine under varying loads, such as the forces resulting from engine-torque reaction upon its mountings. A glance at the figure shows that as the subbase is depressed, the positioning valve opens to supply oil to the large chamber. The diaphragm confining the oil moves upward, compressing the air within its chamber and thus forcing the subbase toward its original position.

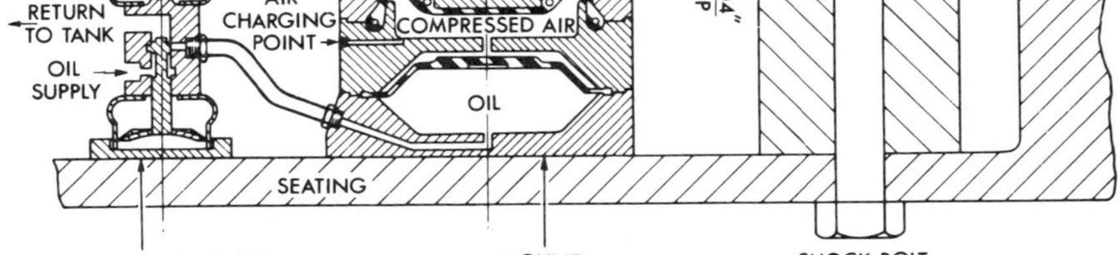

Figure 10.21. Yarrow shock mount [Lindsey (1968)]. From *Transactions of the Diamond Jubilee Meeting,* 1968, published by the Society of Naval Architects and Marine Engineers.

10.5. ALIGNMENT

Alignment is both a design problem and a skilled installation task that is encountered with all propulsion engines and with all forms of mechanical transmissions. The essence of the situation is shown by Figure 10.22: shafting, gears, and engine supported by bearings, mountings, and foundations whose position must be specified by the designer and fixed by the installer, so that the following general requirements are met:

1. Load on each bearing falls within the design range.
2. Bending stresses in the shafts fall within the design range.
3. Design conditions of gear tooth mesh not upset.

In a horizontal plane, alignment is no problem for the designer and a relatively minor one for the installer, because of horizontal symmetry and because there is no component of gravity force. But asymmetry and weight are present in the vertical plane, as pictured by Figure 10.22. The weights of shafting, gears, propeller, and engine place loads upon the supports and produce bending stresses within the elements. The vertical position of the supports determines how the loads and stresses are distributed, and the problems of alignment are those of finding and fixing these positions.

From the designer's standpoint, the problem is fundamentally one of analyzing a continuous beam on multiple supports; this is especially the case for the shafting between gear and propeller. (The analysis can be done by the "method of three moments" that is described in many textbooks on strength of materials in their chapters on beam analysis.) Such analyses show, for example, that raising a support above the straight line connecting all support points increases its loading and distorts the distribution of the load among the supports. Lowering the support reduces its loading, though an increase also occurs if the shaft is forced to follow downward beyond its natural unsupported position. The effect on other supports depends on the length and flexural rigidity (product of modulus of elasticity and moment of inertia about

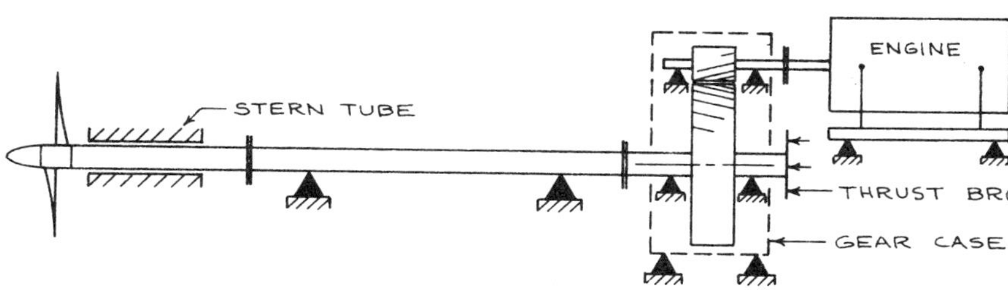

Figure 10.22. Elements of propulsion drive train and its supports.

the neutral plane) of the shaft between supports; the closer the supports, or the stiffer the shaft, the greater is the influence of a support upon its neighbors.

Correct mesh between the bull gear and its pinion requires approximately equal load on their adjacent bearings (i.e., two bearings within the gear case) [Anderson and Zrodowski (1959)]. The alignment of the propulsion-shafting bearings thus affects the gear mesh by determining how the shaft load is distributed to the bull-gear bearings. The designer must specify the alignment so that the desired distribution is attained. In doing so he faces complications such as movement of the gear bearings in operation due to thermal expansion of the gear case. Typically he would specify that the shaft bearings be set high in the cold condition to match the position of the gear when hot.

On the input side of the reduction gear, the problem is the same in principle, but different in detail because it is the engine, rather than a long and heavy shaft, that mates with the input pinion. Engine bearings cannot, of course, be raised or lowered individually, so that it is the engine and its subbase as a unit that must be aligned to the gear. Whatever the original alignment be, however, it is altered in service by changes in bearing position, principally those caused by thermal expansion in the vertical and axial directions. In one sense, the expansion phenomenon makes the alignment problem an unsolvable one, since a perfect cold alignment cannot be perfect when hot if there is any relative expansion between connected components, and there usually is. The practical answer to the dilemma is tolerance and flexibility. Looking back for a moment to the propulsion shafting for an example, the thermal rise of the bull gear means that alignment cannot be the best in both cold and hot conditions, but the phenomenon is made acceptable in practice by locating the first shaft bearing at such a distance from the bull gear that the natural flexibility of the shaft limits the hot-to-cold change on the gear bearings to a range tolerable to the gear-contact specification.

Because the engine is closely coupled to the gear, natural flexibility of shafting is inadequate and must be enhanced by artificial flexibility in the form of a flexible coupling. This is especially true for the problem of axial expansion, which often can be accommodated in no other way. For example, an installation of the Turbo Power & Marine Systems FT4 engine is cited (U.S.C.G. *Hamilton* class cutters [Good and Logan (1969)]) with ½-in. axial thermal expansion measured at the engine-output coupling. The engine also expands in the vertical direction. This expansion is at least partially offset by reduction-gear expansion in the same direction, but equality of expansion would be fortuitous, hence bending flexibility between engine and gear is always required. On the engine side of the gear, therefore, the designer's principal task is determining what relative movements occur and specifying means of accommodating them.

If the resilient mountings are used—as is customary in naval installations—

small movements due to vibration (for instance) are expected in service, giving further need for flexibility between engine and gear. This particular argument is negated, though, if a "common-raft" mounting is used: engine subbase and gear case hard mounted (i.e., without resilience) to a common base (the raft), which is in turn resiliently mounted to ship structure. If resilient mountings are used for sound isolation, the shaft flexible coupling doubtlessly incorporates a resilient member to attenuate sound passing to the sea via the propulsion train.

Since the engine-exhaust elbow usually is interposed between power turbine and reduction gear (exception: if engine output is taken from inlet end, as noted in Section 10.2), an intermediate shaft—furnished by the engine builder—connects the two units. For the large engines (such as General Electric LM2500, or Turbo Power & Marine FT4), this shaft is about 8 ft long and includes the flexible coupling. The LM2500 shaft, for example, can accept an offset of 0.125 in. between its ends and an angular misalignment of ¼° [*LM2500 Marine Gas Turbine Installation Design Manual* (1973)]. Accomodation of thermal growth by this unit requires that it be given a cold tensile displacement of about 0.16 in.

Figure 10.23 is a sectional diagram of the power-turbine end of the FT4 engine, showing the output shaft. The flexible coupling is of the diaphragm type and is located at the turbine (left) end of the shaft. Note that the shaft is unsupported by bearings along its length, so that its weight must be shared by the power-turbine bearings and the reduction-gear pinion bearings. It is the installer's problem during alignment to correct the droop at the initially free end so that on bolting up the load is indeed distributed properly. To do this he must have information from the designer or engine builder, usually in the form of an upward force specified to be applied at the outer end to bring it into the correct line. The builder typically supplies a temporary pedestal with spring scale to facilitate this part of the alignment.

The flexible coupling shown in Figure 10.23 may need to be supplemented by a second such coupling. This was the case with the *Euroliner* installation, where the axial load imposed on the turbine by the output shaft is limited to 1000 lb by a second unit [Kerpestein (1972)]. The installation is pictured in Figure 10.24, this being a diagrammatic view of the *Euroliner* propulsion train from power turbine through bull gear. The two wheels of the power turbine are at the upper left; the two flexible couplings are labeled, with the connecting shaft discussed above showing between them. (An additional intermediate shaft is necessary to reach the pinion in this particular installation because of the gear arrangement. The unit at its right end is a clutch, which is discussed in Section 10.6.) The second coupling is shown in detail by Figure 10.25. In initial alignment it is given a tensile displacement of 0.15 in.; upon heating to operating temperature the turbine expands axially to compress the

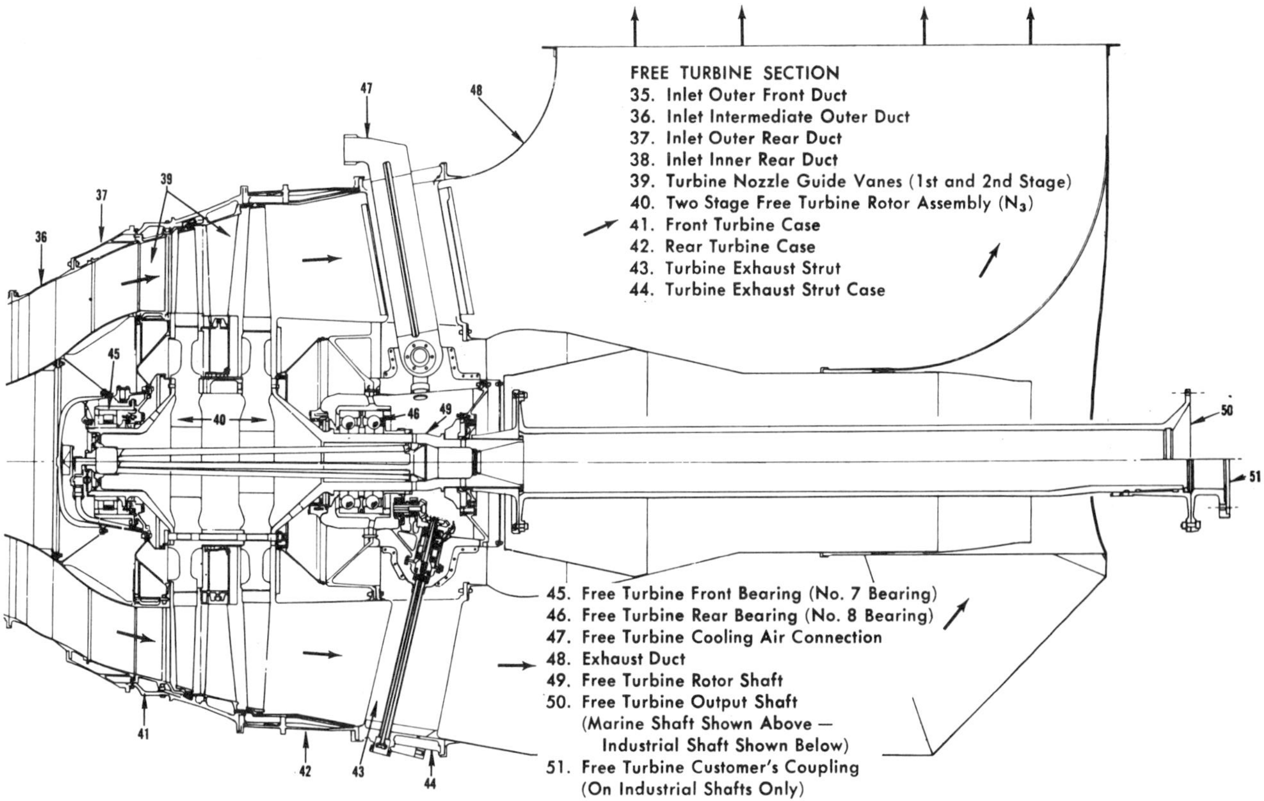

Figure 10.23. Power-turbine section of FT4 engine, showing coupling shaft [Wennberg (1966)]. From *Transactions,* 1966, published by the Society of Naval Architects and Marine Engineers.

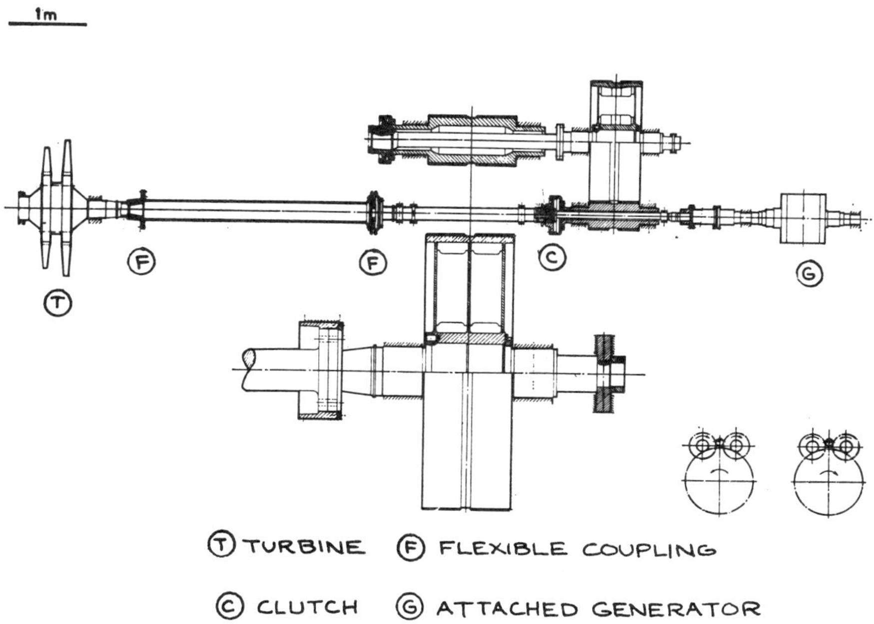

Figure 10.24. Elements of *Euroliner* propulsion drive train [Kerpestein (1972)].

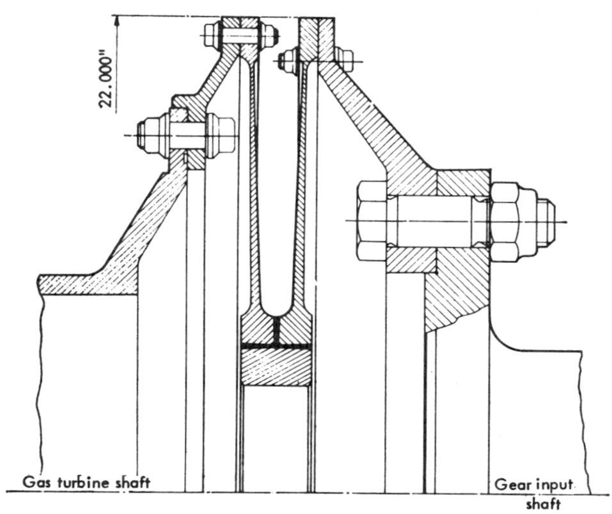

Figure 10.25. Detailed view of flexible coupling from Figure 10.24 [Kerpestein (1972)].

coupling 0.30 in., giving a compressive displacement of 0.15 in. Axial force remains within the 1000-lb limit during this change.

In principle, what is said here applies to both small and large engines, though in practice it may be much simpler with a small engine because of the simpler support. This simplest installations perhaps are found with the small engines that give their output at the compressor end (e.g., Avco Lycoming engines) and have side air intakes. In some instances these engines are cantilevered from the gear case, having no "feet" to be chocked precisely for correct alignment with that gear. A flexible coupling is found between engine and gear, nonetheless.

Our major interest is how the alignment problem as outlined here is unique to the gas turbine. In so far as the propulsion shafting is concerned, there is no uniqueness. Regarding the engine itself, its alignment problem can be compared to that of the steam turbine and is seen to be much the same. The steam turbine also experiences thermal expansions as it heats up to operating temperature, so that a flexible coupling between engine and gear is necessary. The difference is mainly one of degree, since the gas turbine may heat through a range twice as great as that of the steam turbine. The machined chocks that traditionally support a hard-mounted turbine or turbine subbase in the correct alignment to the foundation are the same in both cases. On the other hand, resilient mountings are rarely used in steam practice. Unless resilient supports are involved, the necessity to provide artificial support for a coupling shaft during alignment would probably be the only difference in practice noted by an installer familiar with steam work. There is a little more of a difference between the gas turbine and the diesel engine, mainly because the diesel does not undergo thermal expansion sufficient to make artificial flexibility between engine and gear necessary. This type of engine is often bolted without flexibility to the reduction gear input flange. When flexible couplings are used in this position with a diesel, it is usually to isolate the engine's torsional excitation from the propulsion system.

10.6. CLUTCHES

The clutch is a prosaic industrial device used in innumerable applications to connect and disconnect drivers and loads. It is also a common component in marine propulsion, especially in fixed-pitch geared-diesel plants, where clutches are used to alternately connect the ahead and astern gear trains to the engine. On the other hand, the clutch is quite rare in steam-turbine propulsion practice. From reduction input coupling to propeller, the gas-turbine ship propulsion plant is essentially the same as the steam-turbine plant, except that clutches are typically used with the gas turbine. And when the gas turbine is combined with a diesel engine in the same plant, clutches between engines and reduction gear are also found, but with features that may differ from conventional practice.

Because of the frequent use of clutches in gas-turbine propulsion, this section sketches the clutches and their application. But let's first note the reasons for the difference between gas-turbine practice and that of the diesel and steam alternatives. In general, whenever two or more engines of any type (or of mixed types) drive the same propeller shaft, each must be provided with a clutch so that it can be disconnected when not driving. Many of the applications of the gas turbine have indeed been in multiple-engine plants, often due to its suitability as a boost engine in conjunction with a diesel base-load engine. Diesels themselves often appear in multiple arrangements, but in such cases reverse gearing is nearly always used, making it necessary for the clutches to engage under load from a windmilling propeller during crash reversals. The contemporary CODOG propulsion plants either reverse by means of controllable-pitch propellers, or if reverse rotation of a fixed-pitch propeller is used, rely solely on the diesel part of the propulsion plant. Hence the gas-turbine clutch in a CODOG plant is likely to be different from the typical diesel reverse-gear clutch. In steam turbine practice clutches are rarely seen because multiple turbines are never used (multiple casings, as in the common cross-compound arrangement, yes, but that's still a single turbine) and because it is feasible to build steam turbines with a reversing turbine incorporated. Since it is not feasible to include a reverse turbine within contemporary gas turbines, a gas-turbine propulsion plant with a fixed-pitch propeller requires use of a reversing reduction gear, with its associated ahead and astern clutches. Application as either an element of a multiengine plant, or as the sole engine in a fixed-pitch installation, therefore, requires the clutch between a gas turbine and the propulsion load. And another requirement for clutch use, not at all unique to gas-turbine characteristics, appears in subsequent discussion of the *Euroliner* machinery.

There are many types of clutches, but we shan't attempt a cataloging of them, nor delve deeply into descriptions, but base the discussion on the several functions that clutches are called upon to perform in gas-turbine propulsion. Of course, the general function of all clutches is to connect and disconnect a mechanical power link, but a categorization can be applied as follows, based on the circumstances of the shift from one state to the other:

1. Make or break connection when driver and load are stationary.
2. Add or subtract a driver from a multidriver system without disrupting the load.
3. Make or break connection under heavy opposing torques conditions, that is, driver used to brake and reverse load.

The first of these is exemplified by the *Euroliner* clutch that appears in Figure 10.24. This figure shows the rotating elements of the propulsion drive train, including the rotor of an attached ship service electrical generator (far right). It may be necessary to run this generator in port, requiring that the propulsion train be disconnected by the clutch, which joins the intermediate gear shaft to the primary pinion. But it is quite practicable to stop the engine and generator while the breaking or making of the connection is accomplished, since the standby or emergency generator can be used meanwhile. The clutch, therefore, does not need synchronizing features, nor the ability to handle large torques transiently, characteristic of other clutch types. It consists of two elements—driver and driven—with mating external and internal gear teeth. One of the elements slides axially on splines so that its teeth can engage or disengage the other. Figure 10.26 shows details of the clutch, including the simple manual-lever arrangement used to accomplish the necessary axial motion. Note that the shaft to the attached generator passes concentrically through the clutch and the primary pinion gear, and is not affected by clutch action.

The second clutch function is required when two or more engines drive the same shaft, as is the case with the CODOG plants built by the U.S. Coast Guard and several navies, and with multiple gas-turbine plants, such as the DD963-class destroyers of the U.S. Navy. Since it is undesirable to disrupt

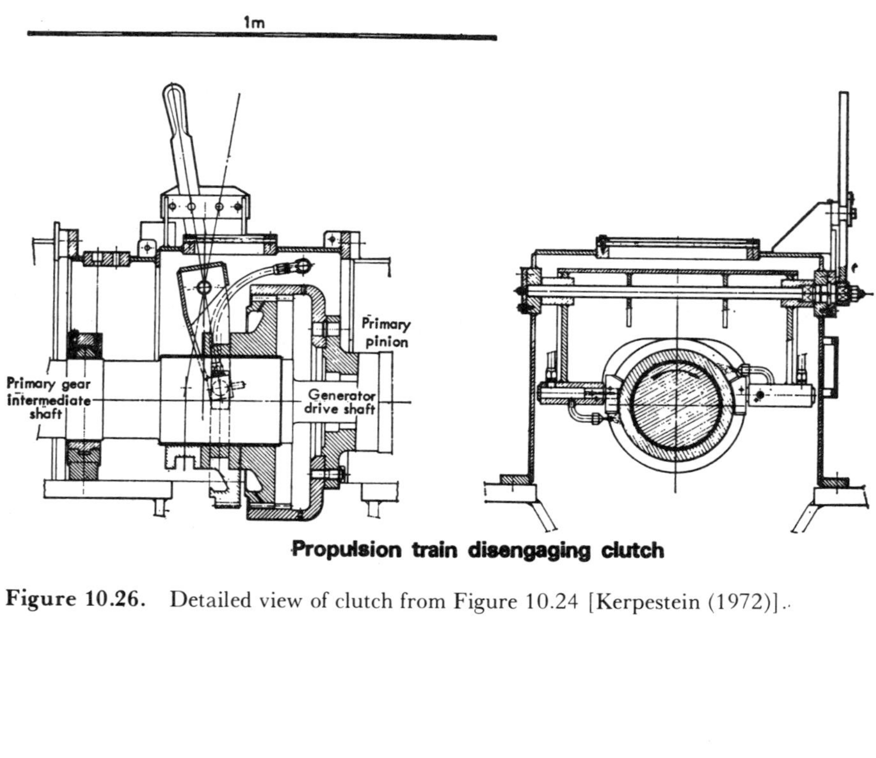

Figure 10.26. Detailed view of clutch from Figure 10.24 [Kerpestein (1972)].

propulsion by stopping the propulsion shaft when switching engines, the clutch that connects each engine to the common reduction gear must incorporate a synchronizing element that brings the shafts to be mated to the same speed before a positive lockup is attempted.

A synchronizing feature is typically accomplished by a separate multidisc clutch that is engaged initially, perhaps under high speed differential, to accelerate the slower shaft to the speed of the faster. The main clutch is typically a dental coupling, with one element sliding axially to engage the teeth of the other to provide the positive lockup, much in the manner of the comparable elements in Figure 10.26. Actuation is pneumatic or hydraulic, though, and must be interlocked so that the permanent engagement is not attempted until synchronization is attained. As an example of this type of clutch, the U.S. Coast Guard *Hamilton* engines (Turbo Power & Marine Systems FT4) are connected to the reduction gear by a clutch with a normal torque rating of 43, 700 lb-ft. The synchronizing element has a static torque rating of 34,220 lb-ft and synchronizes within a period of 5–8 sec from an initial rev/min differential of 2250 [Wennberg (1966)].

An alternative design features self shifting of the permanent dental clutch at the moment the shafts pass through synchronization. Details of such a clutch are diagrammed in Figure 10.27. When shaft speeds are the same, pawls on one element automatically fall into engagement with a ratchet on the other, imparting axial motion along a helical spline until tooth engagement takes place. No external actuation is required and no synchronizing clutch, the speed change necessary to bring the engine shaft up to the speed of the pinion shaft being provided totally by engine-fuel control. If the engine is being taken off the line, the clutch automatically slips out of engagement when the pinion begins to drive the engine. A positive locking feature can be added to prevent inadvertant disengagement during maneuvering. These clutches are used in the CODOG propulsion plants of several of the U.S. Navy PG84-class gunboats and in the last three of the U.S. Coast Guard *Hamilton*-class cutters.

If a reversing reduction gear is used with a fixed-pitch propeller and unidirectional engine, clutches must engage and disengage the propulsion shafting under more extreme conditions than those described above. In a crash reversal, for instance, the engine running in its normal ahead direction must be engaged to the propeller shaft turning in the opposite direction under high torque input from the windmilling propeller. The clutch must, in effect, act as a brake, transmitting, while slipping, the torque necessary to arrest the shaft and to accelerate it in the opposite direction against the continuing windmilling torque. Clutches suitable for this service are widely used in diesel service and are typically pneumatic clutches in which an inflatable toroid presses friction elements against a mating friction drum. Because of the popularity of the controllable-pitch propeller with gas-turbine propulsion plants, these clutches have found

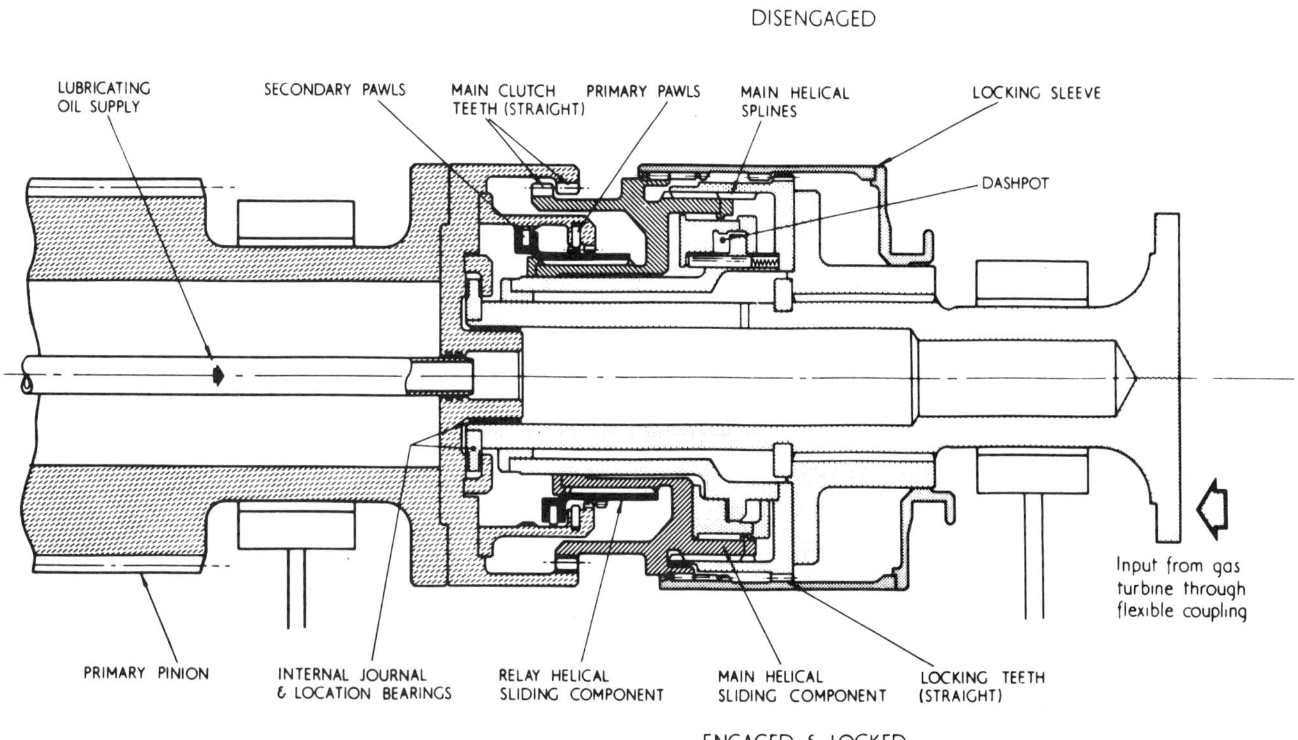

Figure 10.27. Details of synchro-self-shifting clutch (by S. S. S. Gears Ltd.) [Clements (1972)].

limited application with the marine gas turbine, but one outstanding example is the *Admiral William Callaghan*. Figure 10.28 shows the *Callaghan* drive train diagrammatically; the clutches appear in approximately correct scale relative to the other elements. Figure 10.29 shows in section how the clutches are mounted on the intermediate gear shaft and outlines the internal structure in small scale. Figure 10.30 gives a more detailed view of a clutch of this type.

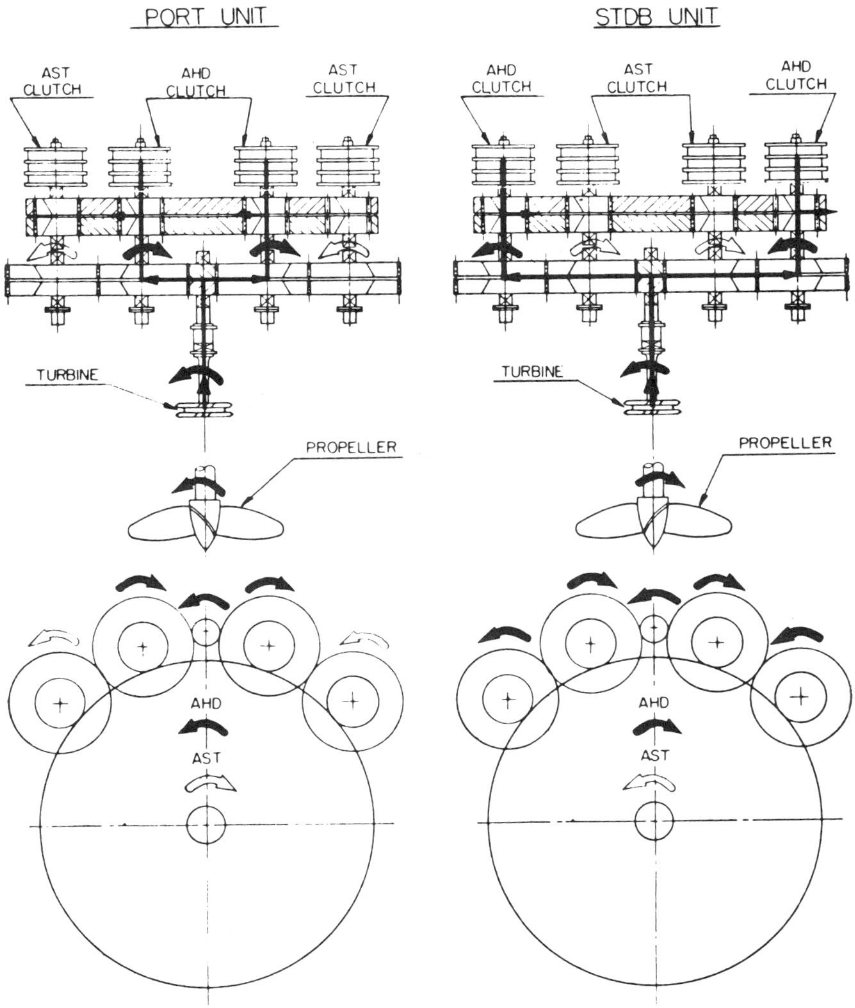

Figure 10.28. Arrangement of *Callaghan* clutches [Richardson (1969)].

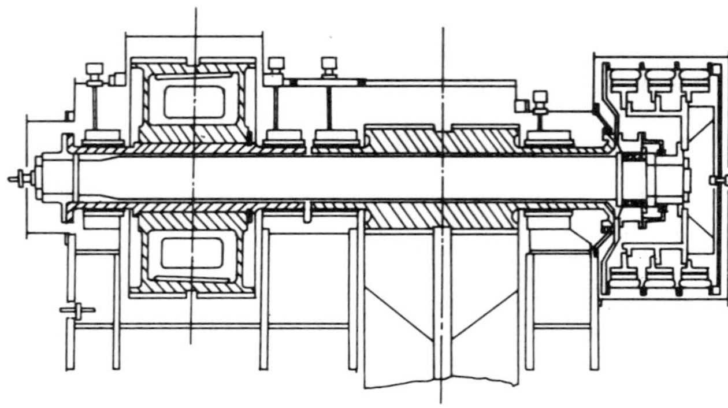

Figure 10.29. Sectional view of *Callaghan* intermediate gearing, showing clutch detail [Richardson (1969)].

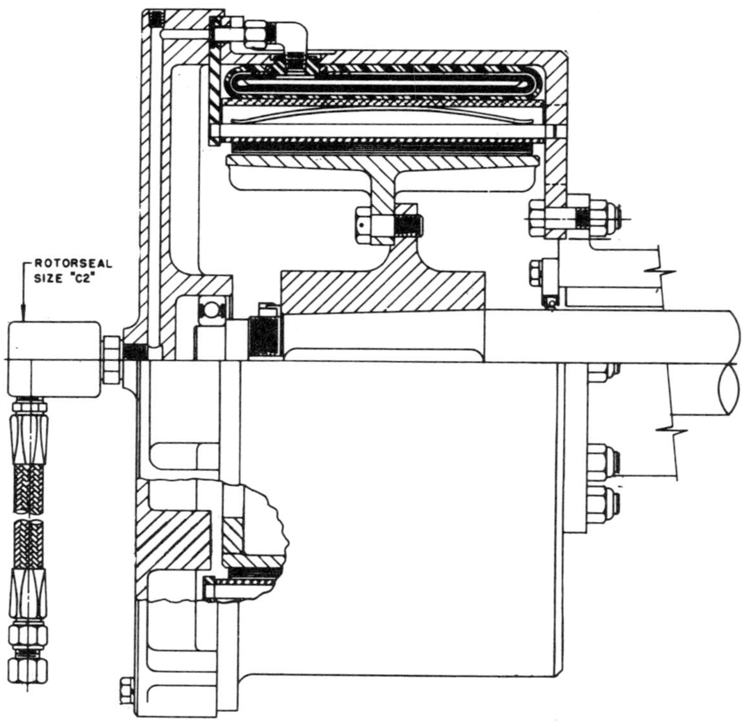

Figure 10.30. Cross section of pneumatic clutch (Fawick Corp.)

10.7. ENGINE ENCLOSURE

The gas-turbine engine is a high-intensity source of both noise and heat leakage. Either of these undesirable outputs is usually sufficient to make the compartment containing the engine uninhabitable unless the engine is fitted within an enclosure. Several earlier figures (Figures 10.12, 10.13, 10.14, and 10.15) show such enclosures. These are for ship-size engines; the more usual solution for small-craft engines is installation within an unmanned space that is lined with sound-absorbing material and has sufficient ventilation to remove the heat at a reasonable temperature. Several figures (Figures 9.18, 10.5, 10.6, 10.7, 10.8, and 10.17.) indicate this alternative. A possible third alternative—encasing the engine in a heavy insulating blanket—is unacceptable because heat escape from the engine is necessary to keep the casing from becoming excessively hot.

Figure 10.31, left, indicates some of the constructional features of the FT4 enclosure. The basic elements are panels constructed of sound-absorbing and heat-insulating materials. Access is provided via doors, and panels are readily removable as necessary to facilitate quick changeout of the gas-generator section of the engine.

Figure 10.31, right, is a sketch of the FT4 engine as mounted within the enclosure and lists the external connections required by the engine. As the figure suggests, the auxiliary connections are brought out via the surface to avoid penetrations of the enclosure bulkheads by piping and wiring. As this also suggests, the engine is virtually a "plug-in" unit. Note from this figure that the power-turbine portion (left end) of the engine is excluded from the enclosure.

The auxiliary connections shown, with one exception, serve the engine itself. The exception is the connection for the enclosure fire-extinguishing system ("FIRE EXT. WATER"). This system is necessary because fuel oil, lubricating oil, and hydraulic oil are present in pressure piping within the enclosure and because hot engine parts provide a source of ignition should any of these fluids escape via a leak or a ruptured pipe. The enclosure makes difficult visual detection of and manual reaction to the resulting fire. Therefore, remote sensing and alarm for the fire and application of a suppressing agent via a fixed system under remote control are essential.

Since Figure 10.31 indicates water for fire extinguishing, a fog system is implied, this being the only means by which water can be effective against oil fires. Other agents, such as carbon dioxide, dry chemical, foam, and halogenated hydrocarbons are, or can be, used for the same purpose.

The *Euroliner* fire detection and extinguishing arrangement is an example of this type of protection within an enclosure [McAllister (1972)]. Abnormality is detected by both heat sensors (set at 350 F) and ultraviolet-flame detectors. If either set of detectors trip, an alarm appears on a control panel in

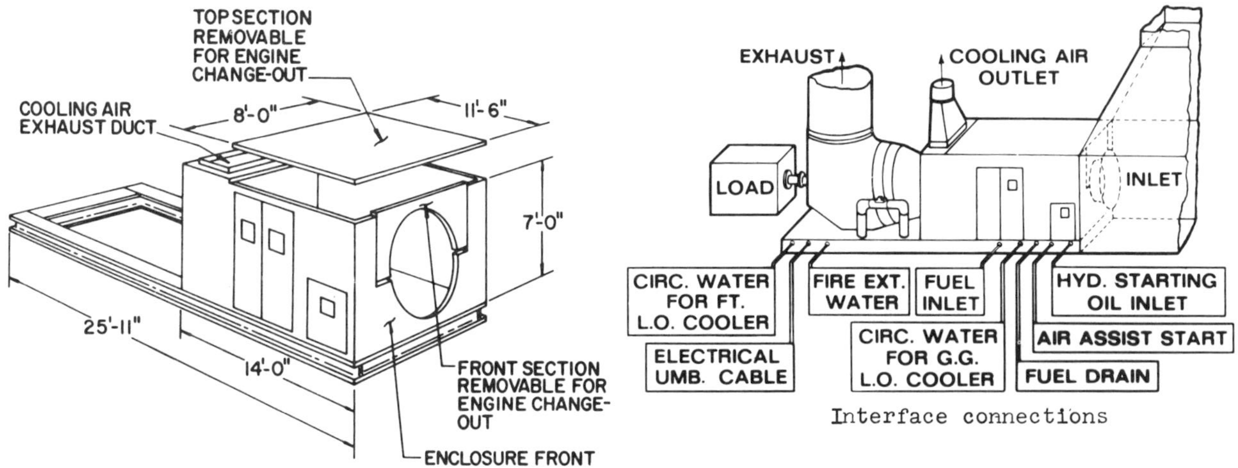

Figure 10.31. Enclosure for FT4 gas-turbine engine (McAllister (1972)].

the machinery control room; if both trip, the extinguishing system is automatically activated. Water fog is distributed throughout the enclosure by 12 fog nozzles supplied from a 1000-liter fresh-water tank. This supply is supplemented by the ship-service sea-water fire system.

Cooling of the enclosure, usually by air, is made essential by the high temperature of the gas-turbine surfaces. A cooling air oulet is evident in Figure 10.31, and several manifestations of cooling supply and exhaust are evident in earlier figures. For example, Figure 10.4 shows a cooling-exhaust duct fitted to an FT4 enclosure similar to that of Figure 10:31, Figure 9.5 shows the ventilation-supply duct branched off the engine inlet for a General Electric LM2500 installation, Figure 9.14 indicates both supply and exhaust for the *Euroliner,* and Figure 9.18 shows cooling of a small-craft compartment via flow induced by an exhaust eductor. Figure 10.32 supplements Figure 10.4, showing how the cooling air is supplied to the enclosure in the same installation (*Euroliner*). A fan (not shown) on the supply side provides the driving head, and the exhausting coolant flows by a short duct into the engine main exhaust, as both the figures mentioned show.

Figure 10.33 plots the surface temperature of the General Electric LM2500 engine at its maximum power, and with a typical cooling air flow. Table 10.1 indicates the additional heat contributions of certain expected leakages from this engine. Its rate of cooling air flow should be about 15%, at the most, of the combustion air supply [*LM2500 Marine Gas Turbine Installation Design Manual* (1973)]. The actual amount depends strongly on the air temperature rise that can be tolerated, since heat removed is proportional to the product of flow rate and temperature rise.

Figure 10.34 shows the air flow as a function of temperature rise for the FT4 engine. The strong dependence upon the temperature rise is quite apparent. Note that this figure is based on a particular enclosure size and that it

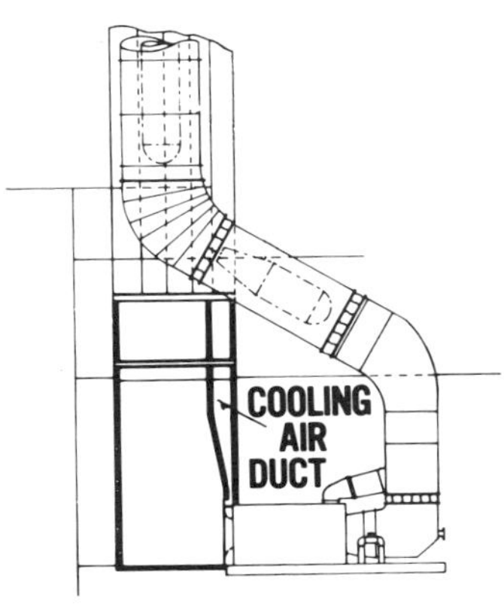

Figure 10.32. Cooling air supply to Euroliner engine enclosure [McAllister (1972)].

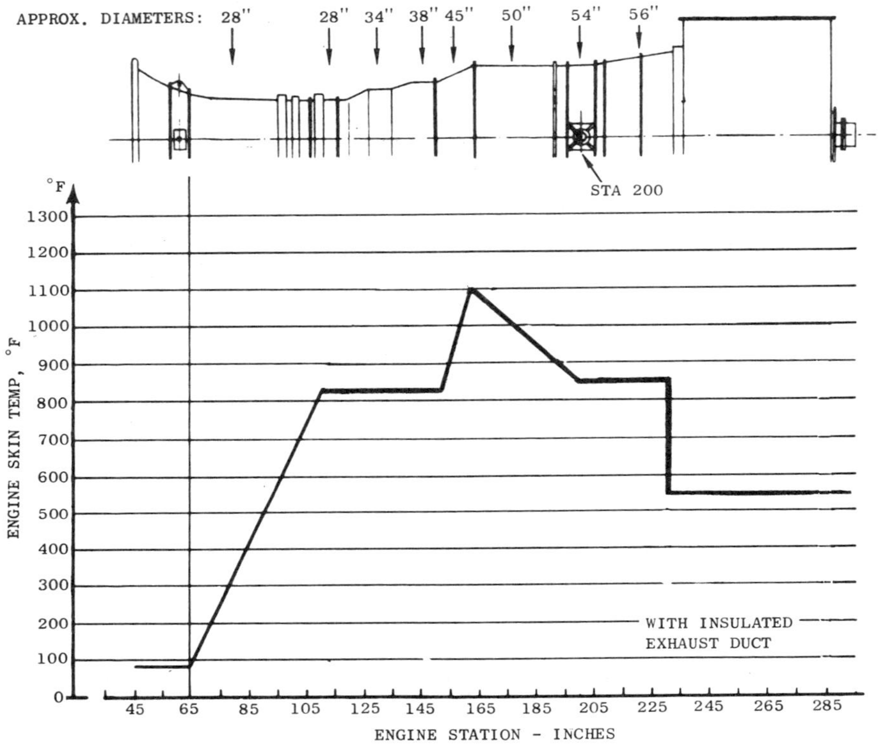

Figure 10.33. Estimated surface temperatures for General Electric LM2500 engine at maximum power, with secondary air flow [*LM2500 Marine Gas Turbine Installation Design Manual* (1973)].

TABLE 10.1. HOT AIR LEAKAGES FROM LM2500 ENGINE [*LM2500 Marine Gas Turbine Installation Design Manual* (1973)]

Engine Source	Max Temp	Max Flow
Flange leakage	800 F	0.17 lb/sec
Compressor rear seal	550 F	0.22 lb/sec
Turbine rear frame	300 F	0.33 lb/sec

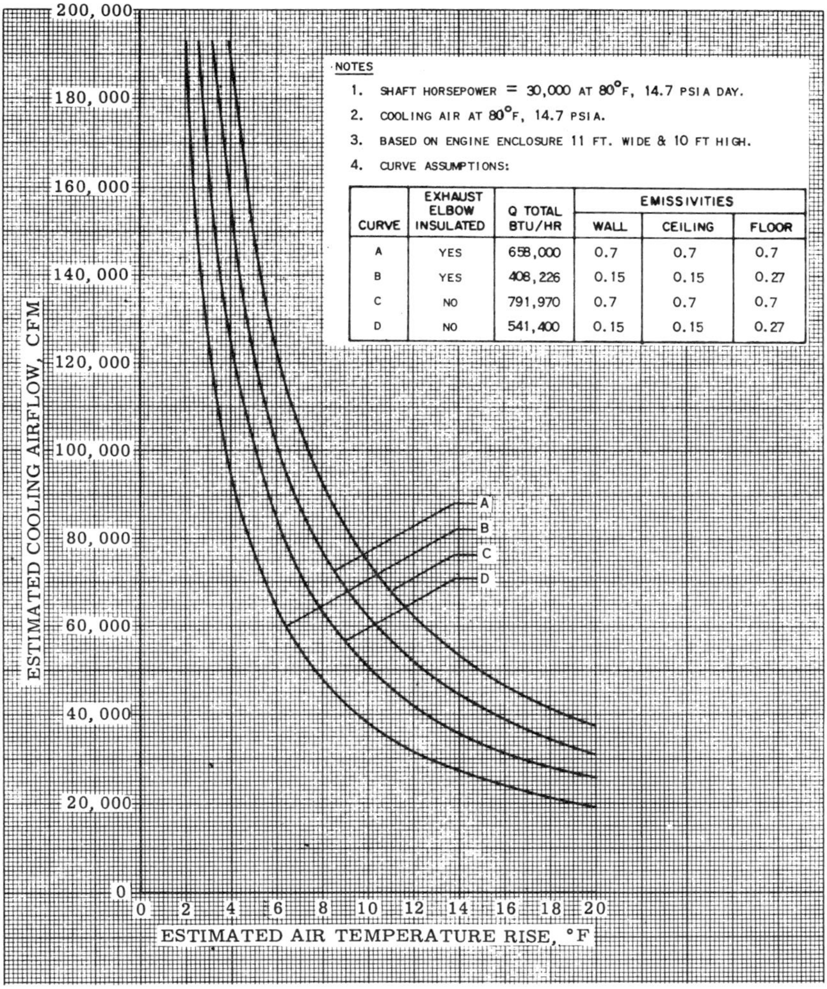

Figure 10.34. Estimated cooling-air temperature as a function of air flow, Turbo Power & Marine Systems FT4 engine [*Installation Design Criteria for Gas Turbine, Application in Naval Vessels* (1968)].

also indicates a dependence on the thermal emissivity of the enclosure inside surfaces.

In the absence of curves such as those of Figure 10.34 (supplied by the engine manufacturer), a detailed calculation of heat transferred from engine surfaces by radiation and convection is necessary in order to determine the rate of heat removal.

10.8. REFERENCES

A Guide for the Selection and Application of Resilient Mountings to Shipboard Equipment (1958), David Taylor Model Basin Report 880.

Andersen, H. C., and Zrodowski, J. J. (1959), "Co-ordinated Alignment of Line Shaft, Propulsion Gear, and Turbines," *Transactions, Society of Naval Architects and Marine Engineers,* **67**, pages 449–523.

Cashman, R. M. (1962), "Design of Marine Machinery Foundations," *Transactions Society of Naval Architects and Marine Engineers,* **70**, pages 723–748.

Clements, H. A. (1972), "Operational Experience of the S.S.S. (Synchro-Self-Shifting) Clutch, Particularly in Naval Propulsion Machinery," American Society of Mechanical Engineers, Paper 72-GT-81.

Daresta, F. G., and Smith, H. F. (1971), "Gas Turbine Propulsion for High Utilization Cargo Ships," American Society of Mechanical Engineers, Paper 71-GT-83.

Fischer, K. E., and Callow, M. J. (1970), "Development and Trialing of HMCS *Bras d'Or*—An Open Ocean Hydrofoil," American Society of Mechanical Engineers, Paper 72-GT-70.

Goode, R. W., and Logan, J. M. (1969), "Propulsion Gas Turbine Experience on the Coast Guard Hamilton Class High Endurance Cutters," American Society of Mechanical Engineers, Paper 69-GT-21.

Installation Design Criteria for Gas Turbine Application in Naval Vessels (1968), NAVSHIPS 0941-038-7010 United States Navy Technical Manual.

Kerpestein, J. B. (1972), "Gear Transmission in the Euroliner Class Vessels, *Marine Engineers Review,* April, pages 17–20.

Lauriat, T. B. (1972), "Riverine Propulsion System Lessons Learned," American Society of Mechanical Engineers, Paper 72-GT-112.

Lindsey, W. H. (1968), "The Marine Gas Turbine," Society of Naval Architects and Marine Engineers, Proceedings of the Diamond Jubilee Meeting.

LM2500 Marine Gas Turbine Installation Design Manual (1973), General Electric Report MID-IDM-2500-2, Marine and Industrial Products Department.

McAllister, P. J. (1972), "Modularization and Installation of the Gas Turbine Propulsion System in Euroliner," American Society of Mechanical Engineers, Paper 72-GT-85.

Rains, D. A., and d'Arcy, R. J. (1972), "Considerations in the DD963 Propulsion System Design," *Naval Engineers Journal,* **84**, pages 65–77.

Richardson, W. S. (1969), "The Friction Clutch Reverse-Reduction Gears for the GTS," *Journal of Engineering for Power, Transactions of the American Society of Mechanical Engineers,* Series A, **91**, pages 245–252.

The Motorship (1971), **52,** 104.

The Motorship (1973), **54,** 190–192.

The Motorship (1973), **53,** 530.

Trewby, G. F. A. (1963), "Recent Operating Experience with British Naval Gas Turbines," *Journal of Engineering for Power, Transactions of the American Society of Mechanical Engineers,* Series A, **85,** pages 46–71.

Wennberg, P. K. (1966), "The Design of the Main Propulsion Machinery Plant Installed in the USCGC *Hamilton* (WPG-715)," *Transactions, Society of Naval Architects and Marine Engineers,* **74,** pages 29–69.

Chapter

ELEVEN

AUXILIARY SYSTEMS

11.1. INTRODUCTION

The gas-turbine engine, like all of the internal-combustion engines, is a self-contained package. Accessory devices such as fuel pumps and lube-oil pumps, as well as major components such as compressor, combustors, and turbines, are integrated by the builder into a whole which the shipbuilder seemingly need only fasten in place and hook up. This last statement is a first approximation—very much an approximation—to the actual process of propulsion plant design and building as interfaced with a gas-turbine engine. As the preceding four chapters illustrate, there is an array of major considerations outside the boundaries of the engine package: integration of engine and propulsion controls, provision of high-quality fuel, design of intake and exhaust ducts in the face of several possible hazards, and alignment of the engine with its foundation and load, to name just one such consideration from each of these chapters. Additional items also demand attention from vessel designer and builder, things which are lumped in this chapter under the heading of *auxiliary systems*.

Of course, the air-supply, exhaust, and fuel systems discussed earlier are rightfully called auxiliary systems also. They have been emphasized by treatment in separate chapters, not because of their greater importance, but because there is enough to be said about them to warrant individual chapters. The systems and items treated in this chapter should not be regarded as less important because they appear to be relegated to a catch-all chapter. The lubrication system is a good example, since there can be no question that it is vital to the engine. Other systems to be discussed, such as anti-icing and fire protection, indeed may play no part in normal operations, but are vital when needs for their services arise.

The auxiliary systems discussed here are those for lubrication, fire protection, starting, and anti-icing. all of these involve interfacing between parts furnished attached to the engine and those to be furnished as part of the shipbuilding job. For example, storage tank, filters, and some piping in the lube-oil system may be ship furnished. The ship designer's main interest is assumed to be in knowing what he must provide on his side of the interface points.

Sections in this chapter are also devoted to electrical and piping considerations, which are elements in auxiliary systems in ways that should not need elucidation at this point.

The lube-oil and fuel-oil pumps are always driven from an accessory-drive gearbox mounted on the gas generator, and usually the starting device connects to the engine in this way also. Of special interest to the vessel designer is the usual provision of one or more spare connections for his use in driving ship-service items. In a small vessel, for example, such a connection might be used to drive the alternator in an automotive-type ship-service electrical system.

The gas-turbine engine as a noise source to structure and to its connecting ducting is discussed in earlier chapters. Discussion is given here of the engine as a source for direct airborne sound within the machinery space. Sound is not, of course, an auxiliary system, but its presence has an influence on vessel design, perhaps calling for sound-isolation measures within the machinery room, and between it and surrounding spaces. These measures may, in a reasonable sense, be called an auxiliary system, though obviously they do not serve the engine itself.

By the end of this chapter, only one commonly required auxiliary system has not been discussed at some place. That one is the rather modest provision usually required for cleaning the engine. Since this feature is so closely associated with maintenance, its discussion is reserved for Chapter 12.

11.2. LUBRICATION SYSTEMS

The lubrication system provides a pumped oil flow to bearings, with filtration and cooling included to maintain the integrity of the lubricant. For the most part, the system is an integral feature of the engine, packaged with it by the engine builder. The necessary pumps, for example, are mounted directly on the engine, and driven by accessory drive shafts. Some items, such as storage tanks, may be furnished by the shipbuilder. Perhaps the shipbuilder's most important responsibility is providing the means of cooling the lubricant. Such responsibilities, and the related design decisions, differ in some respects between the aircraft-derived and heavy-duty engines. The following discussions are segregated, therefore, into sections covering the two types individually.

A. Aircraft-Derived Engines

The most distinctive feature of the aircraft-derived engine is its use of synthetic lubricating oil. The synthetic lubricants are superior to the common mineral-oil alternatives in several properties, but the major incentive for this use is the higher temperatures confronting the lubricating oil in this type of engine. Temperature of oil leaving bearings may be in the neighborhood of 300 F, which is at least 100° higher than is usual with mineral oils.

The major problem presented to the ship designer by the use of a synthetic lubricant is the question of how it shall be cooled. The well-established steam propulsion practice of using cooling by sea water in a simple shell-and-tube heat exchanger is made difficult by the incompatibility of the synthetic oils with copper. The common alloys resistant to sea-water corrosion are copper based; the only feasible choice of an alloy compatible both with the lubricant and the sea water is monel. But even with satisfactory tube material, there remains the possibility of leakage of sea water into the lubricant when the two fluids are present in the same heat exchanger. Therefore, other alternatives are considered, among them cooling by fuel, cooling by air, cooling by seawater via an intermediate fresh-water cooler, and cooling by reduction-gear lube oil (which is in turn cooled by sea water). Each has the advantage of isolating sea water from the oil and of avoiding the material incompatibility problem, but each has its own handicaps.

Cooling by fuel on its way to the combustion chambers is the usual aircraft practice, and it might be expected, therefore, to carry over into aircraft-derived marine work, but the defect that renders it impractical is the comparatively poor thermal stability of the fuels used by contemporary marine engines. The thermal stability of marine diesel oil, for example, is poorer than that of JP4 or JP5, so much so that coking of the former is likely when it is confronted with the high temperature of gas-generator lubricant.

Air cooling is handicapped principally by the larger size of the coolers resulting from the poor thermal transfer and transport capabilities of air. But air cooling has some appeal for engines mounted high in the vessel, since the cooler may perhaps be readily exposed to ambient air, while water coolant piping from below may be eliminated.

The use of fresh-water cooling, with this water cooled in turn by sea water, is common marine practice, being widely used with diesel and spark-ignition engines to avoid passing sea water through their water jackets. Its disadvantage is the necessity for the additional heat exchangers, pump, and piping.

The reduction gear is always lubricated with mineral oil, cooled usually by a sea-water heat exchanger. This oil, after being cooled, can be used as the coolant in the engine-oil heat exchanger. Since the gear-oil heat load is typically much larger than the engine-oil load, the gear-oil cooler is not greatly increased in size (in the sense of a percentage increase). However, the oil-to-oil heat exchanger is larger than a corresponding oil-to-water unit because of the poorer heat-transfer efficiency of oil as a coolant. The thermal-stability problem mentioned for fuel as a coolant appears to be insignificant because the gear oil can be continuously circulated at a high rate through the cooler, whereas the fuel flow must follow the dictates of load demand.

Several of these schemes have been applied in practice. For example:

1. U.S. Coast Guard *Hamilton* class: direct seawater cooling in a monel heat exchanger [Wennberg (1966)].
2. *Admiral Callaghan*: reduction gear oil cools engine oil [Zeien, Smith, and Hirst (1969)].
3. *Euroliner* class: direct seawater cooling in a monel heat exchanger [O'Hare and Holburn (1973)].
4. U.S. Navy DD963 class: reduction-gear oil cools engine oil [Rains and d'Arcy (1972)].

Heat reduction via the lubricating oil is 5000 Btu/min for the General Electric LM2500 engine [*LM2500 Marine Gas Turbine Installation Design Manual* (1973)], and this can be taken as representative of the large (20, 000–30,000 hp) aircraft-derived engines. Note that this is equivalent to about 120 hp, or roughly 0.5% of engine output.

The lubricating oil is consumed at a slight, but perhaps significant, rate. For the LM 2500 the rate is said to be 0.3 lb/hr (7.2 lb, or about 1 gal, per day) [5]. The average rate for the Turbo Power & Marine Systems FT4 is said to be 1.2 gal/day, but with extremes as high as 10 gal/day [*Installation Design Criteria for Gas Turbine Application in Naval Vessels* (1968)]. Consumption by this engine aboard the *Callaghan* was reported to be less than 0.1 gal/hr [Zeien, Smith, and Hirst (1969)].

It is always necessary to vent lubrication systems, gas turbine or otherwise (witness the crankcase breather on the automotive engine), to minimize pressure buildups that would increase rates of consumption. Some oil escapes via the vents, a phenomenon that calls for attention by the marine designer. If the system is simply vented to the engine room—producing an oily atmosphere in that space—several bad things can result. If ventilated clutches are used (as on *Callaghan*), an oily coating may develop on the friction surfaces, thereby interfering with clutch functioning. If the engine room acts as a plenum for engine combustion air, then an oily coating may appear on the compressor blades, promoting the adhesion of other foreign matter. Of course, in many vessels neither of these conditions are found, but there remains the economic consequence of losing the expensive synthetic oil via the vents, and this alone appears to be sufficient incentive for recovery measures to be installed. The typical technique is to use electrostatic precipitators in the vents, these being devices based on the well-established dust precipitators used for cleaning boiler flue gas, and the like. They are used, for example, on both *Callaghan* and *Euroliner* [Zeien, Smith, and Hirst (1969); O'Hare and Holburn (1973)]. On the former ship, with the clutch problem mentioned above, the reduction-gear oil system is vented overboard; the cheaper mineral oil used for the gears does not justify the complication of the precipitator in this instance.

One might also raise the question of human health hazard from oily vapors in the engine-room air. This thought does not appear to have been explored in the marine literature, but if there is any such hazard, precipitation or venting overboard has an additional benefit.

Figures 11.1, 11.2, and 11.3 are diagrams of aircraft-derived lubrication systems; the first of these is the LM2500 system. The other two are the gas-generator and power-turbine systems for the FT4. As the latter separate figures make obvious, separate systems may be used for the respective units of the engine. This feature is a reflection of distinct ancestries, rather than a necessary feature of a marine propulsion engine. Note that these diagrams delineate the boundaries between the engine package and the shipbuilder-supplied items.

The supply tanks are shipbuilder supplied, though certainly their construction is strongly influenced by engine specifications, such as the need to use only those materials known to be compatible with synthetic lubricants.

Tank size is typically specified—or at least recommended—also, and the sizes may surprise readers familiar with steam-turbine lubrication systems. For example, the recommended LM2500 tank is from 15 to 100 gal in capacity [*LM2500 Marine Gas Turbine Installation Design Manual* (1973)], a size that is distinctly modest compared to the several thousand gallons in the gravity tank of a merchant-ship steam-propulsion plant of similar power (though the steam system would also contain the reduction-gear supply). The major reason for the difference is the short storage life of the synthetic lubricants. Because of this, frequent replenishment from fresh supplies is a necessity. In fact, replenishment directly from 10-gal cans is sometimes recommended [*Installation Design Criteria for Gas Turbine Application in Naval Vessels* (1968)] in lieu of replenishment from a large shipboard storage tank in which the oil might deteriorate before even entering service.

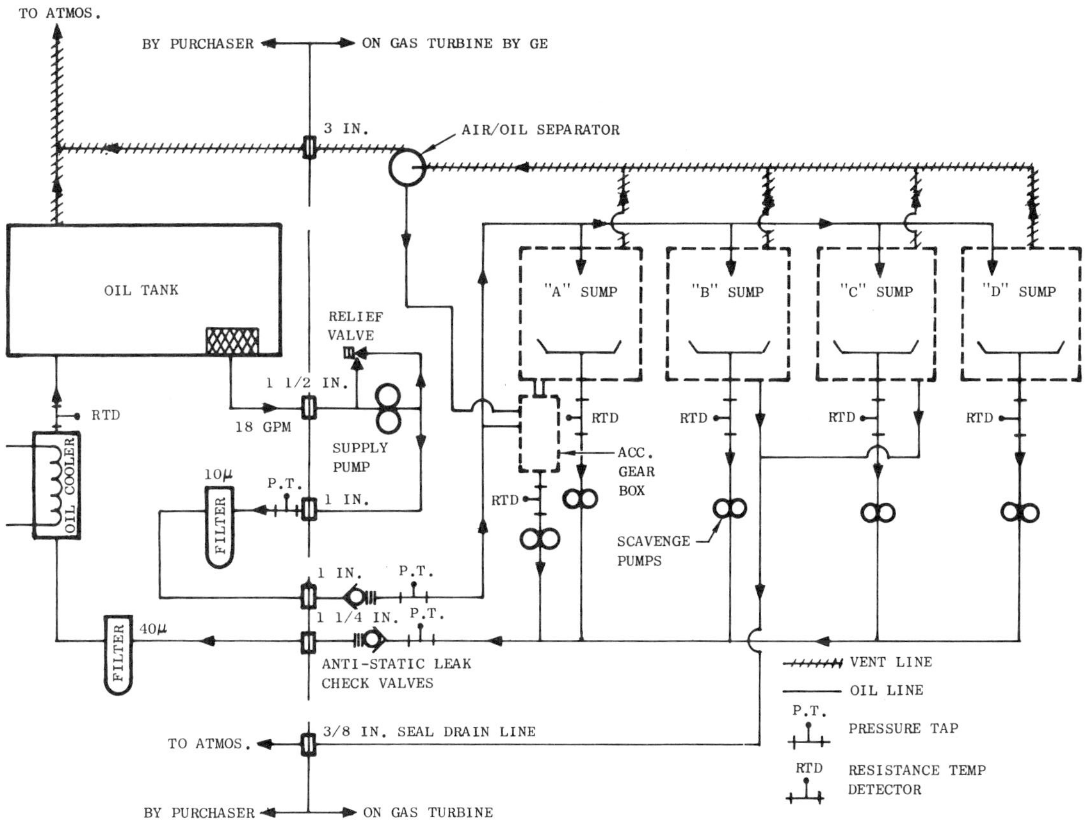

Figure 11.1. General Electric LM2500 lubricating system [*LM2500 Marine Gas Turbine Installation Design Manual* (1973)].

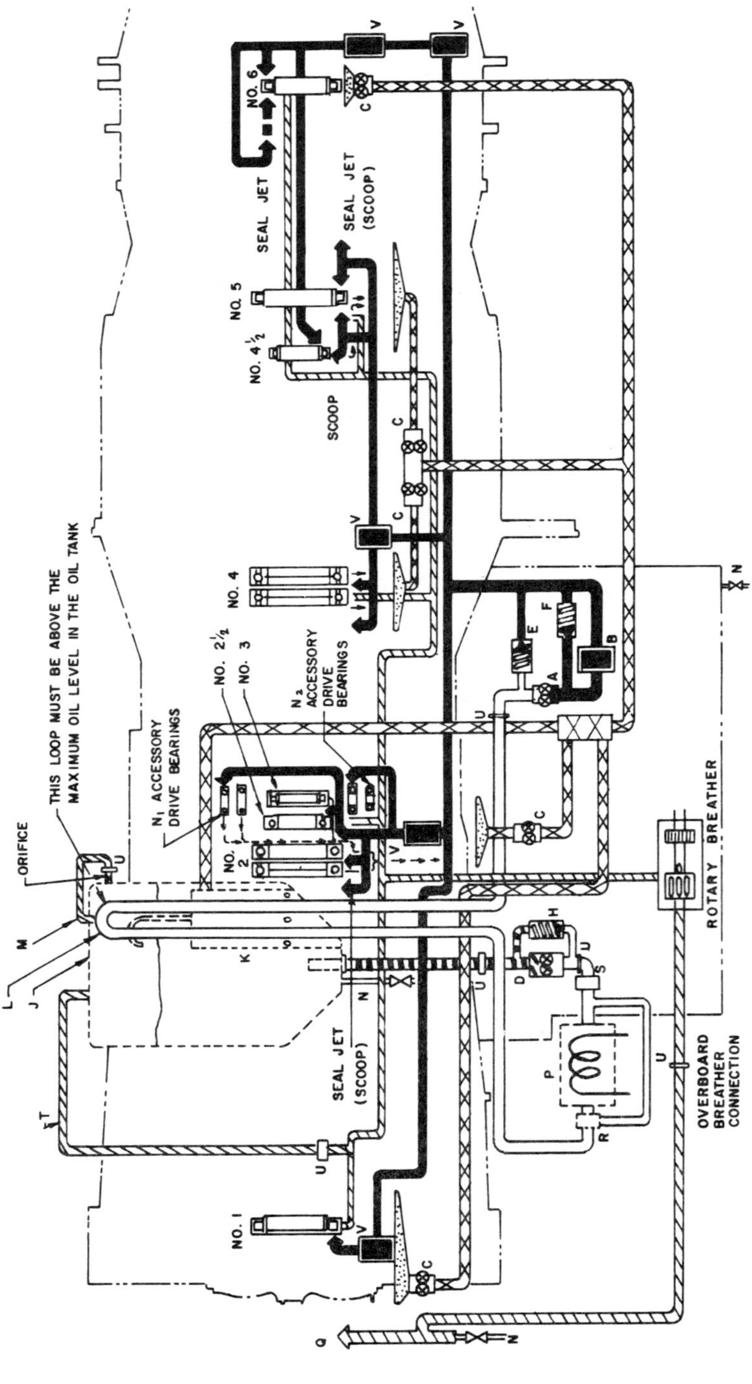
THIS LOOP MUST BE ABOVE THE MAXIMUM OIL LEVEL IN THE OIL TANK
ORIFICE
N1 ACCESSORY DRIVE BEARINGS
NO. 2
NO. 2½
NO. 3
N2 ACCESSORY DRIVE BEARINGS
NO. 4
NO. 4½
NO. 5
NO. 6
SEAL JET
SEAL JET (SCOOP)
SCOOP
NO. 1
ROTARY BREATHER
OVERBOARD BREATHER CONNECTION

LEGEND

A — OIL PRESSURE PUMP (MAIN OIL PUMP)
B — MAIN OIL STRAINER
C — SCAVENGE PUMPS
D — OIL BOOST PUMP
E — MAIN PRESSURE REGULATING VALVE
F — MAIN SCREEN BY-PASS
H — BOOST PUMP RELIEF VALVE
J — OIL TANK (P&WA OR CUSTOMER — SUPPLIED) IF OIL TANK IS P&WA — SUPPLIED, THE OIL TANK BREATHER LINE (T), THE BOOST PUMP SUPPLY LINE, AND SCAVENGE LINE TO TANK ARE ALSO SUPPLIED AND CONNECTED TO THE ENGINE-MOUNTED TANK.

K — DEAERATOR (INTEGRAL WITH OIL TANK)
L — ANTI-SYPHON LOOP LINE
M — ANTI-SYPHON VENT LINE
N — DRAIN
P — OIL COOLER (CUSTOMER-SUPPLIED)
Q — BREATHER VENT & MIST SEPARATOR (CUSTOMER-SUPPLIED)
R — BYPASS VALVE (CUSTOMER-SUPPLIED)
S — OIL FILTER (CUSTOMER-SUPPLIED)
T — OIL TANK BREATHER
U — CUSTOMER CONNECTIONS
V — OIL STRAINERS

BOOST PRESSURE
HIGH PRESSURE OIL
SCAVENGE OIL
COLLECTION POINTS
BREATHER LINES
OIL BOOST PUMP INLET OIL

Figure 11.2. Turbo Power & Marine FT4 gas generator lubricating system [*Installation Design Criteria for Gas Turbine Application in Naval Vessels* (1968)].

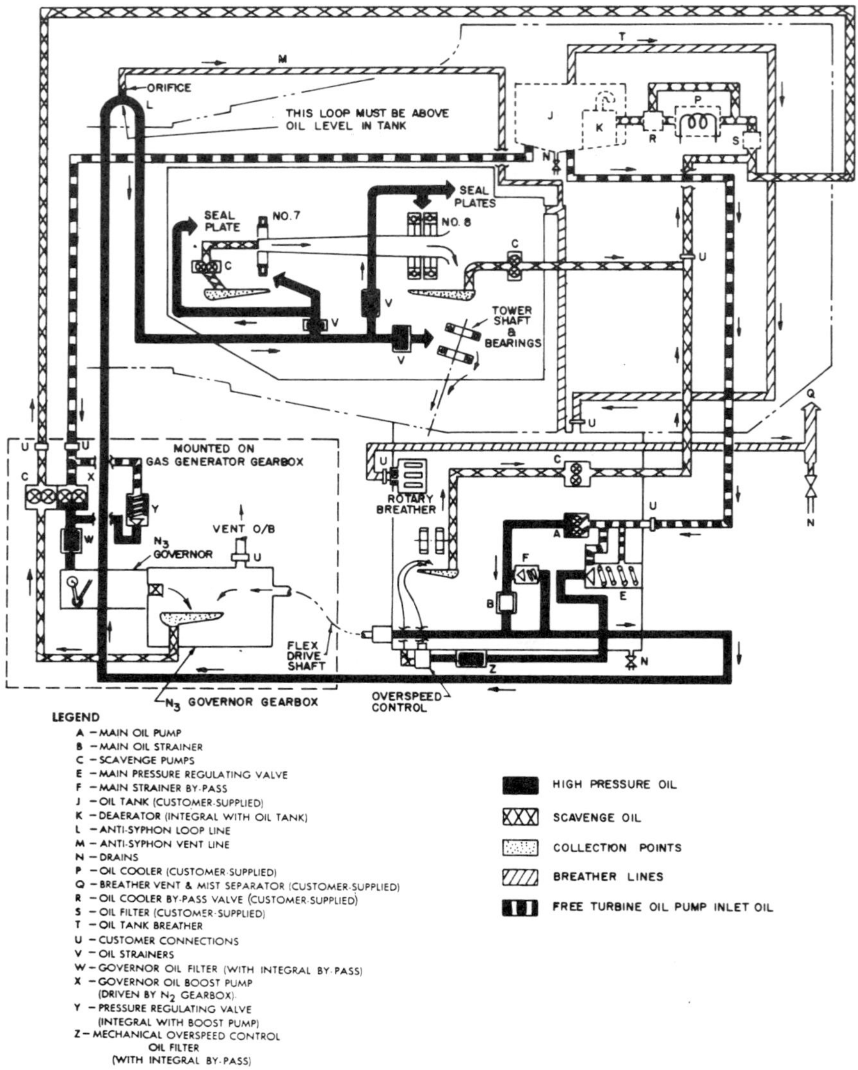

Figure 11.3. Turbo Power & Marine FT4 power-turbine lubricating system [*Installation Design Criteria for Gas Turbine Application* in Naval Vessels (1968)].

The coolers and oil-mist separators are indicated as shipbuilder supplied and have been discussed previously. The filters are also supplied by the shipbuilder, and perhaps his major point of interest is the filter rating (diameter in microns of the largest spherical particle to pass under test conditions). Filters are usually provided by the engine builder at the individual bearings, and the external filter should have a higher rating (stop smaller particles) than the internal filters to avoid general fouling of the latter units. Of course, it is most likely that the engine builder will specify what the external rating must be. The ship designer has the option of specifying either simplex or duplex external filters. In the case of base-load engines, he will probably select duplex, since it permits switching to a clean element without interrupting engine operation. He also may want to add some protection against unrecognized clogging by including external warnings. Some filters are equipped with "pop-up" devices that attract the operator's notice when pressure drop becomes excessive, or differential pressure switches can be instrumented to trip indicators or alarms at control stations.

B. Industrial Heavy-Duty Engines

Perhaps less need be said about the heavy-duty engine, for its lubrication system is more nearly conventional in that it resembles the typical marine steam-turbine system. The bearings are similarly of the sleeve type and operate at essentially the same temperature as steam-turbine bearings. Therefore, the same (mineral-based) oil is used, and because of the suitability of this oil for the reduction gear, gear and engine systems are usually combined.

The lube oil system as furnished with the General Electric heavy-duty engine is pictured in Figure 11.4. Items to be furnished by the shipbuilder are shown dotted. In spite of the remarks of the preceding paragraph, some differences from conventional steam-turbine practice can be noted. One is the absence of the gravity tank that is often used to furnish system head in steam merchant-ship practice. Three service pumps in parallel are provided instead, with reliability equivalent to the nominally infallible gravity system provided by the redundancy of multiple pumps. The main pump is driven hydraulically by the engine; Figure 11.5 gives some detail on this drive. The auxiliary pump is electrically driven by the ship-service power system, and like the main pump, can furnish full rated flow of oil when operating alone. The emergency pump has smaller capacity and is intended primarily to provide lubrication during cooldown periods, but as the name implies, provides emergency service if needed. The emergency-drive arrangements will probably vary from ship to ship according to the preferences of the designer. In at least one instance of

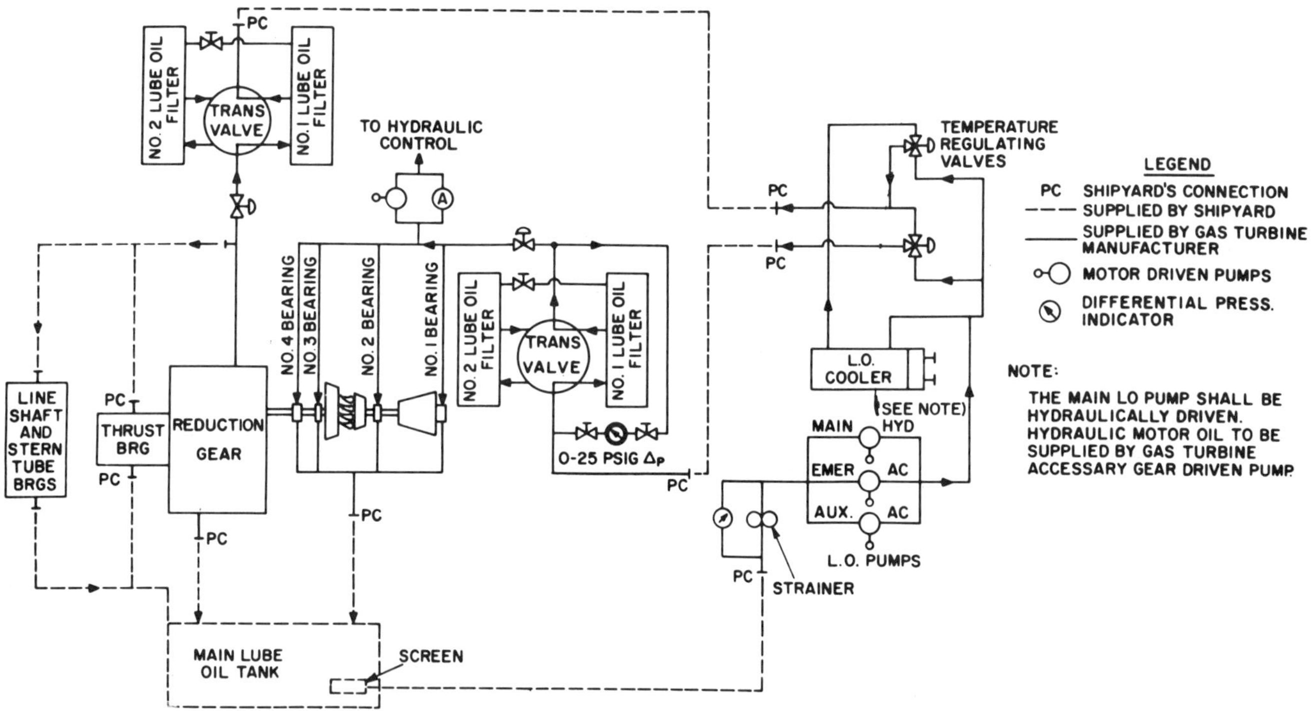

Figure 11.4. General Electric heavy-duty-engine lubricating system [Jones and Hefner (1972)].

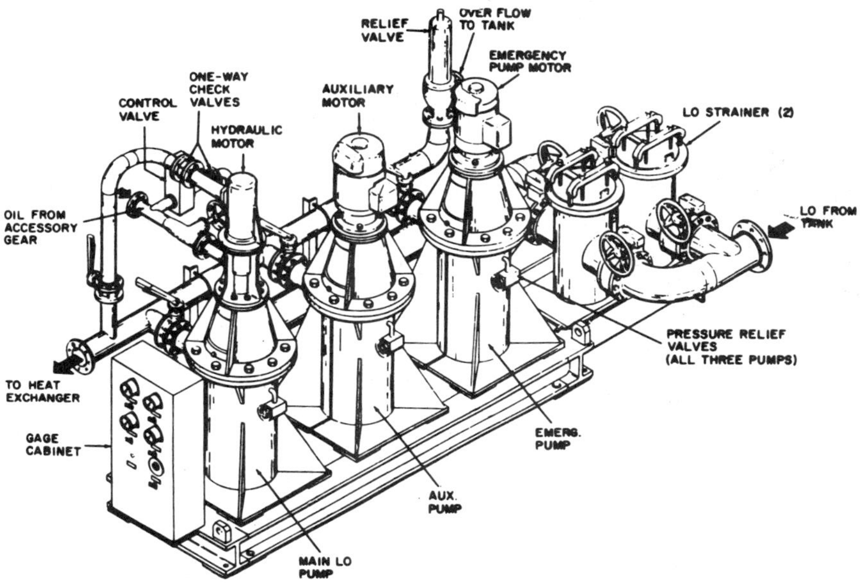

Figure 11.5. Typical pumping unit for lubricating system of General Electric heavy-duty engine. From *Proceedings of the Annual Spring Meeting,* 1973, published by The Society of Naval Architects and Marine Engineers.

contemporary design (Chevron tankers), this pump is driven by a direct-current motor energized by a storage battery. Figure 11.6 is a pictorial sketch of the pumping unit as furnished by General Electric, in this particular instance for the Australian *Iron Monarch.*

Figure 11.6 is a diagram of the General Electric heavy-duty lubrication system and, therefore, covers much of the same ground as Figure 11.4. In addition, it gives some flow and temperature data appropriate to the MS5002 unit (about 25,000 hp). The lube-oil tank capacity is not mentioned in the figure, but should be about 3000 gal for engines of the size stated. The "gravity tank" label near this tank refers to an optional reserve supply.

Another difference between usual steam practice and the system depicted by Figures 11.4 and 11.6 is the absence of a centrifugal oil purifier. Filtration is believed to be adequate to maintain oil quality. If moisture separation is felt to be necessary by the ship designer, a coalescent filter is suggested. A connection for this optional feature is shown on the tank in Figure 11.6.

The coolant medium for the oil coolers is intended to be water from the ship-service fresh-water cooling system, though other choices such as direct sea-water cooling, or even air cooling, are in the hands of the ship designer. As

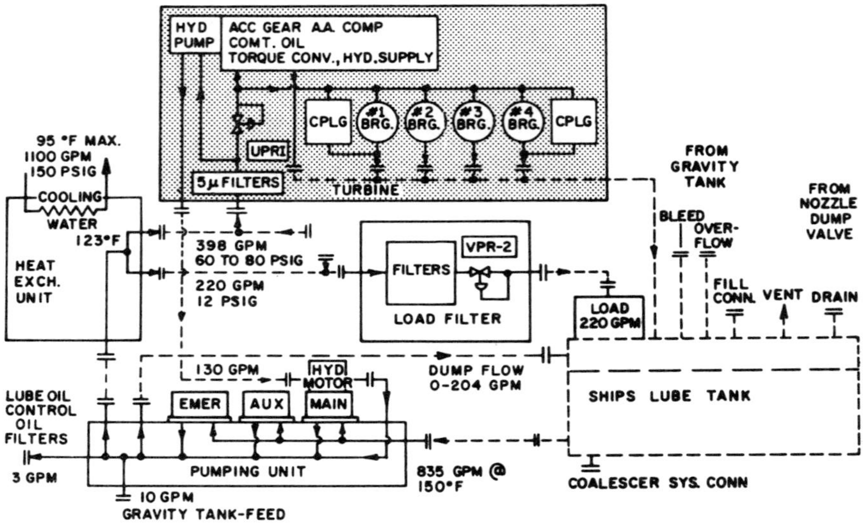

Figure 11.6 General Electric heavy-duty-engine lubricating system Svenson (1973). From *Proceedings of the Annual Spring Meeting,* 1973, published by the Society of Naval Architects and Marine Engineers.

an example of the rather unusual (in marine practice, at least) cooling arrangements that may be contrived, the Chevron tankers tube oil is cooled by a water-glycol mixture that is in turn cooled in an air radiator. Although these radiators are larger than a sea-water heat exchanger would be, their use in these vessels is to some advantage because the engines are mounted high in the ship.

11.3. FIRE PROTECTION

Let's begin by noting significant characteristics of several oils that are found in close proximity to marine gas-turbine engines (Table 11.1). Flash point is the temperature at which vapors form an ignitable air-vapor mixture. Fire point is the lowest temperature at which the vapors continue to burn after being ignited. Auto-ignition point is the temperature at which the vapors ignite spontaneously in air.

Unlike the diesel and steam-turbine engines, the gas-turbine has exposed surfaces with temperatures well above any of the auto-ignition temperatures listed, and further, the piping carrying the fuel and lubricating oils must lie close to the engine to render their necessary services. A leak—combustible fluid spraying on the hot surfaces—and a fire is highly likely. For this reason, the gas turbine requires special attention to reduce the possibility of this most dangerous occurrence and to suppress it should it occur.

The preventive measures consist largely of separating the source of combustible material from the source of ignition. Piping is routed as far from the hot surfaces as its service allows, or it may be shielded from those surfaces with sheet metal. It may be feasible to place a firewall, similar to those used in aircraft practice, within the engine enclosure to completely separate hot and cold areas. Although some of the oil-carrying piping must penetrate this wall, as much as possible should be grouped on the cold side, and perhaps all mechanical joints can be placed there. If piping joints must be close to engine hot surfaces, then the joints can be shielded in such manner that leakage is directed away from these surfaces. The shielding concept can be used also to protect the engine from the bilgewater, should the bilges lie close below the engine. This water is usually oily, and its oil film could be ignited if motions of the vessel cause splashing against the hot areas.

Since gas-turbine engines are usually enclosed or located in unmanned spaces, fixed fire-extinguishing systems and fire detecting systems are necessary.

Several choices of a fire-extinguishing agent are available among common substances that are effective in combating oil ("Class B") fires. The ones usually considered are water fog, foam, dry chemical, carbon dioxide, and the halogenated hydrocarbons. Each has its advantages and disadvantages, which may be briefly summarized as follows [condensed from *Installation Design Criteria for Gas Turbine Application in Naval Vessels* (1968)]:

1. Water fog cools and smothers fires effectively and is available from the sea in unlimited quantity. Sea water is corrosive and, therefore, re-

TABLE 11.1. FIRE CHARACTERISTICS OF OILS (Minimum Temperatures, F)
[*Installation Design Criteria for Gas Turbine Application in Naval Vessels* (1968)]

	JP5	Navy Oil Diesel	Mineral Lube Oil	Synthetic Lube Oil	Mineral Hydraulic Oil
Flash point	140	150	350	400	315
Fire point	190	200	—	490	—
Auto-ignition point	440	440	700	860	—

quires thorough washing of the engine after the fog is used. Recall from Chapter 10 that the *Euroliner* ships have a 1000-liter fresh water tank to supply the fog nozzles, thus obviating the corrosion problem (if 1000 liters is insufficient, sea water is available as backup).

2. Foam is also effective against oil fires by means of its smothering action, and is nontoxic and essentially noncorrosive. Its major drawback is the difficult clean-up job that it requires after a fire. The foam typically consists of a proprietary foaming agent mixed with water. If the water is sea water, it presents its own corrosion problem, as with the fog alternative.
3. Dry chemical (e.g., sodium bicarbonate) is effective by chemically disrupting the combustion process, and is nontoxic and noncorrosive. The action doesn't involve much cooling, so that the danger of reignition may be high after the fire is seemingly extinguished. Dry chemical is a "handy" agent for small portable systems, for example being the most popular agent in hand-held fire extinguishers for small craft. It is rarely used in large-size fixed-pipe systems. The chemical must be expelled by a pressurized gas that is contained in the extinguisher in the small portable units.
4. Carbon dioxide smothers oil fires effectively by diluting the atmospheric oxygen, an action that is usually enhanced by its high density (about 1.5 times that of air). It performs very little cooling, so that a danger of reignition continues after the fire is extinguished. It is nontoxic and noncorrosive.
5. Halogenated hydrocarbons smother fire in much the way carbon dioxide does. They are noncorrosive, but all compounds in this class have some toxicity. An advantage over carbon dioxide is the lower weight of an equivalent system, due largely to the lower vapor pressure, which reduces scantlings of pressurized containers and of piping. A relative disadvantage is the much higher cost of the fluid.

An essential feature of any fire-fighting technique is the immediate stoppage of fuel flow to the engine, since it may be a spray of fuel that is feeding the fire. Lubricating-oil flow cannot be shut off directly, but stopping fuel stops the engine, which in turn stops the lube-oil pumps. A further incentive for stopping the engine is present if it draws its combustion air from the machinery space: such an engine rapidly pumps the extinguishing agent out of the space, as well as drawing in fresh air to supply the fire. In the instance of an enclosed engine with a ducted intake from the outside, the enclosure ventilation performs the same undesirable function (though its flow rate is much less than the main flow) and obviously must be stopped also.

The usual practice of isolating a gas-turbine engine from direct observation by the operators requires that remote fire sensing be used. The sensors are typically either temperature sensors, rate of temperature rise detectors, or flame (visible radiation) detectors. They must be rather artfully located to scan all areas where fire is likely to develop, yet avoid the several situations that might cause false alarms. For example, radiation sensors must not be deceived by normal enclosure illumination, temperature sensors must not be located where temperature can be high in normal operation, and rate-of-rise sensors must be set above the rate produced by engine startup. False signals are of serious consequence: if acted on, they at the very least cause an inconvenient engine shutdown.

The *Euroliner* fire-detection system is outlined in Chapter 10, and the discussion can be amplified here to illustrate an approach to avoiding the false-alarm hazard. Both temperature sensors and flame detectors are installed in the engine enclosures. If either set trips, an alarm is actuated on the machinery-control console. The subsequent course of action is left to the judgement of the operator, who presumably can seek additional input before deciding to actuate the extinguishing system. However, if both sets of sensors trip, the extinguishing system (water fog) is actuated automatically.

11.4. ANTI-ICING PROVISIONS

Formation of ice at the combustion-air weather intake (especially on the moisture separators), and at the compressor inlet (on bellmouth, nose cone, and guide vanes), is a potential problem for gas-turbine engines operating in cold environments. Protection is usually provided by hot air bled from the compressor discharge, or by electrical resistance heaters.

For an example, the General Electric LM2500 engine is optionally equipped with a manifold and valve through which a small part of its compressor-discharge flow is tapped for purposes of heating the incoming air. A flow of 2–3 lb/sec is supposedly sufficient [*LM2500 Marine Gas Turbine Installation Design Manual* (1973)]. Figure 11.7 shows the amount of air available and its temperature. There is, of course, a performance penalty suffered because of the loss of this air from power production; the right half of the same figure demonstrates the consequences to fuel rate. For instance, the 3-lb/sec bleed suggested above would increase fuel rate at full power by about 47%.

It is intended that the hot air be injected into the approaching air stream at a point and in a manner such that it is generally and uniformly heated before reaching the compressor inlet. Injection at the demisters near the duct inlet, with injection distributed across the face of these units should accomplish this objective, as well as protect the demisters. Figure 11.8 indicates in schematic fashion an arrangement of this type. The engine in the figure is the General Electric heavy-duty engine with regenerator; note that the compressor air is tapped off after it passes through the regenerator, a measure that supplies air in the 900–950 F range. A uniform inlet-air temperature of about 38 F is obtained in this way with ambient temperatures as low as −20 F [Svensen (1973)].

An alternative heating scheme is to use electric-resistance heaters applied to the surfaces that are subject to icing. About 45 kW might be required for a large (i.e., 20,000–30,000 hp) engine [*Installation Design Criteria for Gas Turbine Application in Naval Vessels* (1968)]. Or combinations of electric and hot-air heating might be used. For example, the *Admiral Callaghan* as originally built (FT4 engines) used electric-resistance heaters to warm the compressor bellmouth and hot-air bleed to warm the nose-cone and inlet guide vanes. A separate hot-air system protected the moisture separators. Early experience indicated that the latter protection was not needed, even in severe weather of the North Atlantic winter. Meanwhile, the system at the compressor entrance operated about 2% of the time during winter months

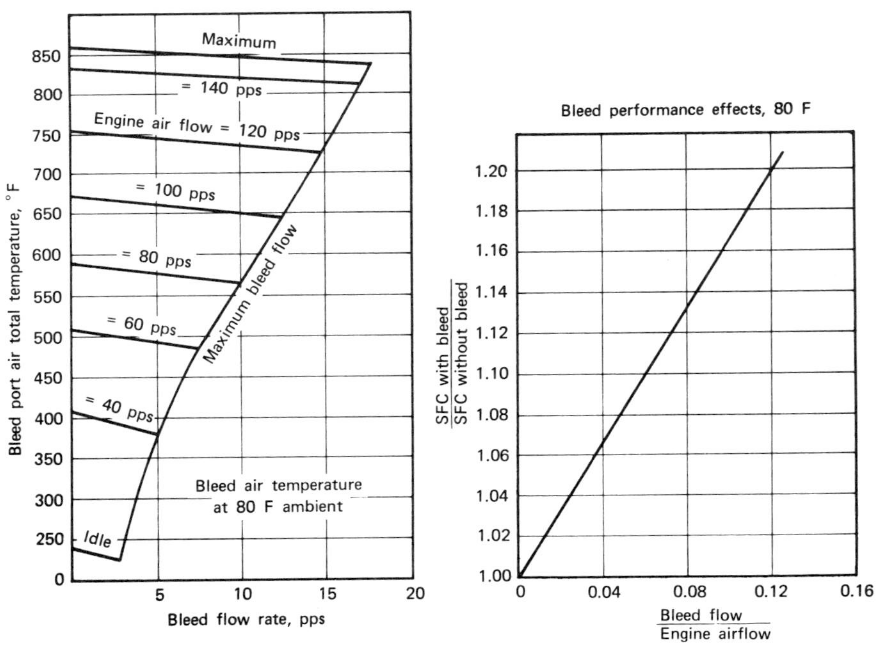

Figure 11.7. Compressor-bleed-flow data for General Electric LM2500 engine [*LM2500 Marine Gas Turbine Installation Design Manual* (1973)].

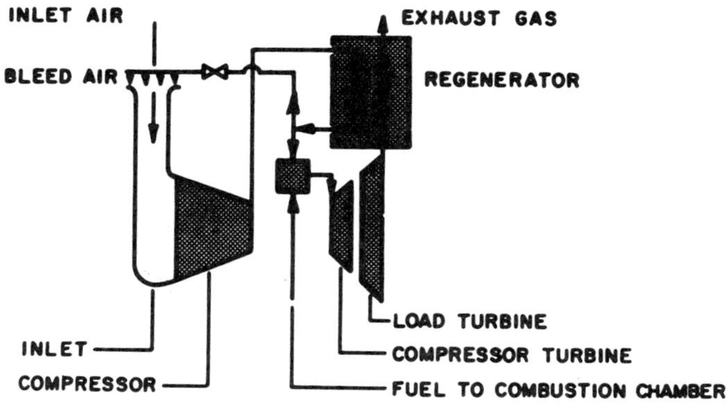

Figure 11.8. Anti-icing schematic for General Electric heavy-duty engine [Svenson (1973). From *Proceedings of the Annual Spring Meeting,* 1973, published by the Society of Naval Architects and Marine Engineers.

[from Zeien, Smith, and Hirst (1969), which covers the first 3 years of service].

11.5. STARTING

The starter is a device that uses an external source of energy to spin the gas generator up to a speed at which it can operate self-sustaining and with sufficient excess torque of its own to accelerate to normal idling speed. The principal function of the starter, therefore, is to supply the torque necessary for this job, but it is also used to motor the gas generator for prestart purging, for rotating cool-down after operation, and for the rotation that must accompany washing of the engine. Typically, the starter is some form of motor or turbine powered either by a stored energy source (e.g., a battery) or by a ship's auxiliary system, or it may be a diesel engine or smaller gas turbine (which in turn have their starters).

The torque required is based on the engine builder's prediction of the gas generator's torque versus speed curve, and on the need of the particular application to attain a specified start-up time. The latter factor determines how much of an excess torque is needed for acceleration, since this excess must be supplied by the starter before firing takes place, and in part even after firing begins. A typical torque prediction is shown by Figure 11.9. The portion of the curve labeled "motoring torque" is simply that external torque required by the starter to spin the gas generator in the absence of firing. If firing occurs at the *light-off speed,* then the external torque required falls until at *self-*

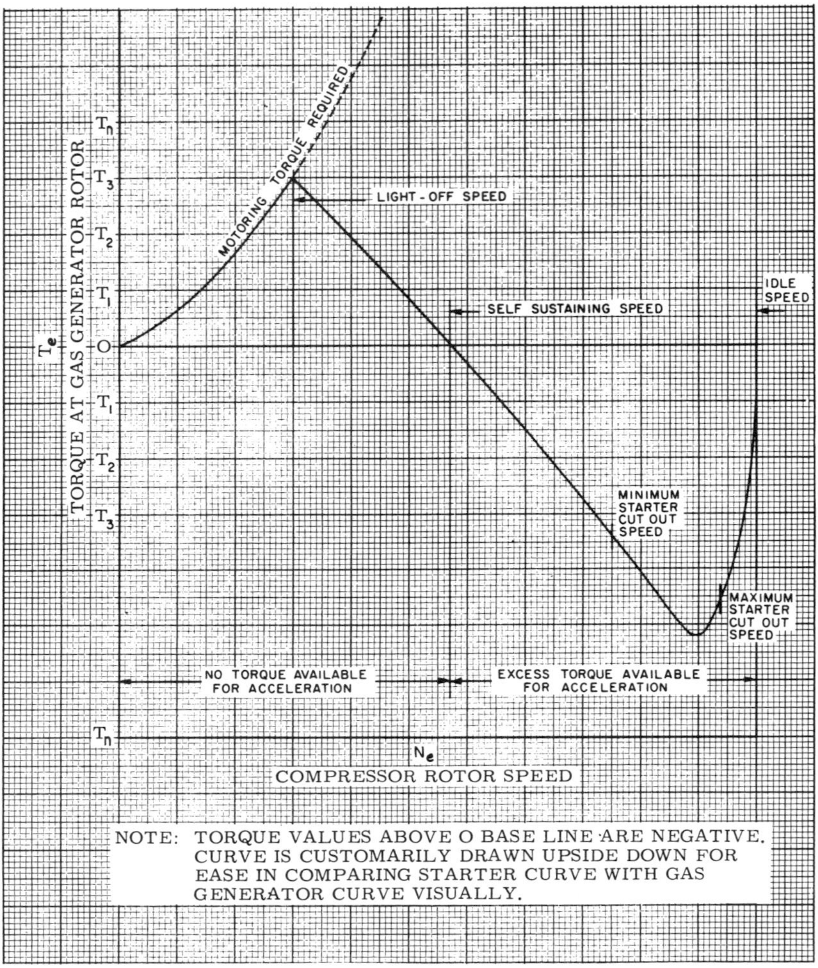

Figure 11.9. Typical torque-speed characteristic of gas generator from rest to idle speed [*Installation Design Criteria for Gas Turbine Application in Naval Vessels* (1968)].

sustaining speed the gas generator turbine is supplying just enough torque to maintain speed without assistance from the starter. The *no torque available* label near the abscissa implies that the engine itself cannot develop the excess torque needed for acceleration over this speed range. At higher speeds, the continued fall shows the negative need for external torque, or the excess of torque generated by the gas generator that causes its continued acceleration.

The abrupt change in the curve—its rapid retreat to zero torque—is a result of the cutback in fuel flow that must be made to avoid acceleration past the idling speed of the gas generator. Note that the starter is kept engaged well beyond the self-sustaining speed, supplying part of the excess needed for acceleration, as noted above.

The greatest demands for excess torque in propulsion service doubtless accompany the quick startup needs of boost engines in military applications, where the engine should be ready to begin assumption of load in less than a minute after the start sequence is initiated. As an example, trials of the FT4 engines in the *Hamilton* class U.S. Coast Guard cutters showed an average of 43 sec required to start and accelerate the gas generator to its idling speed [Wennberg (1966)].

The maximum torque rating of the starter usually can be read from the *motoring torque required* curve of Figure 11.9 at the point where it intersects the highest motoring speed required, whether this speed be for starting or for other duties. As an example, the FT4 starter maximum torque must be approximately twice that needed at light-off speed (i.e., about $2 \times T_3$ in the figure [*Installation Design Criteria for Gas Turbine Application in Naval Vessels* (1968)]), enabling it to run the gas generator for washing at about 50% above light-off speed. An exception to this criterion for establishing maximum torque might occur if the previously mentioned need to accelerate rapidly were to demand an unusually large torque. Another exception could occur if the breakaway torque—the torque needed to start the gas generator from rest—is excessive. This situation is unlikely in an aircraft-derived engine, because of its use of ball bearings, but is possible in an industrial engine which uses sleeve bearings.

Figure 11.10 is quite similar to Figure 11.9 showing much the same information for the Avco TF12 engine. It is of additional interest because it indicates the torque available from the starter and because it shows the torque required at two different temperatures. Note that under very cold (−30 F) conditions, the greatest torque requirement is at 0 rev/min, that is, there is a nonzero breakaway torque, even though this is an aircraft-derived engine. The breakaway torque is not great enough to be the prime factor in fixing starter-torque rating.

Although starter torque may be the first consideration in its design or selection, the total store of energy available to the starter is of the same degree of importance since it determines the length of time over which the engine can be motored. The washing procedure (see Chapter 12) is likely to require the longest motoring run at fairly high rev/min, but repeatability of starts is also an essential consideration; it may be necessary to make false starts for checkout purposes (e.g., to see that lube-oil pressure builds up), or an intended start may be aborted by failure to meet the logic of the starting sequence (see

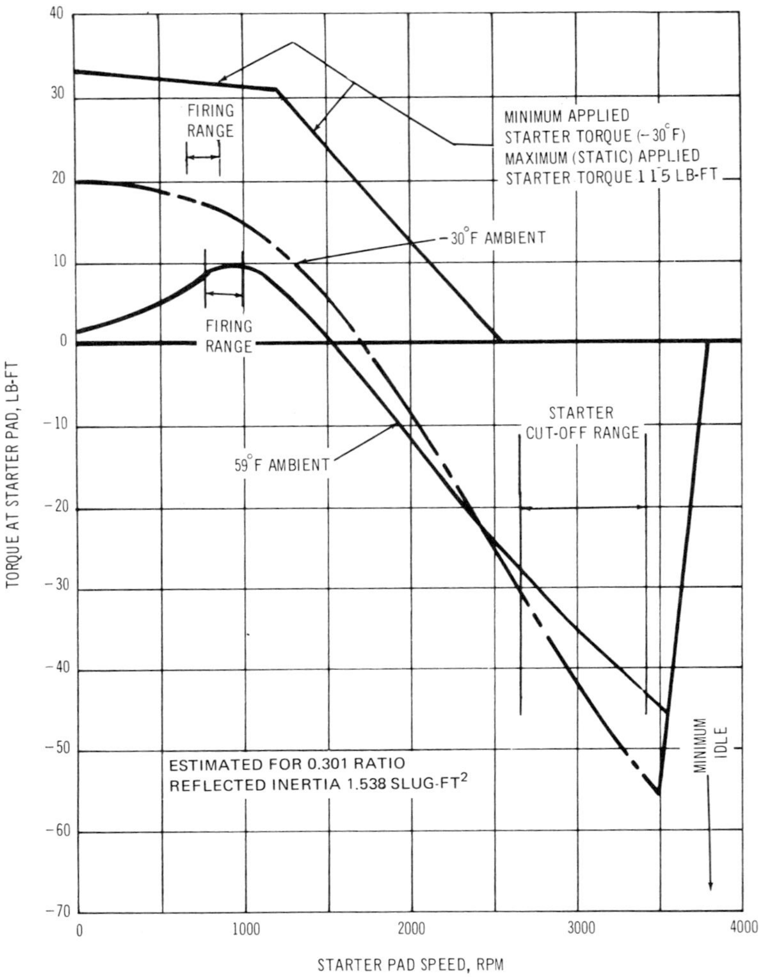

Figure 11.10. Starting torque-speed requirements for Avco-Lycoming TF12/TF14 Engine (from manufacturer's application data).

Chapter 7). Sufficient energy must be stored for several starts, at least. In the case of commercial ships, the classification society may specify how many successive starts must be possible with the energy store immediately available. Of course, if the starter is an engine, or is a motor driven by the ship service electrical system (for instance) without intermediate energy storage, then the direct "energy stored" is that of the ship's fuel supply, and the discussion of this paragraph is superfluous.

Starting systems are classifiable by the type of energy or working fluid that actuates them. The common types for ship propulsion by aircraft-derived engines are hydraulic and pneumatic. Battery-stored electrical systems are used for smaller craft, and even manual starting is feasible at the lowest end of the power scale. Industrial-type engines are more commonly started by diesel engines, steam turbines, or electric motors; in the last instance supply from the ship service electric system, rather than from a battery, is likely.

The electric starting system that may be used with smaller engines is similar to the common automotive or truck system, employing an electric starting motor, battery for energy storage, and attached alternator to recharge the battery. This system suffers the handicap of excessive weight, due largely to the weight of the battery. In a small vessel this handicap may be offset by the advantage of not requiring a pneumatic or hydraulic system that serves no purpose save starting, but a ship may well have such a system for other purposes (e.g., compressed air for use by portable tools and by pneumatic controls). In any case, the increasing weight of batteries as the engine gets larger soon makes this type of electrical starting unattractive. As an example, Avco Lycoming offers electrical starting as standard on its TF series marine engines up through the TF25 (2500 hp), with hydraulic as an option. Their largest engine, TF35, uses hydraulic starting as standard, with electrical as an option. Battery-energized electrical starting seems an unlikely choice for powers greater than this.

A pneumatic starting system is sketched in Figure 11.11, taken from General Electric literature on the LM2500 engine (hydraulic starting is also available with this engine) [*LM2500 Marine Gas Turbine Installation Design Manual* (1973)]. The starter itself is an assembly in a single housing consisting of an air turbine, planetary reduction gear, overrunning clutch, and a splined drive shaft, mounted on the gas-generator accessory-drive gearbox. Note that two sources of the requisite compressed air are indicated. Indeed, there are several options available to the ship designer in arranging the supply, or it might be said that any arrangement that provides air, clean and dry, at the required pressure and quantity, serves.

The following tabulation [from *LM2500 Marine Gas Turbine Installation Design Manual* (1973)] lists several significant features of the LM2500 starter, and Figure 11.12 gives information on air flows and pressures.

Starter weight	32 lb
Duty cycle	Five consecutive starts with 30 sec minimum rest between starts
Motoring	Two minutes motoring, followed by five minutes rest
Start duration	Starter energized 35–45 sec

The several types of air motors, such as the rotary vane type or the reciprocating piston type, may be used as alternatives to the turbine starter mentioned

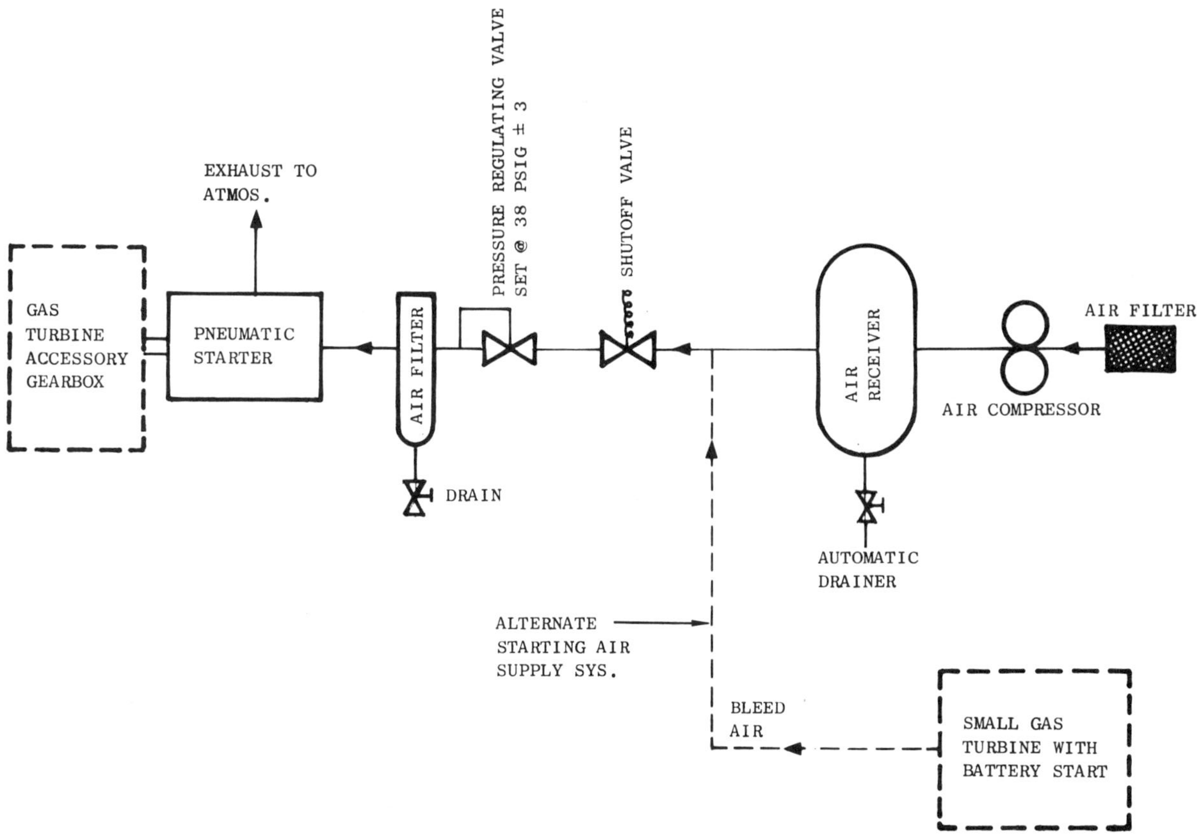

Figure 11.11. General Electric LM2500 pneumatic starting system [*LM2500 Marine Gas Turbine Installation Design Manual* (1973)].

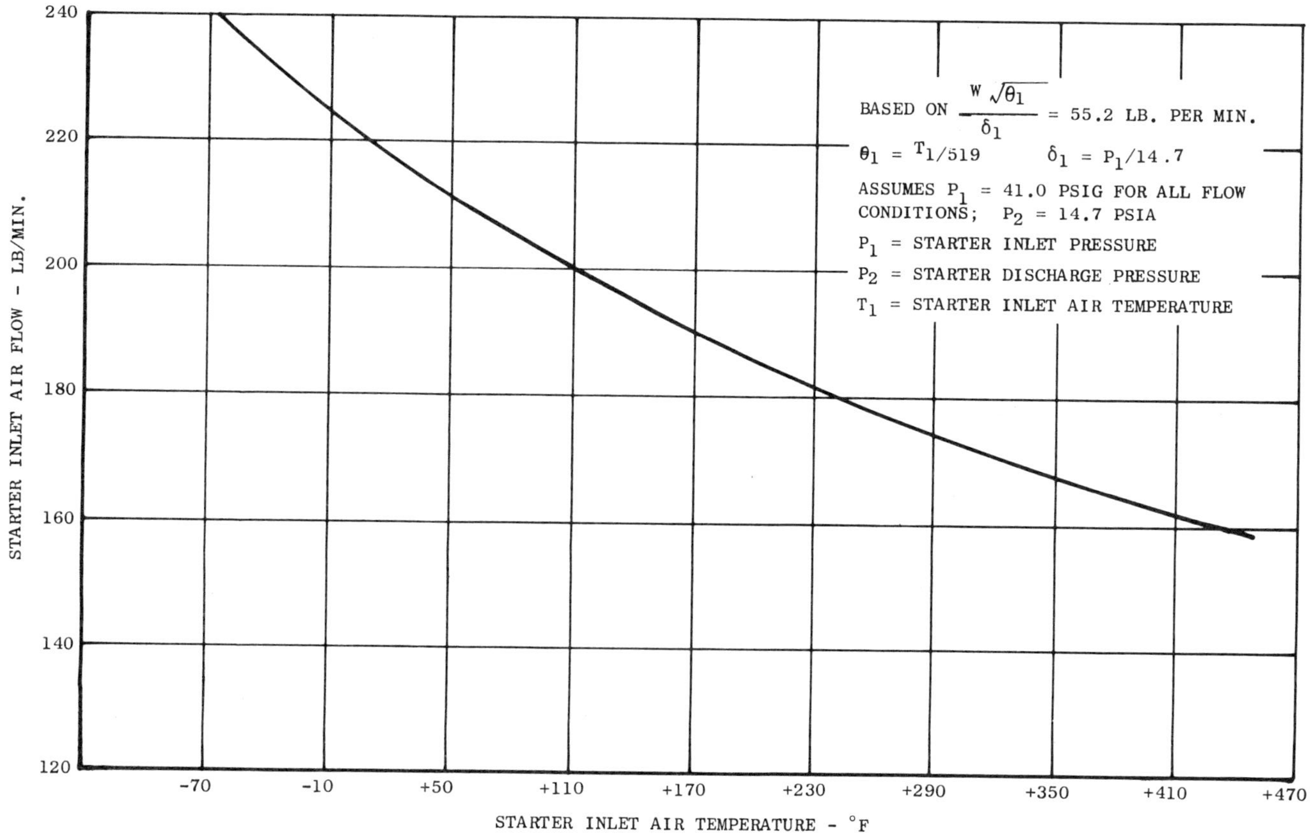

Figure 11.12. General Electric LM2500 pneumatic starting system—estimated air flow as function of temperature [*LM2500 Marine Gas Turbine Installation Design Manual* (1973)].

just above. And as also mentioned, there are options in selecting the source of compressed air. Rotary, reciprocating, and turbine compressors are widely used in compressed air service, and there is the possibility of using air bled from the compressor discharge of an auxiliary gas turbine engine.

A hydraulic starting system is illustrated by Figure 11.13. The starter itself is a variable-displacement hydraulic motor; its energy source is a variable-displacement hydraulic pump driven by the auxiliary-generator diesel engine. Other power sources for the pump are feasible, such as motors energized from the ship service power system. As examples, the *Admiral Callaghan* and the *Euroliner* vessels both use hydraulic starting; the former uses its diesel generator engine to drive the hydraulic pumps [Zeien, Smith, and Hirst (1969)], while the latter use 100-hp electric motors for this purpose [Jones and Hefner (1972)]. Note that the figure includes a heat exchanger, this being required to remove the heat generated by prolonged motoring, such as is necessary for washing cycles (though it would probably be in use whenever the starter is energized). The cooling medium would normally be sea water. The figure also shows the reservoir, relief valves, and filters that are normal components in most hydraulic systems.

A hydraulic starting system is one of the two options available with the General Electric LM2500. The following tabulation [taken from the *LM2500 Marine Gas Turbine Installation Design Manual* (1973)] of data for that engine illustrates some of the magnitudes to be expected in hydraulic systems for large engines (all data approximate):

Hydraulic-pump motor power	225 hp/motor
System pressure	5000 psig
Weight of starter on engine	100 lb
Weight of supply system and piping	1800 lb
Maximum oil flow	65 gpm
Heat-rejection rate	1800 Btu/min

The usual starter for the General Electric heavy-duty engine is a three-phase, 60-Hz induction motor, with torque converter, transfer gear case, and clutch connecting to the engine's accessory gear [Svensen (1973)]. About 100 hp is required for the smallest of these engines (MS1002, 5000 hp) to 750 hp for the largest (MS7002, 55,000 hp). A diesel engine or steam turbine may also be used as starters for this gas turbine. The latter choice would be appropriate when a waste-heat steam system is used with the propulsion engine, since such systems are usually backed up by fired boilers to provide steam when the propulsion engines are off the line. The *John Sargeant,* for instance, was started in this way.

Figure 11.14 is a diagramatic profile of the General Electric heavy-duty engine in an arrangement much like that of the Chevron tankers. The 250-hp

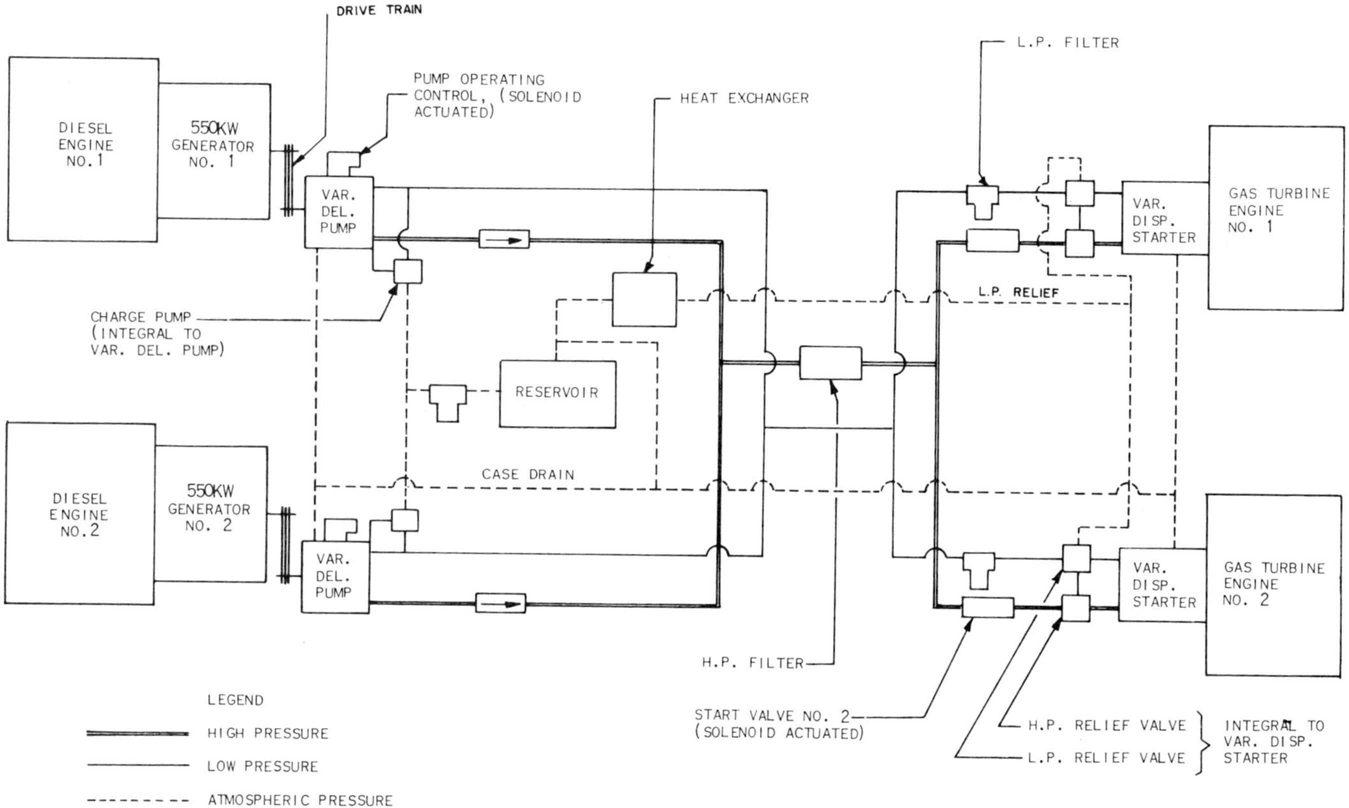

Figure 11.13. Typical hydraulic starting system [*Installation Design Criteria for Gas Turbine Application in Naval Vessels* (1968)].

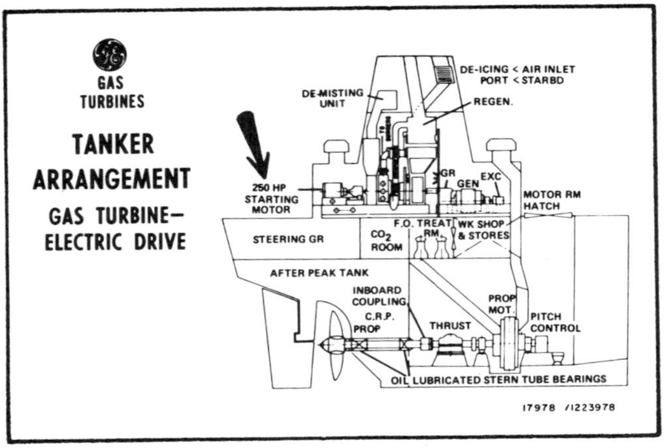

Figure 11.14. General Electric heavy-duty machinery installation profile showing starting motor (from General Electric literature).

electric motor starter used by these ships shows clearly at the aft end of the propulsion engine.

As a final point in this section, let's sketch the overall starting procedure outlined for the *Callaghan* machinery. The following description is quoted from Zeien, Smith, and Hirst (1969):

1. Hydraulic-start system oil-replenishment pump is started from the control console.
2. A main hydraulic start pump is clutched into one of two diesel generators so fitted.
3. The gas-generator crank button is depressed and the gas-generator second-stage compressor rotor brought up to approximately 1800 rev/min. Gas-generator lube-oil pressure gages are monitored to note rise of lube pressure of engine-driven lube-oil pumps.
4. Gas generator stop button is depressed. While the gas generator is coasting down, an inspection is made inside the enclosure to insure that all is in good order.
5. The crank button is again depressed. As the gas-generator compressor passes through 1200 rev/min, the start button is depressed to turn on igniters and then fuel. It is held as the gas generator accelerates to idle speed and the power-turbine rotor speeds up so that its attached lube oil pump develops satisfactory pressure. This sequence takes approximately 30–40 sec.

6. A last check is made of the gas generator and power turbine by going inside the enclosure for about 30 sec.
7. Steps 3 through 6 are repeated for the other engine. These actions, which bring the engines to idle speed ready to assume load, require about 15 min for one man to accomplish.

Recall from Chapter 7 that certain logical steps must be carried out automatically by a starting sequencer. This control device is actuated in step 5 when the start button is depressed. Under normal circumstances, the operator is not concerned with its duties. He is affected by its performance only if some malfunction or improper logic causes the sequencer to abort the start.

11.6. ELECTRICAL SYSTEMS

Some electrical items are normally a part of the gas-turbine engine's accessory equipment and are furnished attached by the engine builder. The most direct concern of the ship designer is to furnish the power required by the equipment. Description of the electrical outfit of a single engine, the General Electric LM2500, serves our purpose here of illustrating this outfit. The following discussion thus applies only to that engine (source is the *LM2500 Marine Gas Turbine Installation Design Manual*), except that a few remarks in the last paragraph are noted as being general, or as applying to other engines.

Wiring is provided to instrumentation sensors, to certain actuators, and to combustion chamber igniters. The sensors include the following items:

1. Chromel-alumel thermocouples that measure the power turbine inlet temperature, there being four sets spaced around the circumference of the engine at the appropriate axial location.
2. Resistance thermometers to measure temperatures of lube oil leaving the four main bearings and the accessory gear box.
3. Two vibration sensors, one on the gas generator and one on the power turbine.
4. Two power-turbine speed pickups.
5. Main fuel-control throttle-lever position switch.
6. Starter-overspeed switch.
7. Gas-generator speed pickups.

The actuators include two fuel-shutoff valve solenoids and the power-turbine topping governor electrohydraulic actuator. The power-turbine topping governor and the power-turbine overspeed switch, which controls the fuel shutoff valves, are mounted off the engine and are connected to the speed pickups, and to their respectively controlled actuators, by appropriate cabling. Power supply, which must be provided by the ship, is 18 V d.c. Maximum current is 3.2 A, drawn by the power-turbine overspeed circuit. It is recommended that this power be supplied via batteries so that overspeed protection will not shut down the engine upon loss of ship-service power; shutdown would occur since the fuel shutoff valves are arranged energize-to-open.

The two igniters are supplied with 115-V, single-phase, 60-Hz power, about 2 A being required during the ignition phase of the starting cycle. This voltage is stepped up by transformers, rectified, and used to charge a capacitor bank which produces a 15- to 20-kV spark in the combustion chamber at a rate of twice per second.

It has been mentioned earlier that the LM2500 engine uses compressor-bleed air for anti-ice protection (if needed) at its compressor inlet, whereas other engines may use electrical-resistance heaters. In the latter instance, the rather modest power requirements typified by the LM2500 would be radically increased. About 45 kW is said to be necessary for the anti-icing function of an engine in the 20,000 hp range [*Installation Design Criteria for Gas Turbine Application in Naval Vessels* (1968)]. This seems to be the only likely deviation of a major nature (major in the sense of power magnitude) from the system described in this section. Of course, it is expected, on the other hand, that engines of different manufacture differ in the details of their electrical systems.

11.7. PIPING

The ship-propulsion gas-turbine engine must be supplied with its fuel via piping from a system external to the engine package itself, and it is noted in Section 11.2 that parts of the lubrication system are often supplied by the shipbuilder. In addition to these major items of piping, there are likely to be certain lesser items that also require piping work by the shipbuilder.

Typical of such piping are drains from several points in the fuel system, with piping led to a waste-fuel tank. This tank, which would normally be supplied by the shipbuilder, need not be large, since leakage rates and the amount of oil to be discharged by a deliberate drainage are typically small. For example, the leakage rate from the General Electric LM2500 is said to be about 5 cm^3/min during full-power operation [*LM2500 Marine Gas Turbine Installation Design Manual* (1973)], with about 750 cm^3 to be drained when the engine is shut down. Drain piping and vent piping from engine-bearing lube-oil pumps may be needed. In this case the likely destination of the drainage is the main lube-oil tank. If oil-mist precipitators are installed in lube-oil vents, these too require drains to return the recovered oil to the system.

Several other piping systems that may not connect directly with attached engine piping are usually provided by the shipbuilder. One has already been mentioned in this chapter: the fire-protection system. In the following chapter, a discussion treats the water-washing arrangements that are required by sea-going gas-turbine engines. A rather simple arrangement to supply water under a moderate pressure is required. Such a system is shown schematically in the next chapter by Figure 12.11. Some of the wash water may collect in the intake plenum and at the lowest point of the exhaust line, making necessary the installation of drains (to the bilges, probably) from these places. The possibility of rain or spray entering the intake and exhaust systems also is a second justification for the provision of these drains.

The need for anti-icing features has been discussed earlier in this chapter, and it was noted that hot air bled from compressor discharge is one of the methods available to meet this need. If used only for the immediate intake area, that is, compressor bellmouth and nose cone, the requisite piping is likely to be an integral part of the engine and consequently beyond the scope of the discussion here. However, if this air is used to protect the moisture separators and inlet screens, then the distribution piping is an item to be designed and furnished by the shipbuilder.

11.8. ACCESSORY DRIVES

Vital accessories such as fuel and lube-oil pumps are invariably driven by the compressor turbine, usually through the agency of an accessory-drive gear box mounted on the compressor casing. This device usually serves also as the input connection for the starter. In addition, spare connections are often provided so that the ship designer may use the propulsion engine as a drive for ship-service items that need run only when the engine is running. An example might be a pump to circulate coolant through the engine lube oil cooler.

Figure 11.15 is a three-view outline drawing of the Avco Lycoming TF25 engine in which the accessory gearbox, with attached pumps and starter motor, is evident on top of the compressor housing. Since this engine is a rather small one, the accessory package appears as a major component in the picture (see also Figure 1.6). On ship-size engines such as the LM2500 and the TF4, the comparable package is dwarfed by the engine itself. Figure 11.16 is a sectional view of the TF4 gas generator on which the accessory gearbox shows, and which illustrates the point just made about comparative sizes.

The power taken from the accessory drives is necessarily limited and usually

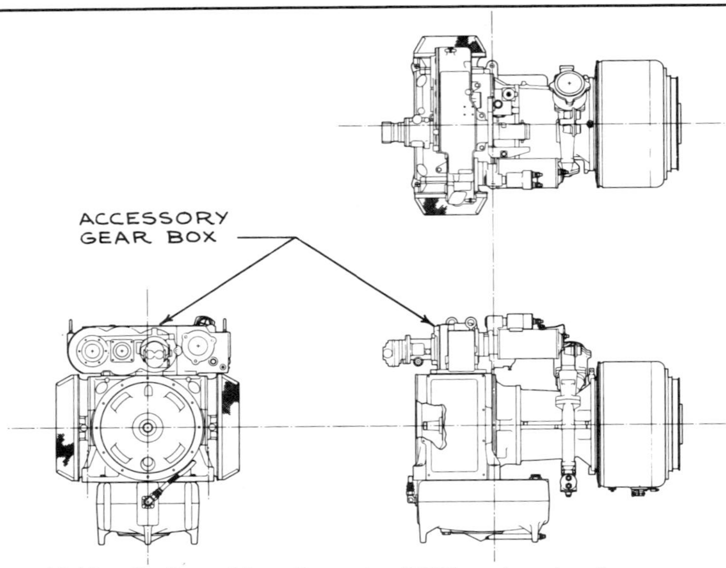

Figure 11.15. Outline of Avco-Lycoming TF25 engine, showing accessory gear box (from manufacturers' application data).

is a small fraction of the engine's main output. For example, the connection provided for the ship designer's optional use on the General Electric LM2500 gas generator is limited to 37 hp at the rated gas-generator speed.

Some engines may provide similar accessory drives from the power turbine. Such an arrangement has the disadvantage that load conditions may slow, or even stop, the power turbine, and consequently do the same to whatever the accessory load may be. An advantage is that the power output of the engine may be less depressed by the auxiliary load if it is taken from the power turbine. For example, the Avco Lycoming TF engines lose about 1.5 output horsepower for every horsepower taken from the gas generator, while they lose one-for-one when it is taken from the power turbine. The explanation lies in the fuel controller: since the accessory load reduces the gas-generator speed, the fuel controller cuts back on fuel to maintain the rated air/fuel ratio. Energy input, and hence output, is reduced in consequence by that extra one-half horsepower for every one tapped off. To circumvent both this and the stated disadvantage of power-turbine accessory drive, Avco Lycoming offers an optional power turbine take-off with a patented cross-over feature: a set of over-running clutches automatically transfers the accessory load to the gas generator when the power turbine slows below a set minimum.

An alternative to the power-turbine drive is attachment of the auxiliary load to an accessory output pinion on the reduction gear. The use of this technique

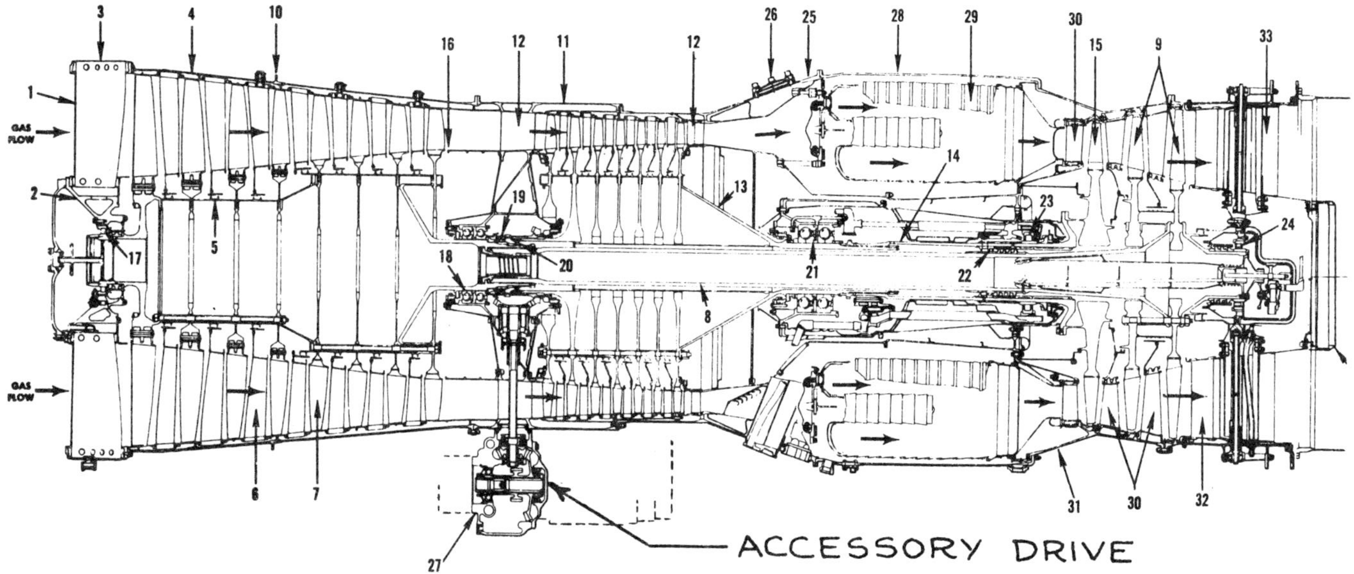

Figure 11.16. Turbo Power & Marine FT4 cross section showing accessory gear box (from Turbo Power & Marine literature).

to drive the ship-service generator is an example. The *Euroliner* generators are driven in this way; see Section 10.6 and, in particular, Figure 10.24. The practice of driving generators from the propulsion-reduction gear is also often used in steam-turbine plants.

11.9. GAS-TURBINE NOISE SOURCE

A gas-turbine engine is a source of high-intensity sound that tends to escape to the vessel's environment via the intake and exhaust ducts, and in the instance of the larger engines, at least, a significant amount of sound energy is radiated directly from engine surfaces into the machinery space. Structure-borne noise can likewise be significant, especially for naval vessels. Some aspects of the noise problem are discussed in Chapter 9 as part of the general treatment of intake and exhaust ducts, and the use of sound-attenuating enclosures and resilient mountings are mentioned in Chapter 10. Here the discussion is mainly aimed at quantifying the gas-turbine engine as a source of concern for airborne noise that affects the machinery spaces and escapes through its boundaries.

Large gas-turbine engines (especially the gas generators) are usually enclosed, with the purpose of the enclosure being, in part, to isolate the sound source that the engine represents. Smaller engines are more usually mounted without enclosures, but within unmanned compartments whose boundaries nonetheless serve as sound-isolating enclosures. Even in the former case, it is likely that the machinery is controlled from a space that is isolated from the

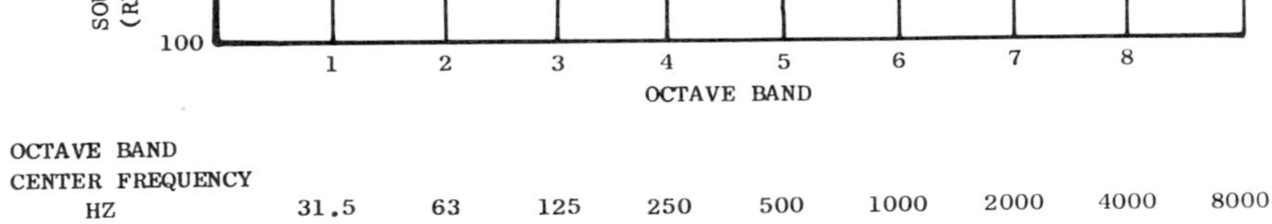

Figure 11.17. Estimated sound levels, airborne from engine casing of General Electric LM2500 engine [*LM2500 Marine Gas Turbine Installtion Design Manual* (1973)].

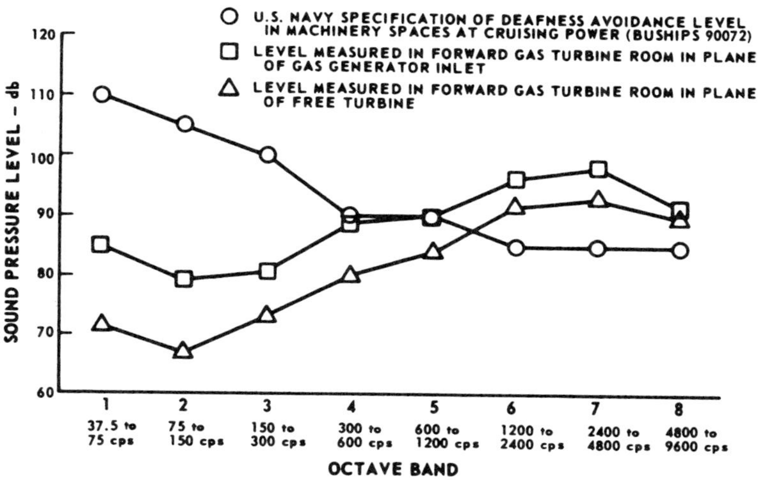

Figure 11.18. Sound levels in *Peder Skram* turbine room [Harper (1969)].

machinery room by a heavy attenuating bulkhead, making the direct exposure of the crew to machinery room noise infrequent. Such measures may create a satisfactory sound situation for the operators, but the specifications applying to surrounding spaces can require that sound-attenuation measures be incorporated in the decks and bulkheads.

The extent and nature of the sound-attenuating treatment is impossible to

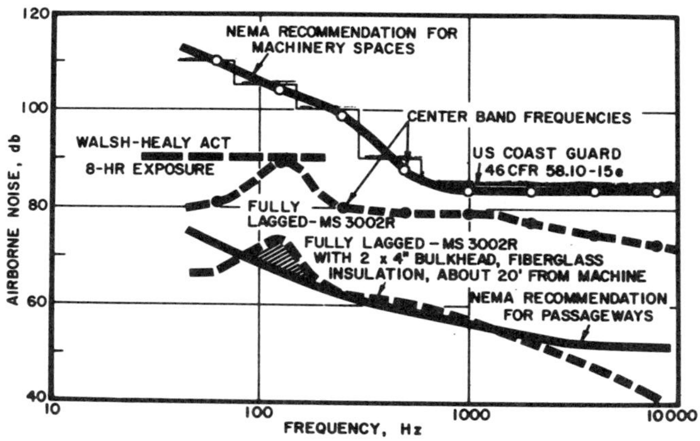

Figure 11.19. Sound levels in machinery space, General Electric heavy-duty engine [Svenson (1973)].

describe adequately in a brief and general treatment, since they depend so heavily on the nature of the particular installation. The location of the machinery space within the vessel and the nature of the surrounding spaces (living quarters, passages, store rooms, cargo holds, etc.) are factors of great importance. Even examples of existing practice, such as can be gathered from the oft-cited *Admiral Callaghan* and *Euroliner* might be misleading and are not given here, since they could not serve as guides to what might be expected in moderately different cases. In fact, noise suppression is usually considered to be something of an esoteric practice beyond the skills of most marine designers, and so it is that acoustic specialists are nearly always brought in to handle this item.

What is presented here is a sampling of sound-source data from several engines to illustrate the levels that are expected in marine gas-turbine machinery spaces, with the purpose of alerting the potential designer to the need for including them in his considerations. Figure 11.17 is the estimated sound level data in the immediate vicinity of the General Electric LM2500 engine casing [*LM2500 Marine Gas Turbine Installation Design Manual* (1973)]. Figure 11.18 is a set of similar measurements made in the machinery room of the Danish frigate, *Peder Skram* [Harper (1969)]. The engine in the latter case is a Pratt & Whitney GG4 gas generator (same as the Turbo Power & Marine FT4 engine) with a Stal Laval power turbine. Note that the figure shows high-frequency noise exceeding the U.S. Navy specification cited, although the engine is enclosed. Figure 11.17, which shows much higher levels, represents data taken without benefit of enclosure. Figure 11.19 is data for a General Electric heavy duty engine, including the effects of surrounding insulated bulkheads [Svesen (1973)]. Levels specified by several regulatory authorities are also indicated in the figure.

11.10. REFERENCES

Harper, R. E. (1969), "Vibration and Noise Characteristics of an Aircraft Type Gas Turbine Used in a Marine Propulsion System," American Society of Mechanical Engineers, Paper 69-GT-8.

Installation Design Criteria for Gas Turbine Application in Naval Vessels (1968), NAVSHIPS 0941-038-7010 United States Navy Technical Manual.

Jones, R. A., and Hefner, W. J. (1972), "Heavy Duty Marine Gas Turbines—A Case Study," American Society of Mechanical Engineers, Paper 72-GT-68.

LM2500 Marine Gas Turbine Installation Design Manual (1973), General Electric Report MID-IDM-2500-2, Marine and Industrial Products Department.

O'Hare, T. L. R., and Holburn, J. G. (1973), "Operating Experience with Gas Turbine Container Ships," *Transactions, Institute of Marine Engineers,* Series A, Part I, **85,** pages 1–23.

Rains, D. A., and d'Arcy, R. J. (1972), "Considerations in the DD963 Propulsion System Design," *Naval Engineers Journal,* **84,** pages 65–77.

Svensen, N.-A. (1973), "What Every Shipyard Needs to Know about Marine Heavy-Duty Gas Turbines," Society of Naval Architects and Marine Engineers, Annual Spring Meeting.

Wennberg, P. K. (1966), "The Design of the Main Propulsion Machinery Plant Installed in the USCGC *Hamilton* (WPG-715)," *Transactions, Society of Naval Architects and Marine Engineers,* **74,** pages 29–69.

Zeien, C., Smith, H. F., and Hirst, F. W. (1969), "The Gas Turbine Ship *Callaghan*'s First Two Years of Operation," *Transactions, Society of Naval Architects and Marine Engineers,* **77,** pages 344–371.

Chapter

TWELVE

MARINE GAS TURBINE MAINTENANCE

12.1. INTRODUCTION

Propulsion machinery of all kinds continuously suffers degradation as its operating hours accumulate; *maintenance* is the continuous effort of the owners and operators to detect the degradation, to minimize it, and to effectively reverse it by repair or replacement of parts that suffer the significant degradations. Thus maintenance consists of inspections, servicing, replacement of operable but somewhat degraded parts (preventive maintenance), and replacement of failed or badly degraded parts (repair). This is generally the way it is no matter what the type of machinery may be, though obviously the details may vary quite noticeably among the types. Each has its unique needs and practices. This is true for the gas turbine no less than for its competitors.

The meaning of *inspection* is nearly obvious. In fact, I feel safe in using the word itself in explanation: for a gas turbine it consists of visual inspection of both engine externals and internals, the latter by opening covers, lifting casings, or by use of borescopes. Simple checking of conditions through the agency of installed instrumentation is also a part of it. *Servicing* is most typically cleaning and replacement of expendable items (e.g., throw-away filter elements). Servicing is a feature of *preventive maintenance*; the latter term also covers repair or replacement of nonexpendable items that give evidence of reaching the end of their ability to perform satisfactorily. And *repair* is any action that restores a part, or restores a larger entity by replacement of a part. By implication here, it occurs after failure (failure = degradation beyond acceptable limits), but actually it overlaps preventive maintenance because some preventive procedures are indeed repairs; you may have noticed the word used just above in outlining the former concept.

This chapter discusses the maintenance aspects of marine gas-turbine propulsion. Unfortunately, it is being written some years too soon for the "true" picture of this maintenance to be presented. Although much can be said about maintenance, based on manufacturers' predictions which in turn may be based on aircraft or industrial service, marine conditions (e.g., sea salt in fuel and combustion air) have already given overwhelming evidence of their ability to cause radically different maintenance problems. The maintenance picture for the marine gas-turbine, therefore, must be based on marine experience, an experience which is still very scant at this time. To be sure, there are and have been many gas-turbine vessels of an experimental nature and a handful of regular applications—all recent buildings. These offer some evidence of what is to come and allow the engine builder to make predictions that are not totally speculative. From discussions of the early experiences and from the manufacturers' predictions, the bulk of this chapter is constructed.

The reader should be strongly cautioned that early experience may appear to be discouraging because of high rates of failure and the appearance of difficulties that were not anticipated in design. These phenomena are not unique to gas turbines, for they plague any new design in the early usage and are also expected in new propulsion plants of well-established design. However, it seems reasonable—even necessary—to discuss the unfortunate occurrences of the short marine-gas-turbine history to illustrate what *can* happen. The bad occurrences will not be diminished by some natural process, but only by the diligent efforts of the designers and builders of gas turbines, and of designers, builders, and operators of propulsion plants. It is especially the latter group that I anticipate as the readers here, and hence a major objective of this chapter is to encourage them to their efforts in bringing about favorable maintenance histories.

No discussion of *reliability* appears in this chapter, even though it is common practice to link reliability and maintenance (or "maintainability") as inseparable twins. Of course, they do have much in common, but my protest and apology that the maintenance story is scarcely beginning to develop apply even more strongly to reliability. Knowledge of reliability (reliability = probability that a failure will not occur over a specified time interval) depends completely on applicable experience; a priori speculation is pointless, for it is usually grossly optimistic. After all, no one expects unexpected failures. There is indeed a considerable record of marine gas-turbine failures, some of which are mentioned in this chapter, but the message that opens the preceding paragraph is repeated here: failures in the nascent stage of the technology—any technology—do not establish the long-term reliability status. The early failure rate is likely to be some magnitudes higher than the eventual steady rate. Readers may recognize this as one of the articles of faith of reliability theory, an article that I see no reason to question in the case of the marine gas turbine. The efforts of all involved to reduce the rates of degradation by improvements in design, manufacture, installation, operation, and inspection, whether intended primarily to minimize maintenance costs, or to increase time between overhauls, or to reduce the incidence of unexpected failures, will have inevitable benefits for all three.

12.2. EXPECTED EXPERIENCE

Certain deteriorations of the gas-turbine engine in service are inevitable, and it is for these that at least a minimum program of inspections and maintenance must be designed. Other deteriorations, while perhaps avoidable under ideal conditions, must nonetheless be expected in marine-propulsion service, and the inspection and maintenance program must be inclusive enough to circumvent them also.

The ideal conditions that, if attainable, would permit an engine to operate with the maximum time between maintenance procedures, are these:

1. Totally clean combustion air: no dirt, no sea salt, no large or small foreign objects, no oily vapor.
2. Totally clean fuel: no solid or liquid impurities, dissolved or suspended.
3. Fuel without composition impurities: no sulfur, vanadium, and so forth (see Chapter 8).
4. Continuous operation at rated load: essentially no thermal cycling due to changes in load, or to starts and stops.
5. Uniform distribution of fuel to all combustion chambers: no temperatures above the mean combustion temperature, as might be caused by an excess of fuel to one of the combustors.
6. No vibration excitation imposed via shaft coupling, mountings, or piping connections.
7. Integrity of lubricating oil.

Under such conditions, the maintenance experience might be the following:

1. If the population of engines were large, a few failures of components would occur in random fashion, under the general principle that perfect reliability is unattainable. Given careful design, manufacture, and inspection, these random failures should be rare.
2. Oxidation of engine hot parts proceeding at a very low rate.

3. Eventual failure of first-stage turbine blades due to creep.
4. If the engine were aircraft derived, and using ball bearings, as is typical for these engines, these bearings would eventually fail according to their recognized life pattern.
5. If the perfection of air, fuel, and lube oil were due to filtration or other treatment, continued maintenance (e.g., cleaning or replacement of filters) would occur throughout the life of the engine. Otherwise, maintenance practices would consist principally of periodic inspections, especially of engine hot parts, to guard against the possibility of the random failure or accelerated deterioration.

The magnitude of the "eventual" stated in items 3 and 4, and hence the life of the engine between major repairs, could perhaps be estimated from fundamental information on loads, stresses, and temperatures, but this would be an exercise of little value because the unavoidable conditions of reality promote earlier degradations. Even if perfection of air, fuel, and lubricant could be attained, practical service is certain to require starts, stops, and major loads changes. Thermal cycling is thus an inevitable feature of engine service experience, and it alone limits life to a value well below the ideal. Thermal cycling is likely to be manifested by cracking and distortions in the light-gage parts of the combustion system components, and by cracking in the first-stage nozzles. Continued service requires that these defects be remedied by change-out of the gas generator (aircraft-derived engine) or by repair or replacement of parts in place (heavy-duty engine); "life" as spoken of here is the time between these major repair events. It is difficult to say what this period would be, because the conditions bespoken are also idealized, but "several years" is a rough approximation. Actual life is more likely to be determined by imperfections of fuel and air.

Sea salt in the combustion air can cause rapid degradation of engine performance by coating compressor blades, though this is a transient degradation not requiring repair, nor reducing engine life (assuming use of corrosion-resistant alloys in the compressor), since salt is removable by periodic washing. More serious and irreversible is the corrosion of turbine nozzles and blades through the phenomenon of sulfidation (Chapter 8), wherein the sea salt acts in conjunction with sulfur from the fuel. If the attack is severe, the nozzles and blades may rapidly lose material to the point when they no longer function. Similarly, small foreign particles in the combustion air contribute to nozzle and blade degradation via erosion.

Larger foreign particles in combustion air—those that might be classed as "objects"—usually cause damage more severe than mere erosion. For instance, it is reported [O'Hare and Holburn (1973)] that a *Euroliner* (actually aboard *Eurofreighter,* the second ship of the class) gas-generator replacement was necessitated by damage to compressor blades resulting from the ingestion of "pop-rivet" shanks. The rivets had been used in construction of an access door to the inlet plenum. The tails, which break off during riveting, apparently fell out of sight, later to be sucked from their lodging places into the engine. Of course, such accidents are not supposed to happen and should not be part of the normal maintenance experience of an engine. But enough such incidents have been reported (another gas generator on the *Euroliner* is said to have been ruined by swallowing a flashlight [O'Hare and Holburn (1973)]) to point up the special hazard to gas-turbine engines from foreign-object ingestion. It would seem that unless great care is taken to keep foreign objects out of the combustion air, damage from this source will be an expected part of an engine's experience. Diesel and steam machinery, on the other hand, are less likely to ingest foreign objects during operation, and their potential for damage in this way is much lower; only the compressor of a turbocharged diesel would seem to be vulnerable to this event.

Dirty fuel can plug fuel nozzles, with consequent local overheating as the remaining nozzles accept the diverted fuel. If the fuel impurity is sea water, it may contribute salt to the sulfidation of engine hot parts. Hopefully, however, both liquid and solid impurities are reduced to a negligible level by the fuel system, so that the maintenance burden (such as changing filter elements) from this source is light and occurs external to the engine. Compositional impurity, such as sulfur, is another matter, since the engine itself must bear the burden. Sulfur from the fuel and salt from the combustion air are the fundamental ingredients of the sulfidation corrosion process.

The maldistribution of fuel to the several combustion chambers may be caused by nozzle plugging, as noted, or by failure of the fuel control to distribute the fuel equally to all combustors. The consequence is over temperature in the hot section served by the over-supplied elements, and in turn the consequence is accelerated oxidation.

Dirty or contaminated lube oil is, of course, detrimental to bearings. Likewise, vibration is detrimental to ball and roller bearings via the phenomenon known as "brineling." Vibration of a magnitude to be significant is usually imposed from sources external to the engine, as in the case of hull vibration transmitted by the mountings. Faulty lube oil is not a normal condition (i.e., not like the sulfur impurity in fuel, or sea salt in combustion air), and the oil is subject to monitoring and replacement should its degradation occur.

Some of the degrading influences mentioned above should not occur, but the ideal conditions are most definitely not to be expected either. Under what might be called "best practical" conditions, a gas turbine serving as a base-load marine propulsion engine is likely to be degraded by (1) thermal cycling,

(2) some salt in the combustion air, and (3) some sulfur and perhaps trace impurities in the fuel. If so, hot section deteriorations proceed slowly. Slow corrosion of nozzles and blades occur, cracking and distortions of light-gage parts develop in the combustion area, and some cracking appears in the first-stage nozzle area. These apparently inevitable processes eventually require replacement of parts; in place in the case of the heavy-duty engine and ashore after changeout of the entire gas generator in the case of the aircraft-derived engines.

So what life between replacements should be expected in the face of this normal degradation? Only a hesitant answer is possible, since a prediction must be based on experience which is yet to be accumulated. This answer is "on the order of 10,000 hr," based on published predictions of expected life [Zeien, Smith, and Hirst (1969); Siemietkowski (1970); Smith (1969); and Wert (1970)]. For example, the initial target time between gas generator overhauls for the *Admiral Callaghan* was 6000 hours, with the expectation of increasing this as experience warranted [Zeien, Smith, and Hirst (1969)]. As this is written, the only other vessels with significant operating experience with base-load gas turbines are the four *Euroliner* ships. According to published accounts [O'Hare and Holburn (1973)], early experience, particularly on the *Euroliner* itself, was distorted by gas-generator changeouts necessitated by abnormal events (like swallowing a flashlight).

The power turbines of the aircraft-derived engines may be changed also, but life is expected to be much longer than that of the gas generator because temperature drops considerably as the gas passes through the latter unit. The question of just what the life should be has almost ignored in the published

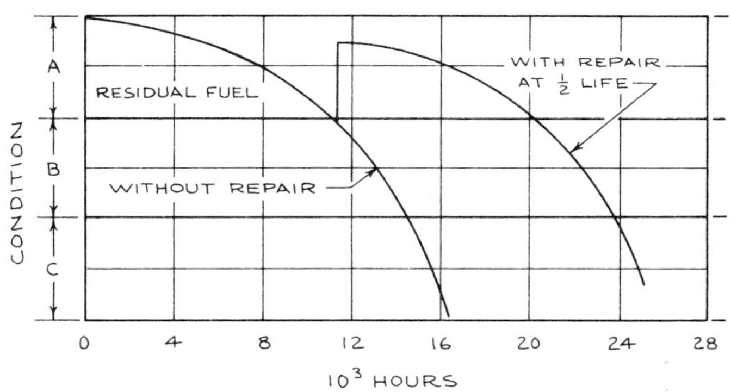

Figure 12.1. Degredation rate of firt-stage nozzle, General Electric heavy-duty engine [Wert (1970)].

TABLE 12.1. NOZZLE REPAIR REQUIREMENTS FOR GENERAL ELECTRIC HEAVY-DUTY ENGINE [Wert (1970)][a]

Condition	Cause	Action
Trailing-edge cracking	Thermal cycling	Repair weld
Side-wall cracking	Thermal cycling	Repair weld
Oxidation	Operating temperature levels	Repair weld/replace
Erosion	Foreign particles in combustion air	Repair weld/replace
Corrosion	Off spec fuel	Repair weld/replace
Diaphragm interstage seal wear	Movement of diaphragm thermal cycling	Repair

[a] Typical repair requirements for second-stage turbine nozzle.

literature. My only public source is a brief mention of an expectation of 20, 000-hr life, found in promotional material of Turbo Power & Marine Systems. In commercial base-load service this would be approximately 4 yr. As with the gas generator, accumulating experience will doubtless reveal a modified figure.

Predictions have been made by General Electric for individual parts life of their heavy-duty engine [Wert (1970)], apparently based on industrial experience. [Sea experience is also scant with this type of engine (in 1973), except for the *John Sargeant* installation.] For example, Figure 12.1 gives a prediction of first-stage nozzle life. The term "condition" used as the ordinate in the figure is rather indefinite, of course, and must be understood to pass from as-new to unserviceable as the curve suggests without intermediate quantification. The likely conditions of degradation to be found over nozzle life, their causes, and the suggested repair action, are listed in Table 12.1.

12.3. EXAMPLES OF EARLY EXPERIENCE

Early experience with gas turbines serving as base-load marine-propulsion engines is perhaps a poor indication of what the maintenance experience should be, for failures have frequently occurred that should not be a part of normal operation. It is nonetheless instructive to discuss some of these, since they point to what can happen and, therefore, must be guarded against by inspections and preventive maintenance. They also point out in some cases the particular spots where care must be exercised in design.

A. *John Sargeant* Experience

The *John Sargeant* was a liberty ship converted to gas-turbine propulsion in the mid-1950s. The propulsion engine was a 6000-hp General Electric industrial gas-turbine that burned low-vanadium residual oil. After 14 voyages, 9270 engine hours, and approximately 1000 starts and stops of the engine, the ship was laid up. Experience with the machinery to that point was reported in an ASME paper [McLean, Tangerini, and Van Cott (1960)]; the discussion following in this section is summarized from that source.

Over the 9270-hr experience there were several failures that caused the engine to be stopped while the ship was underway, but all were due to malfunctions of external equipment, such as the fuel control. The engine itself accumulated essentially no maintenance events, and upon prelayup inspection was found to be in such good shape that operation, if required, would have continued without repair. The major points relative to this inspection, as noted by McLean, Tangerini, and Van Cott (1960), were these:

1. Slight erosion and corrosion of the compressor.
2. Some minor cracks in trailing edges of first-stage nozzle partitions, but not to a degree requiring repair. Apparently due to thermal cycling.
3. Except for the items mentioned in items 1 and 2, hot section of engine in good condition.

The engine had run on diesel oil and on low-vanadium residual oil, with about 75% of the consumption being the residual. The latter fuel was washed by a treatment system essentially the same as that described in Chapter 9. Fuel nozzles were cleaned after every round trip (approximately 700 hr), although there was no indication that this practice was actually necessary. Salt fouling of the compressor was negligible; nonetheless, the engine was washed after each trans-Atlantic passage.

The air intake was located at the top of the deck house, facing outboard. Although there were no salt separators as such, the air followed a baffled path in flowing from the immediate intake chamber into the inlet duct. The engine regenerator through which the air passed may have also tended to screen salt from the engine. On the other hand, no salt fouling of the regenerator was reported.

The generally good condition of the engine hot section appears to confirm the mild impact of salt on the *John Sargeant* engine. However, a more significant factor in engine condition is its operating temperature; the temperature at the first-stage nozzles was approximately 200 F lower than in contemporary practice, and this is probably the major factor in the absence of noticeable sulfidation.

B. *Admiral Callaghan* Experience

A summary of the first two years of *Callaghan* operation was published in 1969 [Zeien, Smith, and Hirst (1969)]. At that time both port and starboard gas generators had been changed once each, one after 3375 hr of operation, the other after 5018 hr. Both changeouts were effected in brief intervals during scheduled port visits; the longer of the two required 30 hr.

The first changeout was made in order to incorporate a new bearing design, this having been found desirable by shore-side experience with other FT4 engines since the building of the *Callaghan*. Upon disassembly for the bearing change, the gas generator was found to be in very good condition, with no noticeable salt buildup in the compressor, and only minor wear on the turbine blades. However, a foreign object was found to have passed through the turbine blades and nozzle guide vanes. The object was one of two bolts that had worked loose from a combustion chamber clamp, having failed to receive the necessary safety wiring at an earlier hot-section inspection. Apparently the damage was not serious and would not have precluded the gas generator from operating without hazard or loss of performance until the damage was discovered at a later hot-section inspection.

The other changeout was necessitated by a failure at sea which caused an engine tripout. The immediate cause was an excessive temperature spread among the six thermocouples sensing gas-generator-discharge temperature. Upon disassembly of the unit after removal from the ship, the defect was found to have been caused by a failure of one combustion chamber, originating with a leak in the fuel manifold caused by chafing between it and a sheet-metal heat shield. Since the heat shield was necessary only in the face of the rigors of the high-altitude aircraft operation for which the engine was originally designed, the fix has been simply to dispense with the heat shield in all marine engines. Except for the damage to the combustion chamber and its immediate environs, the condition of the gas generator was good, with no significant salt deposit or corrosion. It was predicted that had the casualty not occurred, 3000–5000 hr of additional life remained before overhaul would have become necessary. The original target for TBO (time between overhauls) of 6000 hr, therefore, would seem conservative—had the untoward events not occurred.

Although the foregoing indicates that the salt ingestion problem was well under control even in the beginning, it should be noted that this is the consequence, in part at least, of the washing given the engines at the end of every ocean crossing. This procedure is seemingly a fixture of maintenance practice for all sea-going gas turbines, though it is a rather minor burden when downtime (about an hour) is available for the job.

Notable maintenance experience during the first two years of the *Callaghan* service is mostly related to difficulties with dirt in the fuel. The fuel was to some degree contaminated on receipt (e.g., with sea water) and suffered further degradation in the shipboard storage tanks from rusting of the tanks and biological action at oil-water interfaces. Apparently, the problem was of greater magnitude than the designers expected, and the fuel filters were effectively swamped by the job they were faced with. Sufficient fine dirt passed through the filters to clog fuel nozzles on several occasions. Installation of finer filter elements (5 μ) and more frequent cleaning ameliorated the problem.

Another chapter began in the *Admiral Callaghan* experience with the replacement of its original port-shaft engine with a General Electric LM2500 engine in December 1969. The subsequent operating history through September 1973 is summarized in Table 12.2, reprinted from a report by Dupuy (1973). The purpose of the conversion was to gain experience for the Navy with the LM2500 engine; approximately a third of the operating hours summarized in the table were run on a simulated naval operating profile involving about 14 daily fast throttle movements between idling and full power, and about 60% of the hours were run on a heavy distillate fuel similar to that described by MIL-F-24397 (see Chapter 8). A total of three gas generators were involved, designated S/N456-001, 002, 003 in the table. Note from the table that there were five changeouts following failure and five for other purposes (designated F and E changeouts, respectively). The record superficially appears poor, but scarcely so in view of this being the first base-load marine experience of the LM2500 engine, and the first marine experience on heavy distillate fuel of any aircraft-derived engine.

C. *Euroliner* Class Experience

The *Euroliner* class consists of four vessels powered, like the original *Callaghan,* by twin Turbo Power & Marine FT4 engines. The first of the ships was delivered in March 1971, with the others following at intervals of several months. Their experience as of late 1972 was reported in an Institute of Marine Engineers paper [O'Hare and Holburn (1973)], which is the principal source of this summary.

The early record, covering 15–16 months following the delivery of the first ship, is poor. Several engine stoppages at sea occurred, including 12 hr of simultaneous loss of both engines of the *Euroliner* itself. Major cause of the

TABLE 12.2. OPERATING HISTORY OF GENERAL ELECTRIC LM2500 ENGINES IN THE *ADMIRAL CALLAGHAN*, DECEMBER 1969 TO SEPTEMBER 1973 [Dupuy (1973)]

			Hours		
Date	Event	S/N (456-)	Ship	Engine	Remarks
Dec. 69	Installation and trials	001	0	0	Completed installation in 12 days and met all performance requirements.
May 70	Hot Section Disassembly No. 1	001	2,406	2,406	Purpose was to demonstrate in-place maintenance and verify internal condition of engine as previously observed by borescope.
July 70	Accessory drive gearbox changeout	001	3,344	3,344	Damage was result of gearbox overheating during shakedown voyage.
Aug. 70	Hot-section Disassembly No. 2	001	3,929	3,929	Purpose was to install update hardware for evaluation and to obtain engine reference for heavy fuel operation.

Dec. 70	Engine changeout No. 1	001	5,841	5,841	Gas-generator shroud/blade failure.	F
Feb. 71	Engine changeout No. 2	003	7,221	1,380	Scheduled, for reinstallation of engine 001.	E
Apr. 71	Engine changeout No. 3	001	8,181	6,801	Gas-generator stage 2 blade failure.	F
May 71	Engine changeout No. 4	003	8,854	2,053	Scheduled, for reinstallation of engine 001.	E
Dec. 71	Engine changeout No. 5	001	12,535	10,482	LPT flange bolts (37) failed following auto-relights associated with electrical-control problem.	F
Jan. 72	Engine changeout No. 6	002	13,444	909	Scheduled to return engine 002 to NAVSEC-Phil. additional for testing.	E
July 72	Engine changeout No. 7	003	16,052	4,661	For internal inspection following flooding of vent piping.	E
Oct. 72	Engine changeout No. 8	001	17,265	11,695	Evaluation of cracked compressor casing.	E
Jan. 73	Engine changeout No. 9	003	19,052	6,445	Increase in power-turbine vibration levels.	F
Feb. 73	Power-turbine removal	001	19,402	12,046	Evaluation following second incident of unusual increase in power-turbine vibration levels.	E
June 73	Engine changeout No. 10	001	21,198	13,842	Gas-generator thrust bearing failure.	F
Sept. 73	Current status	003	22,900	8,150	Satisfactory operation.	

stoppage was fuel-controller malfunction. One lengthy engine stoppage occurred when its exhaust elbow fractured under the stress of differential thermal expansion and allowed exhaust gas to leak into the engine room.

Since the engines are aircraft-derived, periodic gas-generator changeout was anticipated, but not at the frequency actually necessary during this early operation; sixteen changeouts occurred during the report interval. Maximum engine hours before changeout was 2731 hr. As mentioned in Section 12.2, two of the changeouts were necessitated by foreign-object damage. One changeout followed a leak of sea water from the lube-oil cooler into the lubricating system. Three other changeouts were also the result of casualties. One of the latter events was the failure of a drive-shaft bearing, another was the failure of turbine-nozzle guide vanes, and the third was the failure of a second-stage gas-generator turbine blade. The latter two incidents resulted in severe damage downstream of the initial site as chunks of the disintegrating vanes or blades were swept through the engine. Damage extended to the power turbines, which in consequence were also replaced at the time of the gas generator changeouts.

Ten other changeouts were necessary because of sulfidation corrosion of the gas-generator hot parts. In this instance, the early experience was totally unlike that of the *Sargeant* and *Callaghan,* in which hot-section corrosion was negligible. And such an experience seems surprising since the design of the air-intake system was based on the successful *Callaghan* predecessor. Be that as it may, the inertial separators installed at the *Euroliner* intakes (not identical to the *Callaghan* units) gave inadequate protection from the marine atmosphere. Basically, the faults seemed to be that (1) salt caked on the separator vanes, or salt water draining from these vanes, could easily be reentrained by the air stream, and (2) inertial separators do a poor job of removing fine particles from the air (see Figure 9.10). At the time of the report, some improvement had been made with respect to the former defect by minor modifications to the separators and by frequent cleaning of them. It was apparent, however, that a significant sulfidation problem would continue until additional means were provided to separate the fine particles.

In spite of the rather distressing early experience, it is claimed by the report being summarized [O'Hare and Holburn (1973)] that "The maintenance load is so relatively low it cannot be compared to that obtaining on a conventional propulsion plant, especially at the power levels involved." Perhaps this statement does not reflect the work done ashore in restoring the sixteen removed gas generators and two power turbines. In so far as the shipboard work is concerned, the maintenance work was indeed largely in the changeout procedures and in the hot section inspections that preceded the changeouts. Figure 12.2 shows the breakdown of maintenance hours in the first two ships. An idea of magnitudes may be obtained by considering that a total of eleven

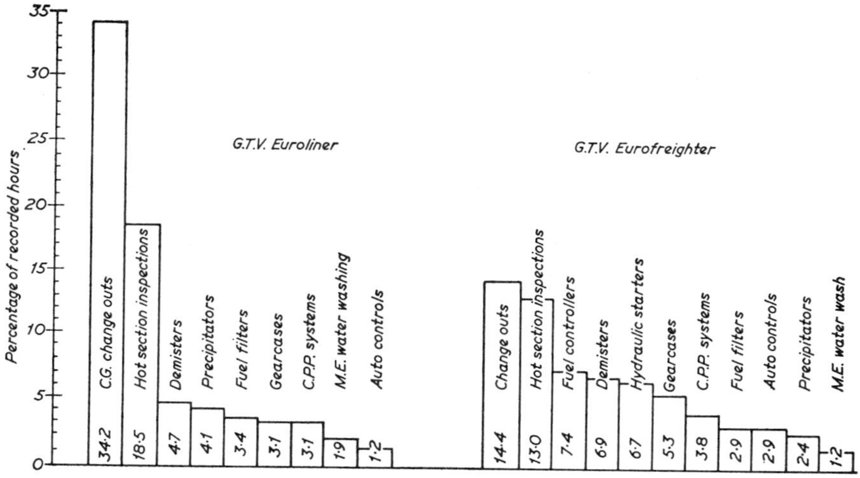

Figure 12.2. *Euroliner* and *Eurofreighter* maintenance experiences [O'Hare and Holbrun (1973)].

changeouts are included and that these averaged about 10 hr each (clock time, not man-hours).

D. Catastrophic Failures

What degree of bad experience should be classified as catastrophic is a matter of opinion. My definition in the present context is any sudden failure that involves sufficient damage to disable the engine immediately and permanently until major repairs can be made in port. Such events have indeed occurred in marine gas turbines; the typical failure begins with fracture of a highly stressed part within the engine, followed by consequent damage as the fragments are swept downstream. A similar event may ensue if a foreign object is ingested. Several such events in the *Euroliner* class have been mentioned in the preceding section.

The purpose of this section is to make the point that catastrophic failures do occur, simply by offering a few illustrations from early experience. The significance of the point is that the gas turbine can react violently, as well as with slow degradation, to inadequacies of cooling, of fuel distribution, of lubrication, of combustion air, and of fuel quality, for the sudden failures have usually been traced to such causes.

Figure 12.3 shows foreign object damage to *Euroliner* compressor blades, a

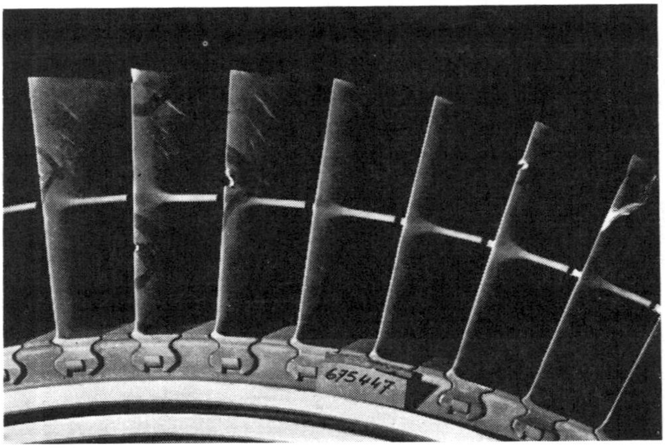

Figure 12.3. Damage to *Euroliner* ninth-stage compressor blades caused by pop-rivet shanks [O'Hare and Holbrun (1973)].

Figure 12.4. Failed and unfailed blades from Proteus engine after severe sulfidation corrosion [Rogus and Weinert (1972)].

result of the "pop-rivet" incident mentioned earlier. This was not quite a catastrophic failure by my definition, since no abrupt stoppage of the engine occurred, but it did trigger an immediate changeout when the damage was discovered.

Figure 12.4 shows the reunited (almost) fragments of a turbine blade from a Bristol Siddeley Proteus engine in hydrofoil service. The failure was the result of severe sulfidation corrosion. A second blade suffering from the same malady is also pictured; the loss of blade material that will lead to mechanical failure is quite evident.

Figure 12.5 is the first-stage turbine disk from an Associated Electric G6 engine from the frigate, HMS *Ashanti*. This failure was perhaps the most spectacular of those described in the marine gas-turbine literature. It is reported [Bowers (1967)] that "men on the upper deck were forced to take cover as red-hot fragments of metal were sprayed from the after funnel." Those fragments were the remains of several of the missing blades; others were found, more or less whole, jammed into downstream blade passages. (Damage

Figure 12.5. Failed turbine wheel from H. M. S. *Ashanti* [Bowers (1967)].

was confined solely to the engine, however, and there were no injuries.) This failure was attributed to prolonged overheating. It occurred in tropical waters under temperature conditions well above those experienced in home waters, or in earlier shore testing.

12.4. EXPECTED MAINTENANCE PROCEDURES

Let's assume that the expected life experience of the gas-turbine propulsion engine indeed occurs as outlined in Section 12.2: deterioration of the hot section proceeds (but slowly) because of thermal cycling and imperfections of fuel and air compositions, and modest salt fouling of the compressor occurs continuously. Further, the possibility of untoward events is recognized. This section then outlines the maintenance procedures that are likely to be instituted to protect against the inevitable degradations and to protect against the untowards. In a broad sense, the whole story can be summarized as one of expected replacement of hot-section parts, but it is organized around scheduled inspections that determine when replacements are necessary. The planned maintenance story, therefore, is largely one of inspection programs. The discussion of these is divided here into separate sections on heavy-duty engines and on aircraft-derived engines. Although much is similar between these two types, the practice of frequently replacing the aircraft-derived gas generator lead to some differences in practice. The discussion is more readily organized into separate section, although there is large overlap.

A. Expected Procedures with Heavy-Duty Engines

Since our only example of the heavy-duty marine gas turbine in U.S. practice is the General Electric engine of this type, the discussion here is essentially a paraphrasing of the maintenance outlined by that company. The immediate source is Wert (1970).

Maintenance of the engine is intended to be organized around four levels of periodic inspection: running inspections, service inspections, minor inspections, and major inspections.

The *running inspection* is basically a countinuing observation of engine conditions as pictured in the normal monitoring instrumentation. It includes, however, not only readings of instantaneous values, but observations of trends that may point to impending trouble. As an example, an increase in exhaust temperature at a constant-power level may indicate fouling of the compressor, with the consequent need for cleaning. This temperature is read by an array of thermocouples spaced around the exhaust hood. If the differences among the individual instruments increases, even though the average temperature remains steady, it may indicate that combustion is no longer occurring uniformly among the combustion chambers (perhaps fuel nozzles are becoming clogged), and accelerated deterioration of turbine nozzles and blades downstream of the hotter combustion is the likely consequence. If fuel nozzles are becoming clogged, then the defect is likely to be evident also in abnormal pressure readings from the fuel system. A continuing increase in vibration level usually indicates a growing imbalance that may be due to compressor fouling, or to significant deterioration of compressor or turbine blades. And recording of the time rate of change of temperatures, pressures, and rev/min during starting cycles is said to be a good check on the condition of instrumentation and control circuits. A change in the time required for a starting event to occur may indicate a drift in instrument calibration, or perhaps a more serious deterioration.

The *service inspection* is principally a check on the combustion system, done at opportunity during shutdown periods. Since the combustion chambers are mounted externally, liners and fuel nozzles are easily inspected by removing lightweight covers. The first-stage turbine nozzles may be seen at the same time by looking through the transition piece between the combustion chambers and the turbine nozzles.

The *minor inspection* is a visual inspection of all hot parts of the engine and requires lifting of the top half of the turbine shell.

The *major inspection* is a thorough inspection of all engine-wearing parts and requires lifting of all upper casings and removal of bearing caps so that these parts are exposed.

The frequency of these inspections depends somewhat on operating conditions and on experience with rates of deterioration, at least after such experience can be accumulated. The running inspection is more or less continuous, or in a manned engine room its frequency is likely to be at least once per watch; in an unmanned room, once a day is likely. At the other extreme, General Electric's estimate of major inspection intervals is a maximum of 30,000 hr. The requirements of classification societies for surveys of machinery may intervene in this instance and require a shorter interval, since these requirements are usually based on calendar time rather than on operating hours.

The story of maintenance actions that may accompany these inspections is summarized by Tables 12.3, 12.4, and 12.5, taken from Wert (1970), which list the repairs or replacements most likely to be needed as a consequence of service, minor, and major inspections, respectively. Note that the suggested in-

TABLE 12.3. ESTIMATED SERVICE INSPECTION SCHEDULE FOR GENERAL ELECTRIC HEAVY-DUTY ENGINE (20,000–30,000 SHP Units) [Wert (1970)]

Cumulative Hours Operation	Estimated Man Hours	Description
4000	6	Remove fuel nozzles for cleaning aboard ship.
6000	6	Inspect combustion liners, transition pieces, and borescope inspection of 1st stage nozzles.
8000	6	Remove fuel nozzles for cleaning aboard ship.
	Repeat after minor inspection	

TABLE 12.4. ESTIMATED MINOR INSPECTION SCHEDULE FOR GENERAL ELECTRIC HEAVY-DUTY ENGINE (20,000–30,000 SHP Units) [Wert (1970)]

Cumulative Hours Operation	Estimated Man Hours	Description
12,000	24	Replace combustion liners and transition pieces. (Repair old pieces for use as spares).
—	64	Replace 1st-stage nozzles, (Repair old nozzles for use as spares).
—	1	Remove F.O. pump for repair.
—	1	Remove flow divider for repair.
—	6	Remove fuel nozzles for cleaning aboard ship
	Repeat after major inspection	

TABLE 12.5. ESTIMATED MAJOR INSPECTION SCHEDULE FOR GENERAL ELECTRIC HEAVY-DUTY ENGINE (20,000–30,000 SHP Units) [Wert (1970)]

Cumulative Hours Operation	Estimated Man Hours	Description
24,000	6	Remove fuel nozzles for cleaning aboard ship.
—	24	Replace combustion liners and transition pieces. (Repair old pieces for use as spares).
—	128	Replace 1st and 2nd stage nozzles. (Repair old nozzles for use as spares).
—	300	Inspect 1st stage buckets, repair as necessary. Lift compressor casings and inspect blading. Inspect bearing.
	Repeat at 24,000-hour intervals	

tervals for the inspections are given, as are estimates of the manhours required for the corrective actions.

The engine is also continually subject to minor external preventive maintenance, most especially the routine replacement of fuel and lube oil filters. Washing of the engine during every port visit is also a possibility, depending on the experience with the extent of fouling, acquired during the lifetime of each individual installation.

B. Expected Procedures With Aircraft-Derived Engines

In principle, the treatment of the aircraft-derived engine is much like that of the heavy-duty engine, being built around a hierarchy of inspections whose most important aim is to detect the need for replacements within the hot section. The greatest difference comes when major replacement is necessary, since the response with the aircraft type is intended to be replacement of the gas generator as a unit, and the same is true for the power turbine should it face the same need. The actual repairs may nonetheless be much the same with both types, but performed ashore when the engine or its gas generator is replaced.

There are some differences in recommended practice among the several manufacturers of aircraft-derived engines, though these differences are minor. To avoid undue repetition, this section doesn't attempt to catalog the near-same procedures of the different builders, but sticks to terms sufficiently general for all. The U.S. Navy also affects the discussion somewhat, since it may organize maintenance programs to suit its unique needs. (The influence of the Navy isn't mentioned in the preceding section since it has not been a user of heavy-duty engines.) But although the organization of programs and the terminology used may differ in some ways from merchant practice, the procedures are in fact much the same for both.

Typical maintenance programs can be described as continuous (or very frequent) monitoring by installed instrumentation and by external visual inspection, minor servicing in port, hot-section inspection at intervals based on the dictates of experience, and eventual replacement ("changeout") of the gas generator for repair ashore.

Monitoring by installed instrumentation is an obvious procedure with an engine of any type, although there may be items of special significance to each type. For example, exhaust temperature is an important parameter in diesel and steam practice, as well as in a gas-turbine plant, but its greatest significance occurs with the gas turbine, since this machine is usually operating close to temperature limits at which accelerated hot-section degradation begins. But between the heavy-duty and the aircraft gas turbines, there is no essential difference with respect to monitoring; the items indicated or recorded are essentially the same, as are the meanings of off-standard readings.

External inspections and servicing while the engine is shut down in port generally cover fuel and lube-oil filters, control linkages, and external piping and wiring. Washing of the compressor is also a common occurrence during port intervals.

Hot-section inspections are carried out at less frequent intervals. The procedure varies in detail among engines of different manufacture; it may require extensive opening of casings, or may rely primarily on internal inspection by means of borescopes (a borescope is an optical inspection device for viewing or photographing the inside of narrow passages). The interval at which hot-section inspections are made is impossible to state generally, since it depends so much on experience which will be scant until gas-turbine propulsion becomes more general than it is now. One source [Smith (1969)] suggests inspections twice a year and an interval that might cover 3000–4000 hr of operation for a base-load engine in a high-utilization cargo ship. This would seem to be an upper limit, attained after some years of experience with the particular installation—or with earlier duplicates—had shown that rates of deterioration permitted an interval of this length. Recall from Section 12.3 that earlier experience with the *Euroliner*-class vessels found gas-generator changeouts necessary in the 1000- to 2000-hr range, indicating that inspections would be appropriate at intervals on the order of several hundred hours. Unless and until base-load gas-turbine propulsion becomes commonplace in the manner of steam and diesel, the history of frequent early inspections, followed by an increasingly long interval up to a limit of several thousand hours seems likely to be repeated in many ships. This should hold for both commercial and naval vessels and may well be the case for the heavy-duty engines also.

Figure 12.6 illustrates some of the steps in a hot-section inspection. The engine is the Turbo Power & Marine FT4; the pictures show the sliding back of a cover to expose the combustion chambers and the removal of these chambers and transition pieces to give access to the first-stage nozzles.

The ultimate consequence of the hot-section inspections—of the one that is eventually unsatisfactory, that is—is replacement of the gas generator. The removed unit subsequently will probably receive a complete overhaul by its builder, then enter a pool of spare gas generators and reenter service at some later changeout. The overhaul typically consists of a complete disassembly of the machine, careful inspection of every part, replacement upon reassembly of any part that does not meet original specifications, and a test run to ensure that standard performance is restored. On the other hand, hot-section inspections may also uncover repair items that do not justify a changeout. Generally these would be faults that are isolated rather than general. For example, if one (or perhaps a few) nozzle-guide vanes are afflicted with trailing-edge cracks, while the gas generator is otherwise in good conditions, the remedy would probably be to make the individual replacements aboard ship, since many such hot-section parts can be replaced without requiring extensive engine disassembly. Whichever tactic is followed, the intention is that shipboard work be accomplished during normal port turn-around time.

The power turbine is included in hot-section inspections, but its rate of deterioration should be much slower than that of the compressor turbine because the hot gas expands to a much lower temperature as it flows through the latter. The power turbine may last through several replacements of its gas generator, but it is also subject to changeout within normal port times should the need appear. Several such events occurred in the early *Euroliner* experience as a result of foreign-object damage [O'Hare and Holburn (1973)].

Boost engines in CODOG or other combined propulsion plants should require inspection and repair much more frequently than base-load engines, if the intervals are measured in operational hours, since their thermal cycling per operational hour is typically at a much greater rate. Of course, if the engine is

Figure 12.6. Hot-section inspection of FT4 engine.

called into service infrequently, the intervals may be longer in terms of calendar time. The experience of the U.S. Coast Guard with the boost engines of the *Hamilton*-class ships can give some idea of how this may be. Hot-section inspections are scheduled at 1000-hr intervals or annually, whichever is the shorter. Changeout intervals are dictated by findings of the hot-section inspections, or by failures detected otherwise. As of early 1974, the maximum gas-generator life had been about 4000 hr. The longest-lived generator was felled, not by performance deterioration, but by that unique bane of the gas turbine, foreign-object damage. (Information on the *Hamilton* class was kindly supplied by the Coast Guard Naval Engineering Division.)

Navy inspection and maintenance are basically the same as in merchant service, for after all, it is the characteristics of the engine, not of the owner, that set the need for certain practices. The most noticeable difference is likely to be in the intervals between events, since naval operating profiles usually differ markedly from those of merchant service. For example, it is only in military service (i.e., naval and coast guard) that boost engines in combined plants are found. Also, the processes may be organized in somewhat different fashion and described in different terms; the U.S. Navy describes its gas-turbine maintenance as being on three levels: (1) organizational, (2) intermediate, and (3) depot [Siemietkowski (1970)].

TABLE 12.6. SCHEDULED MAINTENANCE FOR FT4 ENGINE IN NAVAL SERVICE (from Turbo Power & Marine Systems Literature)

Item	Interval	No. of Men	Total Man Minutes
Condition monitoring	As required while operating		
Waterwash	Recommended on shutdown	2	80
Routine inspection	When shutdown	2	30
Minor inspection	500 hr	2	120
Major inspection (Hot-section inspection)	As defined	3	750
Gas-generator overhaul	As defined	Depot level task	
Free-turbine overhaul	As defined	Depot level task	

Organizational maintenance encompasses the shipboard inspections and the subsequent repair work that can be performed by shipboard personnel, that is, by "the organization." This may include hot-section parts replacement when the task does not require removal of the units from the vessel. *Intermediate* maintenance covers changeouts of gas generators (or complete engines, if necessary) and the repair work that subsequently may be done aboard a repair ship or in a shore facility. Although extensive parts replacement may be involved, the intermediate procedures stop short of complete overhaul. The overhaul is reserved for the category of *depot* maintenance and is performed only at shore facilities specially equipped for this level of work. As mentioned above, an overhaul consists of complete disassembly, inspection of parts, replacements as necessary, reassembly, and testing—a program that restores the unit to essentially as-new condition.

Table 12.6 outlines an expected maintenance schedule for naval service, prepared circa 1968 by Turbo Power & Marine Systems for the FT4 engine. Its major interest is in the estimates of labor effort required for all of the shipboard tasks.

12.5. UNIQUE FEATURES OF GAS-TURBINE MAINTENANCE

Maintenance practices quite obviously must depend on the type of engine. And although there is similarity of practice and procedure among the several types of marine engine, some items stand out as being unique to each. For example, the careful attention to boiler-water chemistry and the periodic repair of furnace refractory (common in older boilers, at least) are unique to steam-propulsion plants. For the gas turbine, two procedures appear to be unique, the frequent internal cleaning of the engine (especially its compressor) and the periodic "changeout" of gas generators—and perhaps of the power turbine—used with the aircraft-derived type. References to both of these are scattered throughout the preceding material in this chapter, but because of their uniqueness, they justify independent discussion, as follows here.

A. Changeout

Changeout normally refers to the process of replacing the gas generator of an aircraft-derived engine as a unit when major repair work is required. On occasion, the power turbine is also included. A major changeout usually means work that requires shop facilities, or that would enforce serious delay of the vessel if done in place. Although applied to both military and commercial vessels, in normal circumstances it is of greater advantage to the latter, since many contemporary merchant ships operate on schedules that call for short port times, times that might be extended with great loss of revenue should engine maintenance times be excessive. Indeed, it might be said that the successful use of the aircraft gas turbine in ships such as the *Callaghan* and the *Euroliner* class has depended on the capability of the engines for short changeout times. Recall the earlier mention of changeouts that occurred during the early life of these ships; to recap, the *Euroliner* changeouts required an average of 10 hr, the longest (both gas generator and power turbine) requiring 18 hr, and the longest changeout time on the *Callaghan* required 30 hr. In all cases reported, the changeouts were made within scheduled port intervals.

Figure 12.7 shows a gas-generator changeout in progress (FT4 engine), though a light deception might be detected by a close look at the figure: it's a changeout being carried out in a shore test facility and is not actually aboard ship.

Figure 12.8 diagrams the scheme for replacement used by the *Euroliner* ships. The unit being changed is moved through a door in the engine room aft bulkhead into the adjacent cargo hold, whence it is lifted by a shore crane. The process requires that the hold be empty and the weather deck hatch not be covered by a load of containers, so that coordination with cargo-handling operations is obviously necessary.

An alternative removal-replacement route for ships that do not have a cargo hold so conveniently adjacent to the engine space is depicted by Figure 12.9, that is, the route via the inlet air duct. (Figure 12.9 is not intended to show a particular ship.) Such routing probably does not enjoy the status of being the preferred way, since awkward handling, removal of silencing splitters, removal of salt separators, and opening of a bolted deck plate are all likely to be required. However, in military vessels, where the spaces surrounding the machinery are usually small and crammed with equipment, the inlet duct may be the only feasible path. If you would glance at Chapter 9, you will note that Figure 9.16 outlines the process via inlet ducts on two naval vessels.

A variation in the changeout scheme is indicated by Figure 12.10. Shown is the Avco-Lycoming TF engine (four marine models, from 1000 to 2500 continuous shp), with the several modules into which the engine can be readily decomposed. The largest model (TF35) weighs 1115 lb (dry); the heaviest of the modules, the power producer (which is essentially the gas turbine itself, less all accessories) weighs 456 lb, and it can be further disassembled into the compressor at 265 lb and the hot-end assembly at 171 lb. The power-producer assembly is approximately 37 in. long and 24 in. in diameter. The prime purpose of the modular construction is to facilitate maitenance through ready re-

Figure 12.7. FT4 gas-generator changeout in progress.

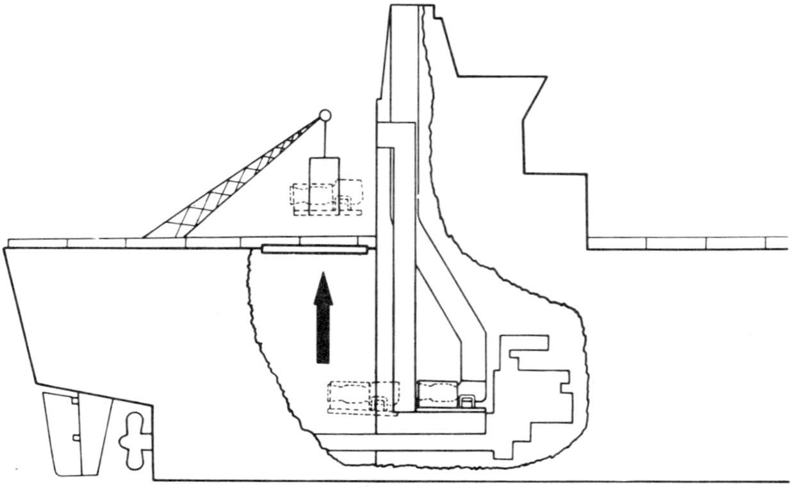

Figure 12.8. *Euroliner* changeout scheme [McAllister (1972)].

moval of modules to shore facilities. With the small individual weights, and the correspondingly small dimensions, the removal and replacement task is a relatively simple one.

B. Engine Internal Cleaning

Dirty gas turbines don't work well, and marine gas turbines do, unfortunately, get dirty. The "dirt" is most often sea salt, borne into the engine by the combustion air. However, other solid or liquid contaminants in the air can be

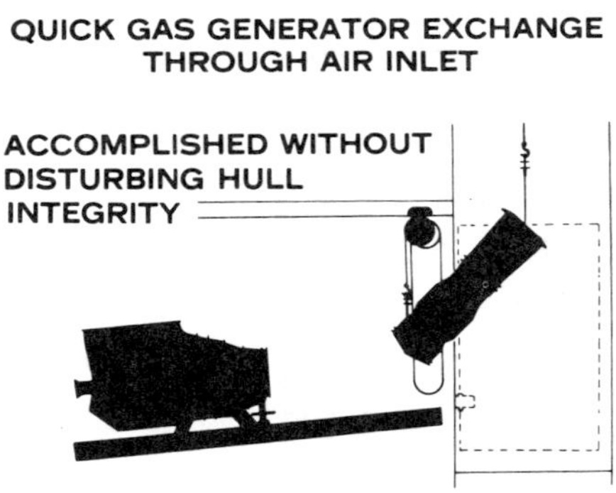

Figure 12.9. Scheme for changeout via intake duct [Smith (1969)].

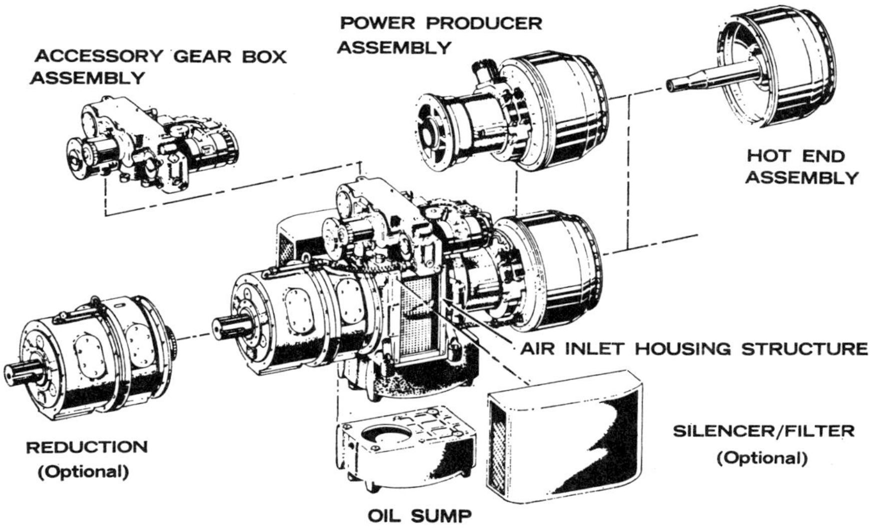

Figure 12.10. Avco-Lycoming modular construction.

equally troublesome. An oily air especially seems to be a key to difficulty, for although an oily film on a compressor blade is perhaps in itself no hazard, it acts as the adherent matrix for dust particles, or for any other solid contaminant in the air. Instances have been reported in which oily vapors originated aboard ship (e.g., from a fuel tank vent close to air intakes [O'Hare and Holburn (1973)]), but oily and otherwise dirty atmospheres may be encountered as the ambient condition in many ports. "Don't work well" means several things that result from deposit build-up on compressor blades. The direct effect is reduction in air flow. The natural consequence is then increase in combustion temperature, since the air/fuel ratio has declined. Since the fuel control prevents much of a temperature rise, fuel is cut back as air flow declines, and, therefore, the end result is a drop in output. The drop may be slow or precipitous, depending, of course, on circumstances. For instance, Figure 9.3 shows what doubtless should be considered a precipitous example—but which is quite possible, for it is an actual case—a decline in power output of about 60% in only 6 hr of running.

The fouling of blades also carries over into the turbine section of the engine. Here, the serious consequence in the case of the usual salt fouling is sulfidation corrosion, implemented in conjunction with sulfur supplied by the fuel. The result may be less immediately obvious than the fall in performance caused by compressor fouling, but is eventually of grave consequence to the engine. Recall Figure 12.4, or the previous mention of the many *Euroliner* gas-generator changeouts forced by rapid sulfidation. Fortunately, however, the compressor takes the brunt of the fouling, and being relatively cool and not exposed to combustion products, should not corrode significantly. Its performance can usually be restored by cleaning, the process that is the topic of this section.

The cleaning is performed either by a liquid wash, or by ingestion of a solid agent that scours deposits from engine sufaces. Engine builders generally recommend one method or the other, and indeed, there is some mild controversy over which should be used. The majority of American builders appear to favor liquid cleaning, though some possible deposits do not yield to the liquid. The solid cleaning also has the advantage of being more readily applicable while the engine is running at power.

The liquid washing agent is usually hot distilled water applied while the engine is motored by its starter. Some details for a typical engine, the General Electric LM2500, illustrate the process; these are the recommendations of the builder [*LM2500 Marine Gas Installation Design Manual* (1973)]:

Engine rev/min	300–2000 (motored by starter)
Water pressure	46 psig
Water temperature	130–180 F
Water flow rate	15 gal/min
Wash quantity	40 gal
Number of rinses	2
Water quantity per rinse	40 gal

This engine is provided with a spray ring built into its inlet bellmouth.

Water suffices for the wash when salt along is the target, but when oily deposits are present, detergent can be added to the water. In severe cases of oily fouling, the use of kerosene in lieu of water has been suggested [Siemietkowski (1970)].

Washing is usually done only after the engine has cooled, and the engine is usually dried carefully afterward by firing for perhaps 15 min at idle speed However, washing while the engine is being fired is possible and has been done. Its advantage lies in the shortened time required for the washing cycle, since the engine is not shut down and cooled as part of the process, and this advantage can be of considerable value to a vessel with short port turn-around times, such as a high-speed hydrofoil or surface-effect ferry vessel. On the other hand, the great disadvantage is the accelerated corrosion of the hot section caused by the flushing of salt from the compressor into that section by the wash medium, all while temperatures are in the corrosive range. Some British experience [Morgan, Lamport, and Smith (1970)] indicates that the corrosion may be greatly accelerated, even though the added exposure to salt is brief.

The frequency with which washings are carried out depends in an obvious way on the rate at which compressor fouling occurs. With relatively low salt concentrations such as those obtained in the *Callaghan* intakes, the engine builder (General Electric) suggests [*LM2500 Marine Gas Turbine Installation Design Manual* (1973)] washing about every 500 hr. For a high-speed merchant ship in North Atlantic service, for example, this would mean at the end of every second round trip. For small vessels, with their intakes closer to the spray produced by the hull at high speed, the interval will almost certainly be much shorter. For small military craft, commuter ferries, and the like, a schedule of washing at the end of every day's service is likely to be appropriate. An engine in boost service has special requirements, depending on the frequency of its use. An example is the U.S. Coast Guard *Hamilton*-class ships; their schedule of water washing calls for this to be done after every shutdown of the engine, or weekly if the engine has not been operated.

As mentioned earlier, the General Electric LM2500 engine is equipped with a washing spray ring built into the inlet bellmouth, and such a permanently installed washing feature is typical of many engines. Figure 12.11 shows an alternative scheme involving spray nozzles mounted on the inlet plenum bulkhead, facing the compressor inlet. This figure, taken from [*Installation Design Criteria for Gas Turbine Applications in Naval Vessels* (1968)], is based on recommendations of Turbo Power & Marine Systems for their FT4 engine. Another alternative is the use of a hand-held hose with suitable spray nozzle. This scheme is awkward to apply, except for engines that take their air from the machinery space, rather than from a closed duct system. The Avco-Lycoming TF engine is an example. Yet another scheme is injection of water via the combustor fuel nozzles. This reportedly [O'Hare and Holburn (1973)] has been used aboard the *Euroliner* ships (in addition to normal injection via the compressor inlet) in an attempt to improve the removal of deposits in the engine hot section.

Cleaning by solid agents is intended to remove deposits that do not yield to liquid washing. The necessary characteristics of an agent are good scouring action without abrasiveness that might erode the engine internals. These desiderata are best met by hard vegetable organic matter such as ground nut shells. The most commonly used material appears to be "carboblast," said to be a finely ground mixture of walnut shells and apricot pits. Although the word is a trade name (Turco Products), one often notes "carboblasting" being used as a generic term for the solid cleaning process. Other agents have been used, such as wheat hulls, rice hulls, or the grain itself. Typical quantities for a cleaning of a large gas-turbine engine should be in the range, 10–20 lb [Wert (1970)].

In constrast to the water washing, this method of cleaning is most frequently used when the engine is operating, so that the agent is destroyed by combustion in the hot section and swept out as a fine ash. However, unfortunate

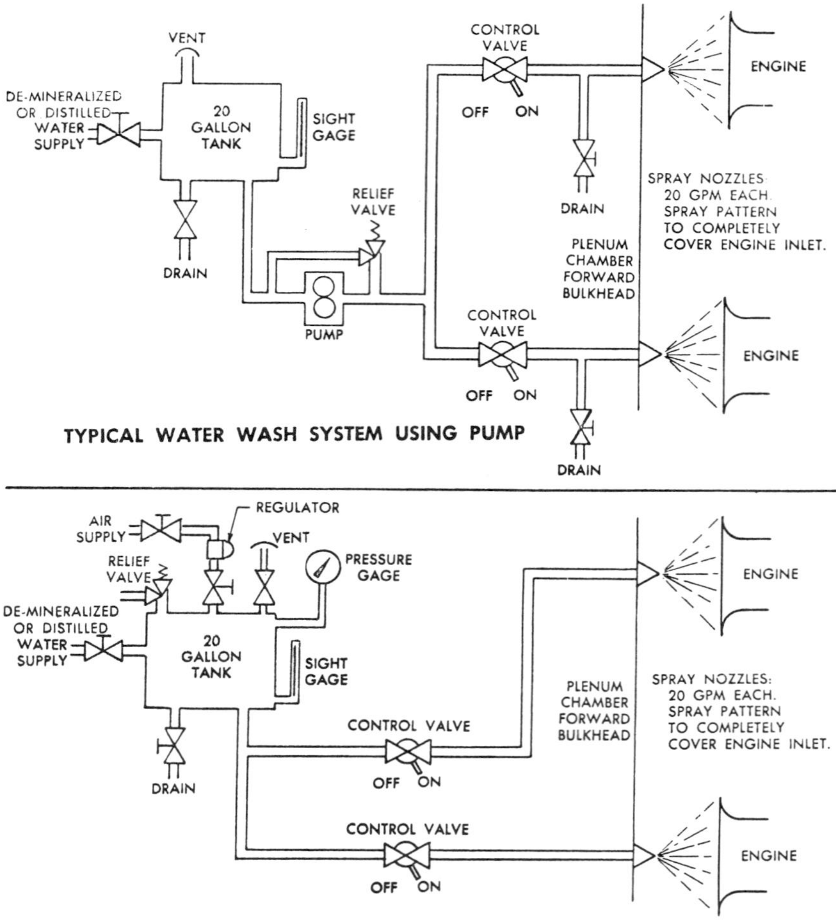

Figure 12.11. Typical system for injecting wash water into compressor inlet [*Installation Design Criteria for Gas Turbine Applications in Naval Vessels* (1968)].

consequences have been reported from clogging of either lubricant passages or cooling air passages by residue of the cleaner: an example is the failure of a compressor shaft in a British hovercraft engine, attributed to oil starvation following excessive use of solid cleaning [Woodward (1972)]. Also, the accelerated corrosion of hot sections apparently caused by compressor salt being swept into those sections by the cleaning process, as mentioned earlier, may occur with either solid or liquid cleaning when they are applied to an operating engine. Solid cleaning more often receives mention in this regard, since it is the one usually used with the engine under power. In sum, it would seem that solid cleaning should be used sparingly; if frequent cleaning of this type becomes necessary, the much-preferred remedy is doubtless to attack the cause of the fouling, rather than to increase the frequency of cleaning.

Table 12.7, taken from Trewby (1963), summarizes some early British experience with cleaning agents of several different kinds. Figure 12.12, taken from *Installation Design Criteria for Gas Turbine Applications in Naval Vessels* (1968), pictures permanently installed injection equipment for solid agents, as recommended by Turbo Power & Marine Systems.

12.6. CLOSURE

This section is intended to reiterate certain key points that can be distilled from the discussions of this chapter. The biggest point that might be summarized is the maintenance position of the marine-propulsion gas turbine: how it compares in down time and expense of repair to its competitors, the steam turbine and the diesel engine. Unfortunately, this topic cannot be treated at all, for this history of the gas turbine as a base-load engine at sea is scarce begun as this is written—the vessels in service are only a handful and early in their careers. Early experience with machinery of a new type is always peppered with failures that are due both to inadequacies in design attributable to the very lack of that experience and also to the "early-failure" defects that are unfortunately inevitable even in new examples of proven machinery. You probably have observed that this chapter is constructed largely on examples taken from reports of early experience, plus predictions (rather than records) of necessary maintenance by the engine builders. In fact, the introduction promised you essentially this.

Therefore, we must wait for some years from now for a clear pattern to emerge from the obscuring aspects of early failures and of builders' hopeful predictions. Meanwhile, let's summarize the points relative to maintenance that should be carefully noted by the marine engineer designing a gas-turbine propulsion plant. The factors that most directly influence engine degradations, and hence maintenance, are (1) metal temperatures, (2) thermal cycling (especially starts and stops), (3) fuel quality, (4) combustion air quality, and (5) extraneous forces and moments. Some aspects of these are beyond the control of the propulsion-plant designer, but others are definitely under his control and call for his best attentions.

Temperatures within the engine are established by the designer of the engine itself, but one important ingredient of internal temperatures, the ambient temperature, falls more within the ken of the plant designer. Of

TABLE 12.7. EARLY BRITISH EXPERIENCE WITH ENGINE CLEANING METHODS [Trewby (1963)]

Installation	Engine	Cleaning agent	Quantity	Rate of Injection	Frequency, hr	Total hr	Load kW	Effect on Machine during Injection	Condition of Machine Examination
H.M.S. Cumberland	Ruston, TA, 750 kW	Kerosene	3 gal	—	50	670	—	Nil	Compressor badly fouled.
H.M.S. Cumberland	Ruston, TA, 750 kW	Bran	1½ lb	As convenient for tipping bucket	Daily	250	250	Slight surge	Convex sides of stator and rotor blades fouled. Bran deposit on first four rows.
H.M.S. Cumberland	Ruston, TA, 750 kW	Canary seed	1½ lb	As convenient for tipping bucket	Daily	—	Nil	Slight surge	Slightly better than bran but deposits left on later stages.
H.M.S. Cumberland	Ruston, TA, 750 kW	Ground walnut husks	1½–2 lb	As convenient for tipping bucket	8 and 12	500	125 and 250	Nil	Rotor blades clean, stator blades slightly fouled. Slight erosion of stator and rotor blades of first few rows.
Naval Marine Wing	Ruston, TA, 750 kW	Wheatings	1 lb	Rapid by air pressure	24	500	400	Little effect	2½% restoration of compressor efficiency. Slight loss of blade metal in first few stages. 50% blockage of turbine cooling passages.
Naval Marine Wing	Ruston, TA, 750 kW	Ground walnut husks	1–1½ lb	Rapid by air pressure	4 cleans	345	50–400	One slight explosion due to rate of injection.	Very clean. Slight blockage of turbine cooling passages.

Naval Marine Wing	Allens, 500 kW	Distilled water	1½ gal L.P. 4½ gal H.P.	1½ gal/min approx.	4	190	200	Nil	H.P. and L.P. compressors fairly heavily coated with dry black deposits. Later H.P. stages deposits baked on.
Naval Marine Wing	Allens, 500 kW	AVCAT/water emulsion followed by water/	1½ gal L.P. 4½ gal H.P.	1½ gal/min approx.	12	210	200	Governor controlled adequately with slight variations in H.P. and L.P. line conditions.	L.P. compressor lightly coated with black oily deposit. H.P. compressor early stages very clean, later stages fouled.
Naval Marine Wing	Allens, 500 kW	DIESO/water emulsion followed by water/ detergent	1½ gal L.P. 4½ gal H.P.	1½ gal/min approx.	4	150	200	Governor controlled adequately with slight variations in H.P. and L.P. line conditions.	L.P. compressor very clean except for some inlet guide vanes and last few stages. H.P. compressor early stages very clean remainder deteriorating toward compressor outlet.
Messrs. Allens Ltd.	Allen, 500 kW	i. Wheatings	2 lb	As convenient for tipping into inlet	One injection only	—	Nil	Nil	High percentage of performance regained after one clean, but turbine cooling passages becoming blocked.
		ii. "Carboblast"	2 lb	As convenient for tipping into inlet	One injection only	—	Nil	Nil	High percentage of performance regained after one clean, but turbine cooling passages becoming blocked.
A.E.I. shore trials unit	G.6	AVCAT/water emulsion followed by	20 gal	4 min	Daily at least	—	4,000/5,000 gas generator rpm	Slight speed rise	Early stages of compressor clean. Improvement in performance slight, later cleans making little difference as final stages foul up.
		water/ detergent	20 gal	3 min	Daily at least	—			
A.E.I. shore trials unit	G.6	Distilled water	20 gal	3 min	Daily at least	—	4,000/5,000 gas generator rpm	Nil	Ineffectual in removing industrial grime but expected to remove salt deposits.

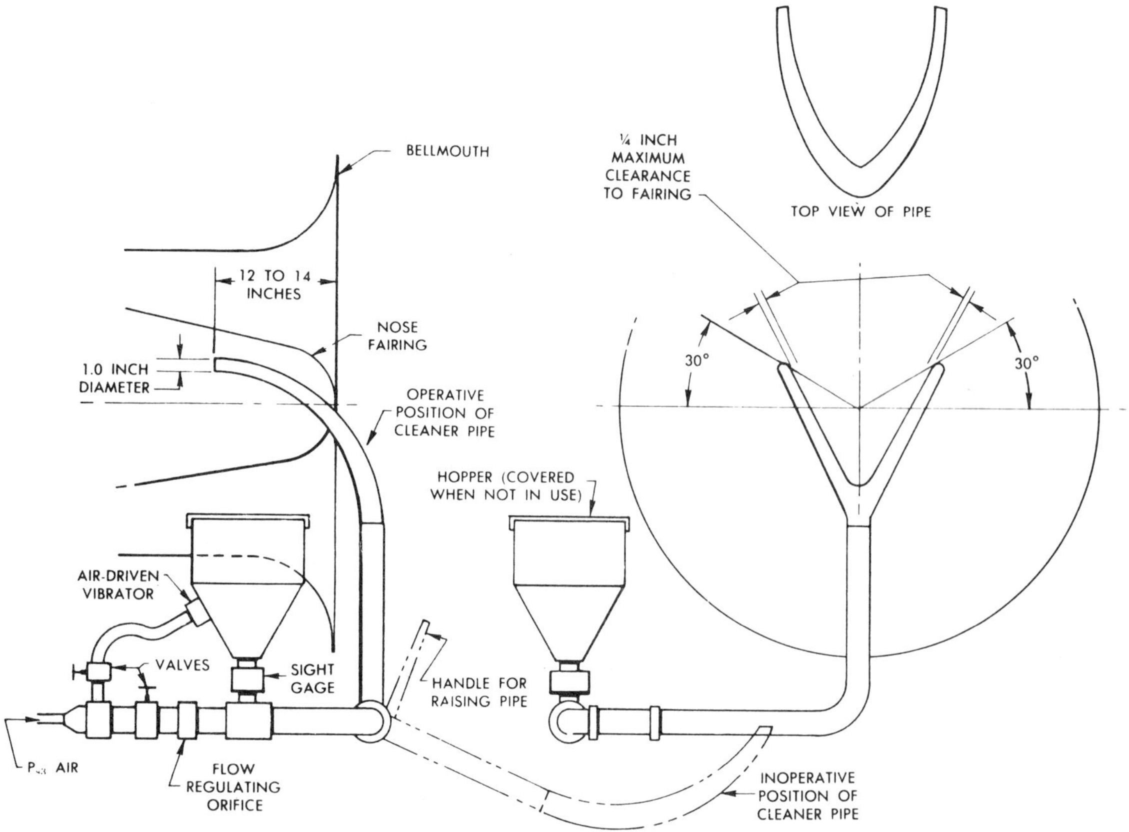

Figure 12.12. Typical system for injecting solid cleaning agent into compressor inlet [*Installation Design Criteria for Gas Turbine Applications in Naval Vessels* (1968)].

course, he cannot control this temperature, but he can investigate the climatological conditions under which the vessel will operate and thereafter set his power rating at a level appropriate to the temperatures to be encountered.

Thermal cycling is perhaps beyond the realm of the ship designer's control, for it depends on the operating profile to be followed in service. In usual circumstances, this profile is, in effect, a specification handed the designer at the outset of his work. Nonetheless, the designer should be aware that a maintenance benefit can be had from reduced thermal cycling, should unusual circumstance give him the opportunity as his design progresses.

It might also be said that fuel quality is beyond the reach of the designer's influence, since he probably won't choose the type of fuel to be used, nor will he do the buying and inspecting throughout ship life (obviously not). But nonetheless, his decisions in this case do have a major influence, for they determine what happens to the fuel stored aboard the vessel, and in its passage from tank and to engine. Choices of tank materials or coatings, similar choices for piping, provisions of settling tanks and clean fuel tanks, choice of strainers, centrifuges, and filters, all affect the cleanliness of the fuel as it reaches the engine. In opting for alternatives that produce the cleanest possible fuel within the constraints of rational engineering decisions, the designer has a significant impact on subsequent maintenance history.

And after changing the word "fuel" to "air" in the sentence just past, it may be repeated. Perhaps the ship designer's impact on maintenance is strongest of all here, for excess sea salt reaching the engine can most quickly cause serious degradation via the sulfidation phenomenon, in addition to the rapid loss in performance. He usually faces many decisions relating to location and protection of air intakes, choice of separating inputs (such as oily vapors or heated air), and by no means least, the careful avoidance of design details that could increase the possibility of foreign-object ingestion.

Extraneous forces and moments can occur because of loads that might be imposed as a result of thermal expansions of engine or of attached ducts and pipes. The engine manufacturer may well provide the engine mounted on a base in a way that minimizes the expansion problem of the engine relative to its foundation, but the attachment of inlet and exhaust ducts, and of piping, is accomplished as part of the ship building process. Provision of the necessary features, such as expansion joints, to avoid the distorting forces and moments that those things can place upon the engine, therefore, is usually within the scope of the vessel designer's work.

To summarize, the maintenance history of a marine gas turbine appears to be in a number of ways dependent on the decisions of the marine engineer designing the propulsion plant. For a good history, he must strongly slant these decisions toward providing a favorable environment for engine operation.

12.7. REFERENCES

Bowers, N. K. (1967), "Gas Turbines in the Royal Navy," *American Society of Mechanical Engineers Transactions, Journal of Engineering for Power,* **89,** pages 95–124.

Dupuy, P. A., (1973), "LM2500 Marine Gas Turbine Improvements Derived from Operational Experience Aboard the GTS *Admiral Wm. M. Callaghan,* The Marine Engineering Society of Japan, paper presented at the ISME Conference, Tokyo.

Installation Design Criteria for Gas Turbine Applications in Naval Vessels (1968), NAVSHIPS 0941-038-7010 United States Navy Technical Manual.

LM2500 Marine Gas Turbine Installation Design Manual (1973), General Electric Report MID-IDM-2500-2, Marine and Industrial Projects Department.

McAllister, P. J. (1972), "Modularization and Installation of the Gas Turbine Propulsion System in *Euroliner,*" American Society of Mechnical Engineers Paper 72-GT-85.

McLean, H. D., Tangerini, C. C., and Van Cott, W. H. (1960), "Report on 9000 Hr Operation of Marine Propulsion Gas Turbine in the *John Sargeant,*" American Society of Mechanical Engineers, Paper 60-GTP-5.

Morgan, S. G., Lamport, A. W., and Smith, A. J. R. (1970), "Gas Turbines in the Royal Navy, 1965–1969," American Society of Mechanical Engineers, Paper 70-GT-10.

O'Hare, T. L. R., and Holburn, J. G. (1973), "Operating Experience With Gas Turbine Container Ships," *Transactions, The Institute of Marine Engineers,* Series A, Part 1, pages 1–23.

Rogus, B. J., and Weinert, E. P. (1972), "Generalized Outlook on Marine Gas Turbine Hot Section Deterioration," American Society of Mechanical Engineers, Paper 72-GT-95.

Siemietkowski, J. S. (1970), "Maintenance of Marine Gas Turbine Engines," American Society of Mechanical Engineers, Paper 70-GT-51.

Smith, H. F. (1969), "Gas Turbine Propulsion for LNG Tankers," American Society of Mechanical Engineers, Paper 69-GT-47.

Trewby, G. F. A. (1963), "Recent Operating Experience With British Naval Gas Turbines," *American Society of Mechanical Engineers Transactions, Journal of Engineering for Power,* **85,** pages 46–71.

Wert, W. (1970), "Marine Maintenance," Paper 7, General Electric Northwest Regional Marine Gas Turbine Seminar.

Woodward, M. L. (1972), "British Experience Using the Aero-Derived Gas Turbine for Main Propulsion in Air Cushion Vehicles and Fast Patrol Craft," American Society of Mechanical Engineers Paper 72-GT-76.

Zeien, C., Smith, H. F., and Hirst, F. W. (1969), "The Gas Turbine Ship *Callaghan*'s First Two Years of Operation," *Transactions, Society of Naval Architects and Marine Engineers,* **77,** pages 344–371.

Chapter

THIRTEEN

GAS-TURBINE ENGINES IN OCEAN ENGINEERING

13.1. INTRODUCTION

There are indeed many power applications in the several subfields of ocean engineering. If "ocean engineering" may be defined to include the traditional transportation uses of the seas, then one of these applications is the powering of conventional ships, the topic to which this book is largely devoted. It is, of course, the predominant use of power on the oceans, but others are also important and are becoming increasingly so as new utilizations of the ocean resource are discovered and developed. This chapter sketches those power applications—other than conventional transportation—in which the gas turbine engine is making an important contribution.

By far the biggest nontraditional use of the sea is the extraction of minerals from the sub-sea crust of the earth, which in the 1970s, at least, means petroleum and natural gas. The extraction industry (i.e., oil companies) is characterized by a variety of mobile "drill rigs" from which exploration is carried out, by stationary "platforms" from which the long-term extractive operations are carried out, and by vessels that serve their needs for personnel, supply, and construction services. The power requirements are met for the largest part by diesel engines, and were met almost exclusively by diesel before about 1965. Over the 10-yr span that has elapsed from then until this is written, however, the gas turbine has come into appreciation for the advantages that acrue to offshore installations from its lightness and compactness, and it has made a noticeable penetration into offshore services. The major use has been in supplying the power needs of the production platforms. Several of these applications are discussed in this chapter to illustrate these services.

Commercial fishing is also a major ocean-extractive industry. Its power needs are mostly those of conventional vessel propulsion, plus some auxiliary power for handling the fishing gear, and in some cases for processing the catch. As yet the diesel engine reigns almost totally unchallenged in these services. Therefore, without attempting explanations of why this may be, or attempting predictions of when and how it may be challenged by the gas turbine, this chapter regretfully ignores the fishing aspects of ocean engineering.

Some underwater power needs are developing. Of course, naval submarines are nearing their one-hundredth anniversary of development and use, but I refer rather to the research and industrial applications that were almost nonexistent before the 1950s. Since that time, a number of small research submarines have been built, as well as a number of experimental stationary underwater "habitats." Their power needs have been met mostly by storage batteries, but the habitat needs, in particular, may grow beyond the economic reach of storage battery systems if they expand into underwater industrial (e.g., mining) installations. The gas turbine is, of course, ordinarily a combustion engine for which immersion in the atmosphere is essential. On the other hand, it can operate as a closed-cycle, external-combustion engine, with the oxidant needed for combustion supplied from a stored source, or the heat source can be noncombustive, such as a nuclear source. Thus it is that a gas turbine engine can be adapted to underwater use to provide a high-density source of power. Developments of gas turbine power for use in space missions have contributed much of the necessary technology. Although the day of this application appears to be not yet at hand and faces potential competition from other sources (e.g., fuel cells), development has been accomplished and reported to the extent that some features can be illuminated in this chapter.

13.2. THE GAS TURBINE AS AN UNDERSEA POWER SOURCE

If the gas turbine is arranged to run on a closed cycle, so that the working fluid is continuously recycled, noncombustion heat sources can be applied to produce an atmosphere-free engine. Possible heat sources could be a nuclear reactor, radioactive material, or stored heat in some form such as molten salt. Or it is quite possible to provide an isolated combustion heat source through the agency of stored oxidizer. But even in the latter instance, a closed cycle would also be used in order to conserve working fluid.

The closed-cycle gas turbine is virtually unknown in the United States as a stationary or mobile power plant for earthbound services, although it is used in Europe in some stationary applications. On the other hand, small closed-cycle engines have been developed in this country for space-mission power plants, with the prospect of a variety of heat sources. Figure 13.1 shows a compact closed-cycle engine developed for electric-power generation in space by the AiResearch Division of Garrett Corporation. This is a single-shaft machine with an alternator mounted on the shaft between compressor and turbine. Net electrical output is 50 kW at a frequency of 400–3200 Hz. The unit is hermetically sealed with gas bearings lubricated by the working fluid, which typically would be a mixture of helium and xenon [Phillips (1969)].

Note that the figure shows no combustion chamber; as the arrows indicate, the compressed gas leaves the unit for heating in an external heat exchanger. The engine itself is therefore adaptable to any situation—such as space or under water—in which a closed cycle can be provided to heat the gas and to cool it after expansion through the turbine. Figure 13.2 diagrams a complete cycle of this type, incorporating the engine in an underwater cycle. Cryogenically stored hydrogen and oxygen are combined in a catalytic reaction chamber to produce a hot gas that heats the working fluid to 1600 F in a heat exchanger. Heat rejection is to the sea via another heat exchanger. The combustion products must be condensed, since exhausting overboard against a large hydrostatic pressure would not be feasible, and the condensate would probably be stored. With 29 F sea water as the heat sink, the overall efficiency

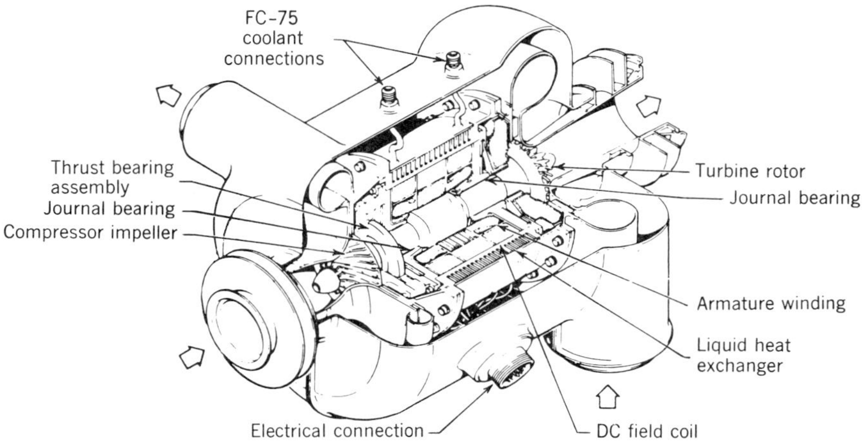

Figure 13.1. Compact gas-turbine electric generator for space or undersea use [(Garrett Corp.) Phillips (1969)]).

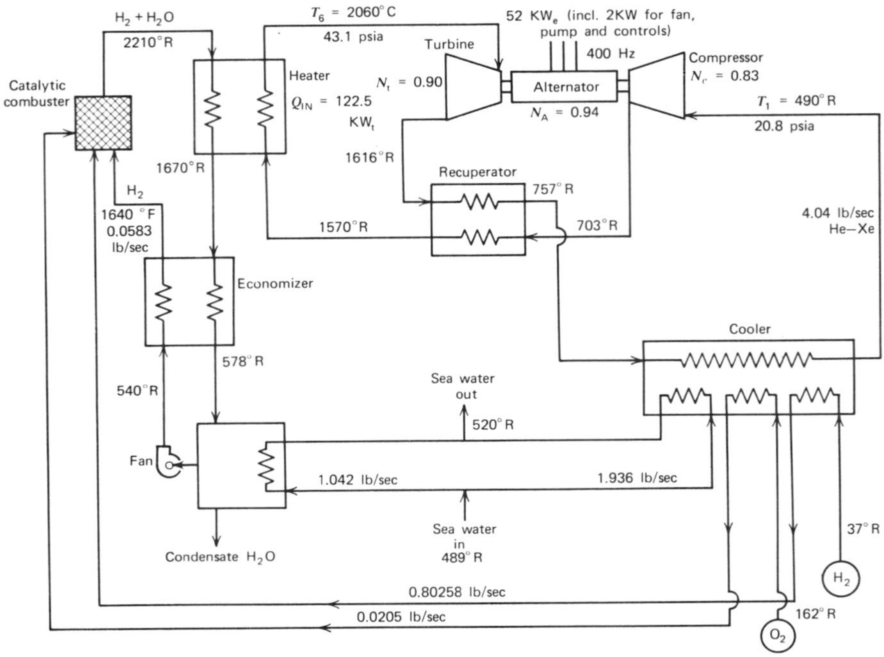

Figure 13.2. Cycle diagram for power plant using the ending of Figure 13.1 [Phillips (1969)].

should be 42.4%, with specific reactant consumption of 1.66 lb/kW-hr [Phillips (1969)].

The appeal of a gas-turbine power source such as this for underwater use lies in the compactness that is usually associated with gas turbines in any situation, plus the compactness of a combustion energy source (as contrasted to batteries, for example). Table 13.1 [taken from Phillips (1969)] lists the prospective weights and volumes. The rather small weight shown for the hydrogen and oxygen tanks are for the Dewar inner wall only; the sea pressure vessel that encloses the machinery is assumed to furnish the outer wall.

Figure 13.3 is a plot of specific energy (lb/shp-hr) as a function of endurance (hr) for several possible underwater energy sources. The data from Table 13.1 (3.2 lb/shp-hr for 20 hr; 50 kW converted to 67 shp) is spotted on this figure, indicating its relatively good position. Note that the figure shows a spread of weights, with the highest line including the weight of buoyancy material necessary to offset the weight of machinery and energy storage; the buoyancy weight is not included in the added gas-turbine point.

TABLE 13.1. WEIGHTS AND VOLUMES FOR A 50-KW, 1000-HR UNDERSEA GAS TURBINE SYSTEM [Phillips (1969)]

	Weight, lb	Volume, ft:
Power-conversion unit	1666	23.5
Hydrogen tank (dry)	281	52.0
Hydrogen reactant (including residual)	190	—
Oxygen tank (dry)	183	26.0
Oxygen reactant (including residual)	1505	—
Condensate storage required	—	26.5
	3825	128.0

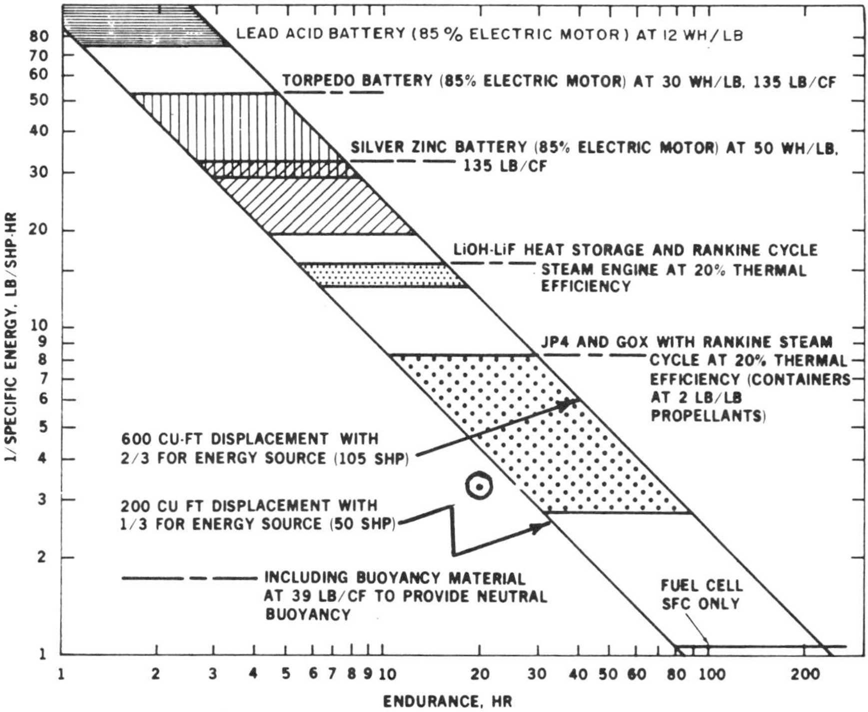

Figure 13.3. Submerged vehicle endurance with various energy sources [De Vries, Karig, and Drage (1969)].

Several other choices of heat source may be feasible, such as radioisotopes, nuclear reactors of several types, and stored heat (e.g., molten salt). Table 13.2 lists several significant parameters with values that might be obtained with the gas-turbine engine of Figure 13.1 used in a cycle heated by decay of plutonium 238 and by a zirconium hydride reactor. The left column of data in this table represents the cycle shown in Figure 13.2. Also, with any heat source, a wide range of flow rates, temperatures, and pressures are usable, giving a designer many choices in seeking the best combination of parameters for a particular application. Questions relating to these choices are explored in the rather limited literature on undersea gas turbines. For a reference that does so for the engine and cycles mentioned here, I suggest a paper by Balukjian and Rackley (1971).

TABLE 13.2. SIGNIFICANT PARAMETER VALUES FOR SEVERAL UNDERSEA CLOSED-CYCLE GAS-TURBINE SYSTEMS [Pietsch and McCormick (1970)]

Power output, kW	50	10	50
Heat source	$H_2 + O_2$	Pu^{238}	ZrH reactor
Working fluid	HeXe	HeXe	HeXe
Turbine inlet temp, R	2060	2060	1730
Compressor inlet temp, R	490	530	510
Shaft speed, rev/min	24,000	36,000	24,000
Recuperator effectiveness	95	90	95
Pressure loss factor	0.92	0.92	0.92
Compressor inlet pressure, lb/in^2	20.8	14.1	33.9
Compressor pressure ratio	2.16	2.10	1.90
Compressor efficiency, %	83.0	81.0	84.4
Turbine inlet pressure, lb/in^2	43.1	28.6	61.9
Turbine pressure ratio	1.99	1.94	1.75
Turbine efficiency, %	90.0	88.4	90.7
Generator frequency, Hz	400	1200	400
Generator efficiency, %	94	92	94
Electric, windage, and bearing losses, kW	4.110	0.930	3.00
Gas flow rate, lb/sec	4.04	1.01	4.91
Cycle heat input rate, kW	122.5	31.6	147.5
Cycle efficiency, %	42.4	33.2	33.9

13.3. APPLICATION BY THE OFFSHORE PETROLEUM INDUSTRY

A. Typical Uses

As readers are probably aware, a significant part of world petroleum and natural gas extraction takes place in the shallow seas surrounding the continents and islands. The extraction sites are commonly marked by platforms that house equipment required for extraction, for certain processing, and for pumping the product ashore. See Figure 13.4 for an overall view of a typical platform. The energy requirements are large for the diverse services; from ten thousand to up to several tens of thousands of horsepower is common. Although diesel engines are the usual prime movers, gas turbines began to be used for this service in the late 1960s, and their use increased rapidly in the early 1970s.

An example of an offshore platform powered totally by gas turbine engines is the Union Oil Company's Grayling platform in Cook Inlet, Alaska [Anonymous (1970)]. In 1970 this platform employed 16 Solar Saturn* (about 1200 hp each) gas turbines for all of its power requirements. Two drove compressors that delivered natural gas ashore, one drove a crude oil delivery pump, three furnished electric power via 800-kW generators, four drove gas-lift compressors, and six drove water-injection pumps.

The last two items, which may sound unfamiliar to readers with strictly a ship background, are measures to increase the recovery of petroleum. Gas lift involves injecting gas into the underground formations to reduce the density of the liquid through intermixing of gas bubbles. The water injection serves the same function through displacing the less dense petroleum upward.

An inspection of gas-turbine and oil-industry trade press for the years 1970–1973 shows similar installations being made in the Gulf of Mexico, Lake Maracaibo, off southern Australia, and in the Persian Gulf.

B. Advantages of the Offshore Gas-Turbine Engine

From its beginning through the writing of this chapter, the gas turbine has had a higher first cost than a diesel of comparable rating, and also the fuel economy has yet to equal that of the diesel. However, advantages offsetting these handicaps can be identified, and they are strong enough to make the gas turbine a strong—and often successful—competitor for the diesel aboard off-

* Saturn is the registered trademark of Solar Division, International Harvester.

Figure 13.4. A typical offshore petroleum production platform (Shell Oil Company, Brent North Sea Field).

shore installations. The advantages are the same as those that give the former its appeal in several vessel-propulsion applications, namely compactness and lightness.

Lightness may not be an obvious asset in a stationary application, but it is indeed significant in this one since the structural scantlings of the platform are a function of the weight that must be supported above the sea. The unit weight of a gas turbine generator (as an example) should be in the range of 13–25 lb/kW, while the figure for a diesel generator should be in the 30–60 lb/kW range [Wittner and Culp (1973)]. Using the Cook Inlet example of the preceding section, the sixteen gas turbines there may be saving a machinery weight of about 200 tons (based on sixteen engines at 800 kW, and using the highest lb/kW figure quoted just above).

The same installation also takes advantage of gas-turbine lightness in a special way. The four gas-lift compressor units were an add-on at some time after the original platform construction and nominally would have required enlargement of the unit. However, the entire gas-lift compression outfit could be fitted into a 115-ton package [Anonymous (1970)], a package that was light enough to cantilever off the side of the platform.

Lightness of structural support is also enhanced by the comparative freedom from vibration excitation originating in the gas turbine engine.

Space saving is also an important advantage in the expensive off-shore environment. Construction costs as of 1973 have been cited as \$800–1500/ft^2, and the area density of a gas-turbine generator as 10–18 kW/ft^2, compared to 4–8 kw/ft^2 for a diesel generator [Wittner and Culp (1973)].

Weight and space advantages may allow a major part of the construction and testing processes to take place ashore, rather than in the more expensive off-shore environment. The add-on of four gas-lift compression units to the Cook Inlet platform, mentioned three paragraphs above, is an example; the 115-ton package was assembled ashore before being barged intact to the offshore site. And other examples are described in the trade press. One, for instance, describes the construction in Texas of a complete gas-compression station for service in Lake Maracaibo, Venezuela. The compressor and electric-power generator are driven by an engine consisting of a Rolls Royce gas generator and a Dresser Clark power turbine, 13,500 hp at NEMA rating [Reddy (1973)]. The entire platform is a double-deck assembly weighing 365 tons that was shipped as a complete unit from its assembly site in Texas. Field work consisted only of mounting on pilings, hook-up, and the necessary field testing.

Figure 13.5 shows a similar unit afloat at its building site. This one features two tandem sets of Elliot compressors, each set being driven by a Turbo Power & Marine engine of 22,500 hp. There is also aboard an 800-kW Solar Saturn generator.

C. Operation in the Offshore Environment

The offshore environment is, naturally, much the same as that aboard ship. A platform surely lacks the motions of a ship hull, but in the most important question of atmospheric sea-salt contamination, the situation is essentially the same. Although a fast-moving vessel may generate spray that contributes to the burden on its gas-turbine engines, natural processes also contribute enough

Figure 13.5. A typical offshore compressor barge with gas-turbine prime movers (Turbo Power & Marine).

spray to produce a problem of about the same magnitude for a stationary engine at sea. For this reason, the platform gas turbines usually are equipped with inlet moisture separators, and the compressor inlets are fitted with nozzles for the injection of washing water.

Since petroleum is produced in all the earthly climates (well, in a great variety), gas turbines that are employed therein may be required to operate in either very hot or very cold conditions, whether ashore or at sea. This might be regarded as a handicap to the gas turbine because of its sensitivity to ambient temperature. A gas-turbine engine being selected for service in Lake Maracaibo, for example, is likely to require more derating for high temperature than a diesel engine of the same nominal rating. Going to a cold climate, on the other hand, means a greater gain for the gas turbine, but it may not be feasible to take full advantage of this phenomenon; limited capacity of the fuel system, load limitations on transmission parts, and possible shift of the compressor operating point into the surge region are factors that can limit such output increase [Kindl (1973)]. In cold regions, inlet icing may be a problem, especially associated with the marine environment because of the high humidity. The problem and its solutions are much the same as aboard ship (Chapter 11). An additional option that is available to the platform engine if it is burning natural gas (which is often the case) is recirculation of a small fraction of the exhaust for the purpose of preheating the incoming air. With the usual liquid-petroleum fuels burned aboard ship, the likely fouling of compressor surfaces by the combustion products precludes this practice.

Figure 13.6. Total-energy-packaged plant for offshore platform (North American Turbine Corp.) [Wittner and Culp (1973)].

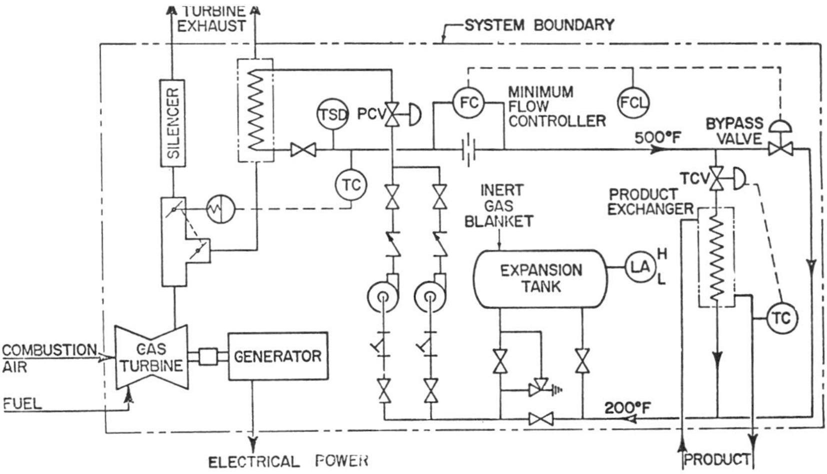

Figure 13.7. Diagram of the waste-heat use for the unit shown in Figure 13.6.

D. Total Energy Offshore

The "total-energy" concept involves supplying all of the several energy needs by a single source, especially where high quality (mechanical) and low quality (low-temperature thermal) energy are needed simultaneously. The appeal is high efficiency resulting from maximizing the usage of energy available following its combustion. The necessary tactic is to use in low-temperature services that heat normally exhausted to the surroundings. Although this is frequently and reasonably done with diesel engines, the practice is the more attractive with the gas turbine because it rejects more of its energy at high temperature—roughly half the heat rejected from a diesel engine is usually in the form of water at less than 200 F.

Use of the total-energy idea is not common in the marine gas-turbine field, but examples can be found. One such is the Canadian DDH-280-class destroyer, wherein exhaust from the ship service generator gas turbine produces steam that is used for galley, heating, and distillation of fresh water [Breene and Warner (1973)].

Similarly, use of the total-energy gas-turbine concept in the offshore industry is not yet common, but has an appeal where large amounts of thermal energy are required in addition to the mechanical or electrical energy that the engine normally provides. A package unit for such service is shown being assembled ashore (North American Turbine Corporation unit [Wittner and Culp (1973)]) in Figure 13.6. The mechanical function of the gas-turbine engine here is drive of petroleum-delivery pumps, and the thermal function is dehydration and viscosity reduction of the petroleum product. The heat usage is effected through an intermediate heat-transfer fluid (a liquid of hydrocarbon base). Typical conditions are shown in Figure 13.7, this being a diagram of the heat-use system, including typical parameter values. Note that temperature in the intermediate fluid is controlled by bypass dampers in the engine exhaust line and that a minimum flow control is used to keep heat-transfer-fluid flow from being so low that coking would occur on the gas-heater surfaces.

Efficiency of a total energy-packaged unit such as that of Figure 13.5 is said to be in the 45–60% range. First cost as completed ashore is $240–300/kW (circa 1972), of which $25–35 is the added cost of the heat-recovery system [Wittner and Culp (1973)].

13.4. GAS TURBINES IN THE VESSELS OF THE OFFSHORE INDUSTRY

The previous section has outlined uses of gas-turbine power by the offshore petroleum industry aboard its stationary installations. This type of power also

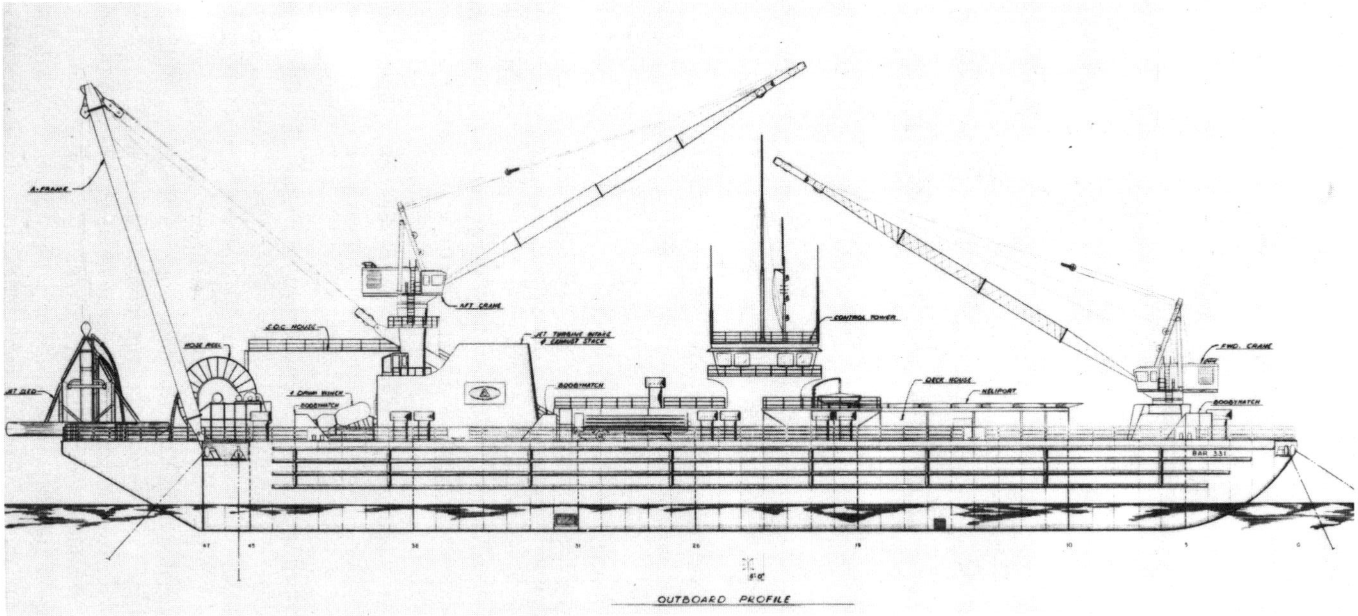

Figure 13.8. Outboard profile of the pipe bury barge, BAR 331 (Brown & Root, Inc.).

Figure 13.9. Gulf Coast offshore crewboat powered by Ford gas-turbine engines.

has application aboard floating equipment in situations where much the same advantages appear, namely, large concentrations of power in light and compact packages.

An excellent example is the application of two Turbo Power & Marine FT4 engines to a pipe bury barge of Brown & Root, Inc. (The designation comes from its principal function, which is the laying of underwater pipelines in a sea-bed trench.) This vessel, named BAR 331, entered service in 1974 from Levingston Shipbuilding and requires power for a number of services. Four 1500-kW generators are driven by Alco diesels; two 1250-hp Detroit Diesel Allsion engines drive the air compressors and pumps of dredging units. The big power, however, is required for jet pumps that deliver power to the jetting sled. This sled moves along the sea bottom, jetting a pipe trench with flow of water of up to 20,000 gal/min at 2500 lb/in^2. A single pass of the sled can cut a trench from 4 to 12 ft wide, up to 10 ft deep. The jet pumps are driven by two FT4 gas turbine engines rated at 19,750 hp each for this service [Peacock (1974)].

An outboard profile of the barge is shown by Figure 13.8. The jetting sled appears on the stern. The barge is intended for service in the rather difficult environment of the North Sea, in water of depths of 400–500 ft. Its principal dimensions are 350 ft overall length by 92 ft beam by 28 ft depth of hull. Its size is effectively determined by the deck area needed for machinery and the volume needed for fuel, stores, and crew accommodations. The saving in machinery deck area made possible by the compactness of the gas-turbine plant is translated into a smaller, hence cheaper, barge.

The offshore-oil industry makes use of a large number of supply vessels, some of which are the high-speed "crew boats" that shuttle personnel between shore and the offshore sites. High speed in a small vessel seemingly offers the gas turbine an advantageous application, although as this is written its penetration into this service has been very small. Most of the vessels built with gas-turbine power have been experimental or demonstration projects. Figure 13.9 pictures one such boat, crew boat built by Teledyne Sewart Seacraft in 1971. Engines are twin Ford truck gas-turbine at 450 hp each. Passenger capacity is 100 persons, with reported speed of 26 mph.

13.5. REFERENCES

Anonymous (1970), "Alaska's Grayling Platform Goes All-Gas-Turbine," *Gas Turbine International,* **11,** pages 24–25.

Balukjian, H., and Rackley, R. (1971), "A Closed Brayton Cycle Power System for Deep Submersible Vehicles," *Journal of Hydronautics,* **5,** pages 5–17.

Breene, G. T., and Warner, C. E. (1973), "Total Energy Applications for Ships Service Power," American Society of Mechanical Engineers, Paper 73-GT-69.

De Vries, G., Karig, E., and Drage, G. (1969), "Eutectic Molten Salt Thermal Storage System," *Journal of Hydronautics,* **3,** pages 191–195.

Kindl, F. H. (1973), "Gas Turbines in the Arctic Environment," American Society of Mechanical Engineers, Paper 73-GT-83.

Peacock, D. (1974), "Gas Turbines and Diesels Combine on Deep Water Pipe Bury Barge," *Diesel and Gas Turbine Progress,* **24,** pages 43–44.

Phillips, T. W. (1969), "Closed Cycle Gas Turbines as a Proposed Undersea Power Source," *Journal of Hydronautics,* **3,** pages 184–190.

Pietsch, A., and McCormick, J. E. (1970), "Development Status of Closed-Brayton Cycle System for Space Power Applications," (Intersociety Energy Conversion Engineering Conference, 1966), excerpts reproduced in *Gas Turbine International,* **11,** pages 30–31.

Reddy, R. (1973), "Commissioning an Offshore Compressor," *Gas Turbine International,* **14,** pages 44–48.

Wittner, B. R., and Culp, R. E. (1973), "Turbine Total Energy for Offshore Rigs," *Gas Turbine International,* **14,** pages 14–18.

SUBJECT INDEX